ANTIQUE **PACK**AGING

GINGKO
PRESS

First Published in the United States of America, September 2008

Gingko Press, Inc.
5768 Paradise Drive, Suite J
Corte Madera, CA 94925, USA
Phone (415) 924 9615 / Fax (415) 924 9608
e-mail: books@gingkopress.com

Author: JOSEP Mª GARROFÉ
Art Direction: Estel Roca

ISBN: 978-1-58423-318-3

To **Francesc Serra**.
This book would not have been possible without his kind and patient collaboration. Because we never tell him how much we appreciate him, we want to take the opportunity to do it now.

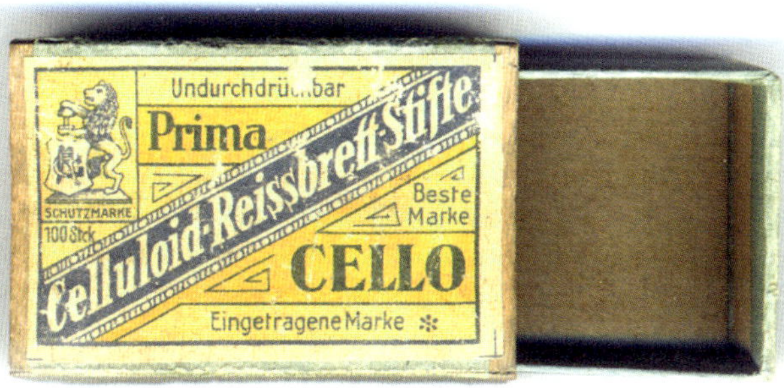

ANTIQUE **PACK**AGING

Sulfoguayac
Fosfato de

ADULTOS:

NIÑOS: La

la dosis,

PR

LOOKING TO THE PAST TO FACE THE FUTURE

Antique Packaging is a healthy exercise for designers and creative minds -especially those who work in the packaging industry- who would like to look to the past to face the future with an added sense of confidence and an eye for innovation. The collection of boxes included in Antique Packaging represents an interesting review of the major trends from other eras, especially in terms of the types of images used and the graphic resources employed in their design. They are also a pleasant, nostalgic look at days gone past.

The boxes selected for inclusion in Antique Packaging where chosen first and foremost for their design and aesthetic aspects. Then, other aspects like theme and industry were also taken into consideration. Our most basic yardstick for inclusion, however, was that the packages had to be made of cardboard. This means that we had to rule out a large number of metal packs, which, though beautiful and historically significant, bear no direct relation to the project at hand.

Born of the private collection of Tribu-3, Antique Packaging will undoubtedly become an interesting reference work and source of inspiration for design professionals and a cult object for those who love the purest forms of retro style.

I hope you like the packs featured in this book. I love them.

JOSEP MARIA GARROFÉ

The packs included in Antique Packaging are perhaps some of the last remnants of the days of letterpress printing and the handcrafted production of containers, an era that began with the Gutenberg press around 1450 and carried on for centuries with little in the way of technological advances. For nearly 400 years, printers worked using the same process: awl carving, the preparation of matrices, typeface molding, typesetting and printing. All this was carried out entirely by hand, and any corrections were done manually, making it extremely slow process.

Typography

The first printing presses worked by means of pressure on a raised service, a system common to all the various models that have replaced it over time: Minerva printing presses, flatbed presses and rotary printing presses.

At any rate, until the relatively recent appearance of offset printing and, later, digital printing systems, printing any image or text required considerable effort from a range of specialized professionals: typographers, compositors, correctors, collators, typesetters, designers, illustrators, editors, printers, etc.

Perhaps the most important role of the whole lot was that of the printer. Considered by their contemporaries to be highly learned persons, a veritable social reference, printers were often respected, even worshipped, because of their profession. Later, printers began to lose their extreme influence in society as the general level of education among the population began to grow.

The contribution of the draughtsman, or illustrator, is also worthy of mention. The illustrator, an artist in all senses of the term, manually prepared the decoration and illustration for all sorts of print publications. So-called "typographic ornamentation", which was originally done with woodcuts, experienced a noteworthy revival at the end of the 19th century and beginning of the 20th century as a result of the Art Nouveau and Art Deco movements.

This is how things were done in the days of job printing, the branch of typography dedicated to the production of print documents for business, industry and the private sector, that is to say, non-book printing. The pieces that form the Antique Packaging collection belong to this type of typography.

Towards offset printing

The end of the 19th and beginning of the 20th centuries saw the birth of pioneering machines allowing the mechanical placement of more than 7,000 characters per hour. This was the era of the linotype and the first rotary printing presses.

In addition, the development of photography opened the door for direct engravings or autotype in the case of typographic machines and photolytes in the case of offset, a system still in its infancy at that time.

This was the start of the decline of heavy printing, which gave way to other systems that were lighter in terms of materials, very fast and cheaper to produce, such as wood engraving, print and lithography. These, in turn, eventually yielded to offset printing, the true precursor to typesetting and computer-age production, which would revolutionize the graphic arts industry from the 1970's on.

But that is no longer the story of hand production. It is history of technology.

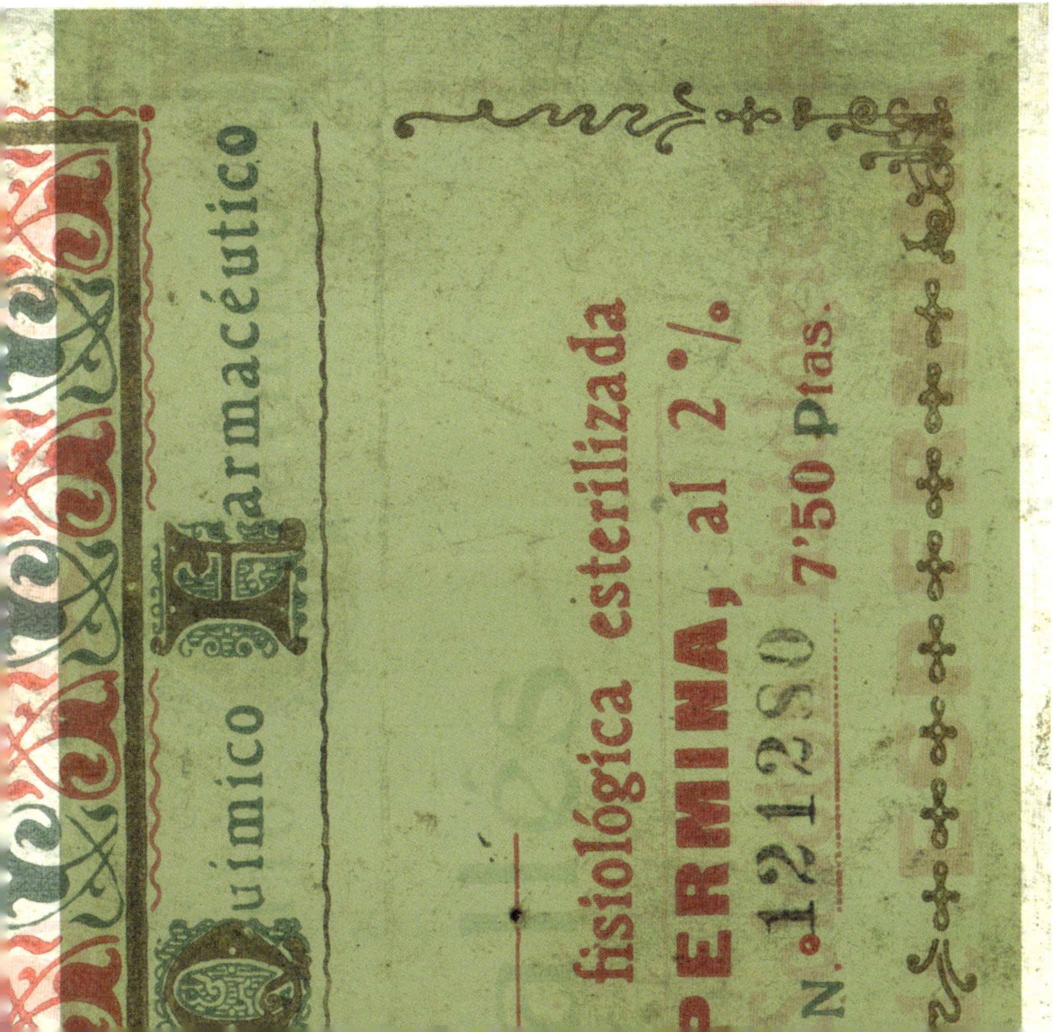

CONTAINER = A receptacle that wraps or contains business good or other objects for their storage or transport.

This apparently simple definition masks a complex universe dominated by design and marketing, where a truly effective container, be it for personal use or mass consumption, must meet a series of conditions. It must:

BE SIMPLE

COMMUNICATE THE CHARACTERISTICS OF A PRODUCT AT A SIMPLE GLANCE

BE BETTER THAN THE COMPETITORS'

BE PRINTED CORRECTLY

SHOW ATTRACTIVE IMAGES OF THE PRODUCT

AND IN THE CASE OF REDESIGNS, MAINTAIN VISUAL TIES WITH ITS PREDECESSOR SO THAT CONSUMERS STILL RECOGNIZE IT

Nevertheless, these characteristics, which seem so basic today, have not always been the norm for containers. Although the need to store and transport food and other products has existed for millennia, nothing was subject to rules or conditioned by market requirements in the early days of Humanity.

In prehistoric times, human beings ate fruits and other types of foods that were protected by natural containers. They soon learned to imitate these containers, adapting them and improving them little by little. The first human attempts to create packaging were based on interwoven vegetable fibers and unfired clay pots and date back to around 8000 B.C. Later, Greek and Roman cultures used cloth wineskins, wooden barrels, glass bottles and fired clay pots.

YESTERDAY, TODAY, ALWAYS

From then on, packaging has gone through an incomparable, dizzying evolution based on the changing, increasingly more demanding needs of different societies and cultures. Today, any product worth its salt must be presented to the consumer in a functional container that is both attractive and original, and the power of packing is so great that at times, it even overshadows product inside.

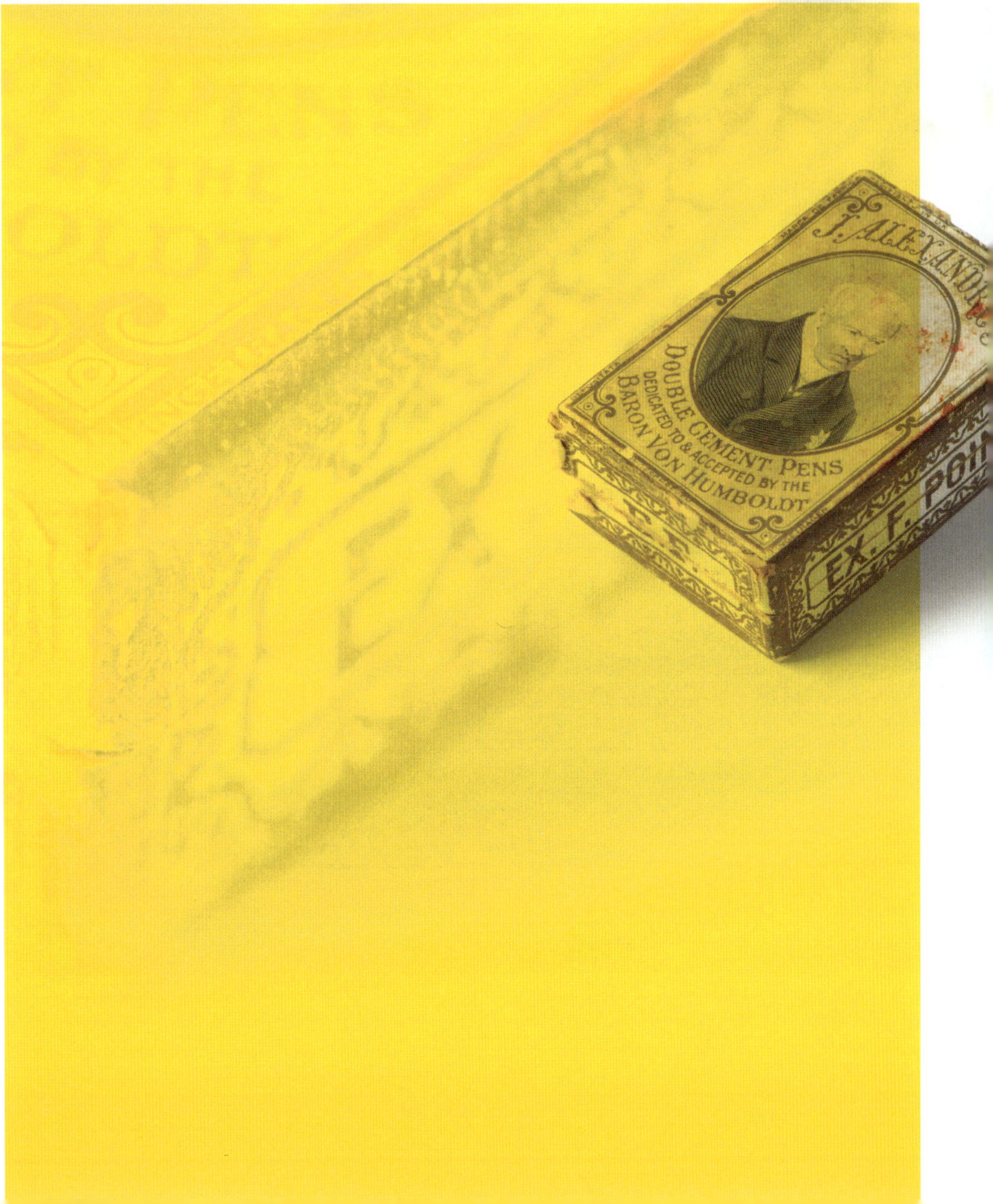

In the origins of modern-day packaging, paper was the simplest means used to wrap products. For several decades, paper's place was taken by plastic, but it is now seeing a strong comeback due to the ever-growing need to recycle materials and develop sustainable human activities.

Cardboard, produced by placing several layers of paper together, is stiff enough to store and transport specific products. The cardboard box is an American invention dating back to the 1870's, when Robert Gair, a printer and paper manufacturer in Brooklyn, was printing seed bags. During production, a metal ruler used to crease the seed bags lifted up by accident and cut a bag. With this simple cutting and creasing operation, Gair began to develop different types of boxes and packages.

Throughout history, human beings and product packaging have gone hand-in-hand, first with natural products such as leaves and vegetable fibers and later with paper and other paper-based products.

PAPER AND CARDBOARD PACKAGING. A BRIEF CHRONOLOGY

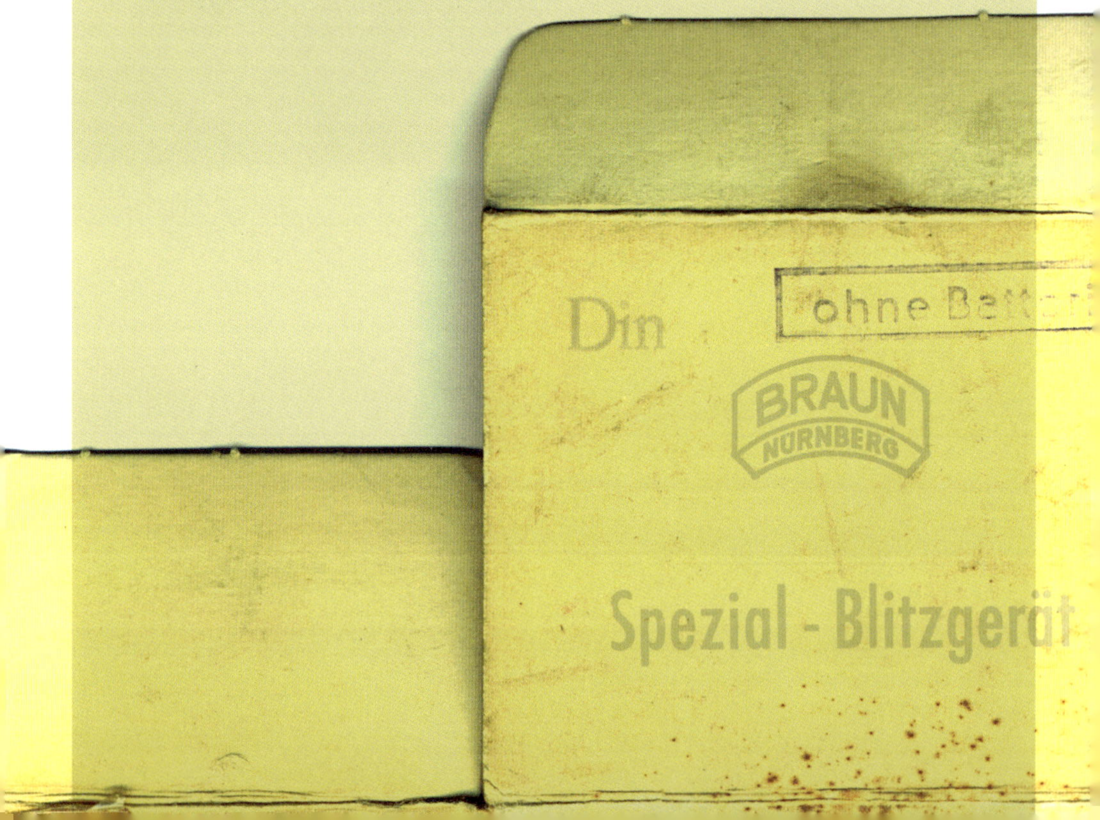

The first modern containers appeared when the industrial and economic advances of the 19th century made it possible for the farmers and manufacturers of the day to send already packaged products to business establishments. This was the birth of packaging and -with it- marketing.

A lot has happened since then, and our society has changed so much that we now have very little in common with that far-off 19th century. First, there was the mass society, where we seemed alike, and recent times have brought about the global society, where we all want to be different in a macro-universe of global production and consumption.

In today's world, products are still wrapped in packaging, but the fundamental difference is that wrapping them is no longer enough. Since the 20th century, especially in the 21st century, packaging is no longer just the skin of a product; it is one of its best weapons for presentation and one of the major factors in a consumer's decision to purchase it. With its packaging, consumers take hold of a product and come to identify with a brand and participate in its success.

For this reason, the development of packaging is not a simple task of decorating a product or dressing it up with a label. It implies creating a strategy of needs and maintaining a constant of innovation that is both effective and aggressive in business terms, using formal and graphic resources that match a process of marketing and global design, taking into consideration new materials and the legal environment of the market.

And all this… for what? Because the market is filling up with more and more products that are increasingly more similar. This generates a non-stop war where competition is skyrocketing, turning packaging into an effective means for attracting customers and differentiating oneself from competitors. Packaging is becoming more important when it comes to recognizing a product, positioning it in the market and differentiating it from the competition.

Together with the brand image, packaging is the cover letter for any type of product. As such, good packaging must attract the attention of consumers, create a desire and need for consumption, reinforce the brand and explain the characteristics of the product inside in such a way that it stands out above the competition and gains the trust of consumers.

In a not-so-distant future, packaging will even talk to consumers. It will allow them to see, smell and touch its product. It will include integrated chips, decodable in any television set or computer, that will explain a product's particular advantages. And if we are ready for all this, it is thanks, in part, to the legacy of packaging -nowadays classic- that was truly innovative in its day.

PACKAGING TODAY

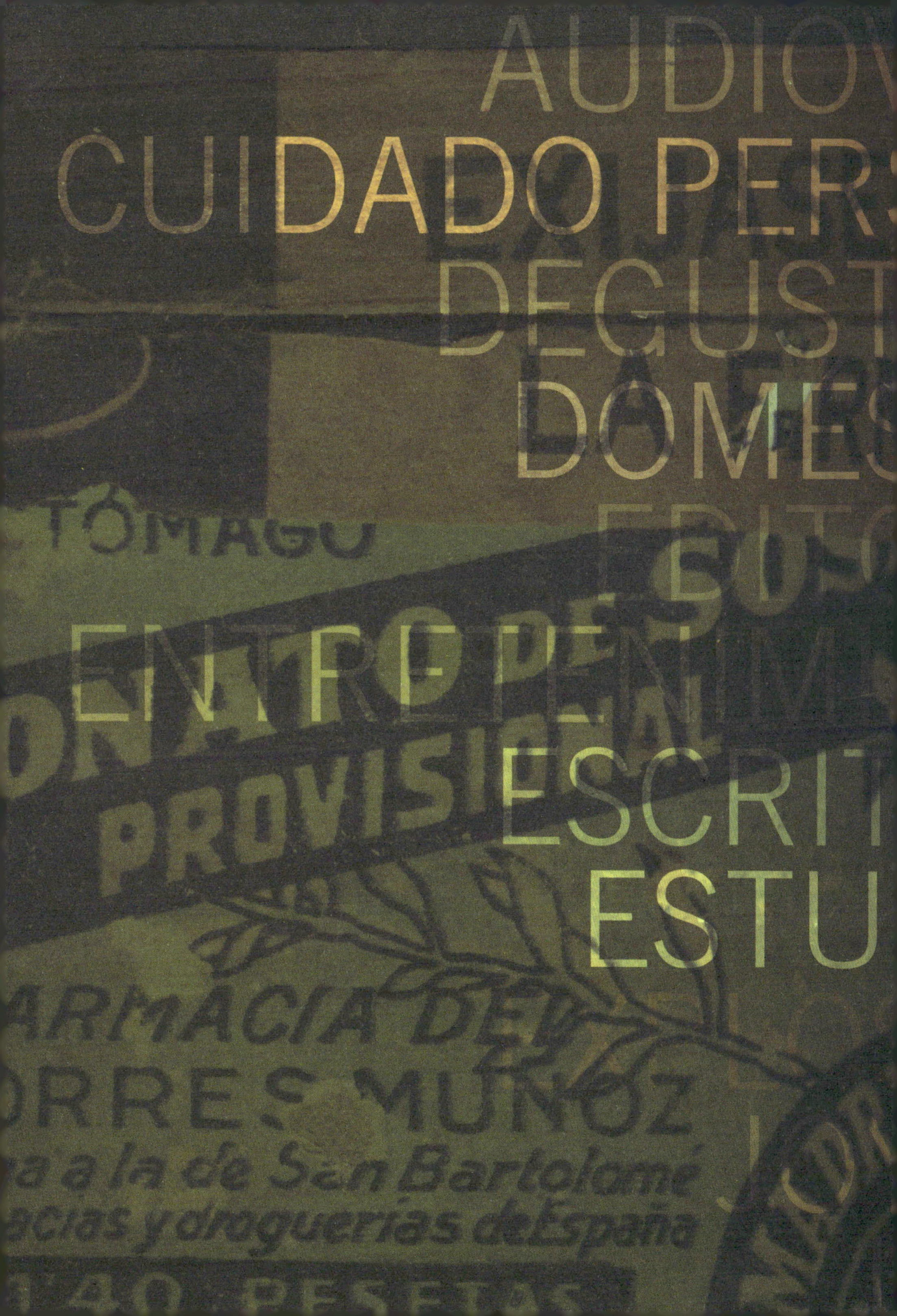
AUDIOV
CUIDADO PERS
DEGUST
DOMES
TÓMAGO
ENTRE
PROVISI
ESCRIT
ESTU
ARMACIA DE
ORRES MUÑOZ
a a la de San Bartolomé
acias y droguerías de España

ARMACY
MACYPHARMACYP

PHARMACY

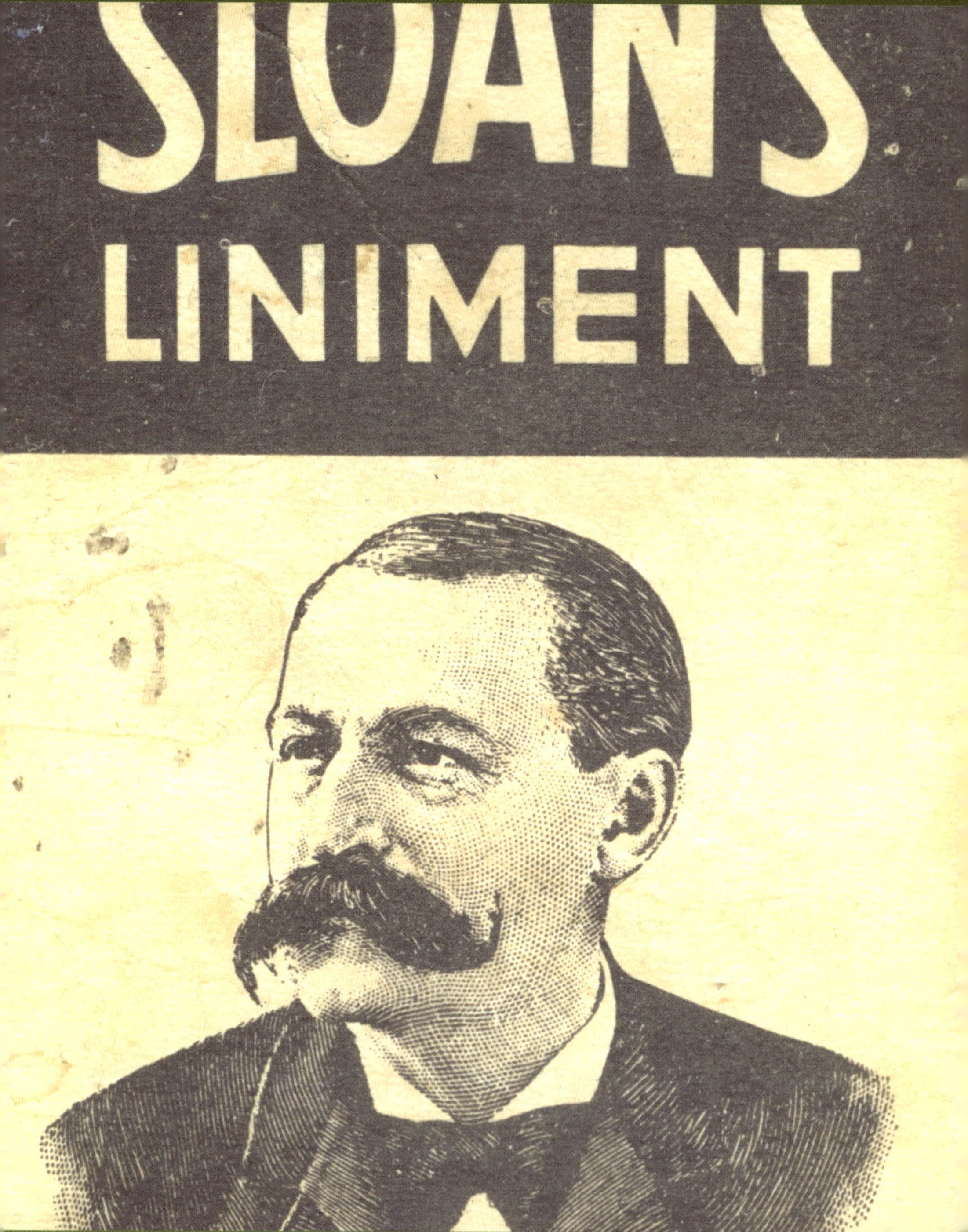
SLOAN'S
LINIMENT

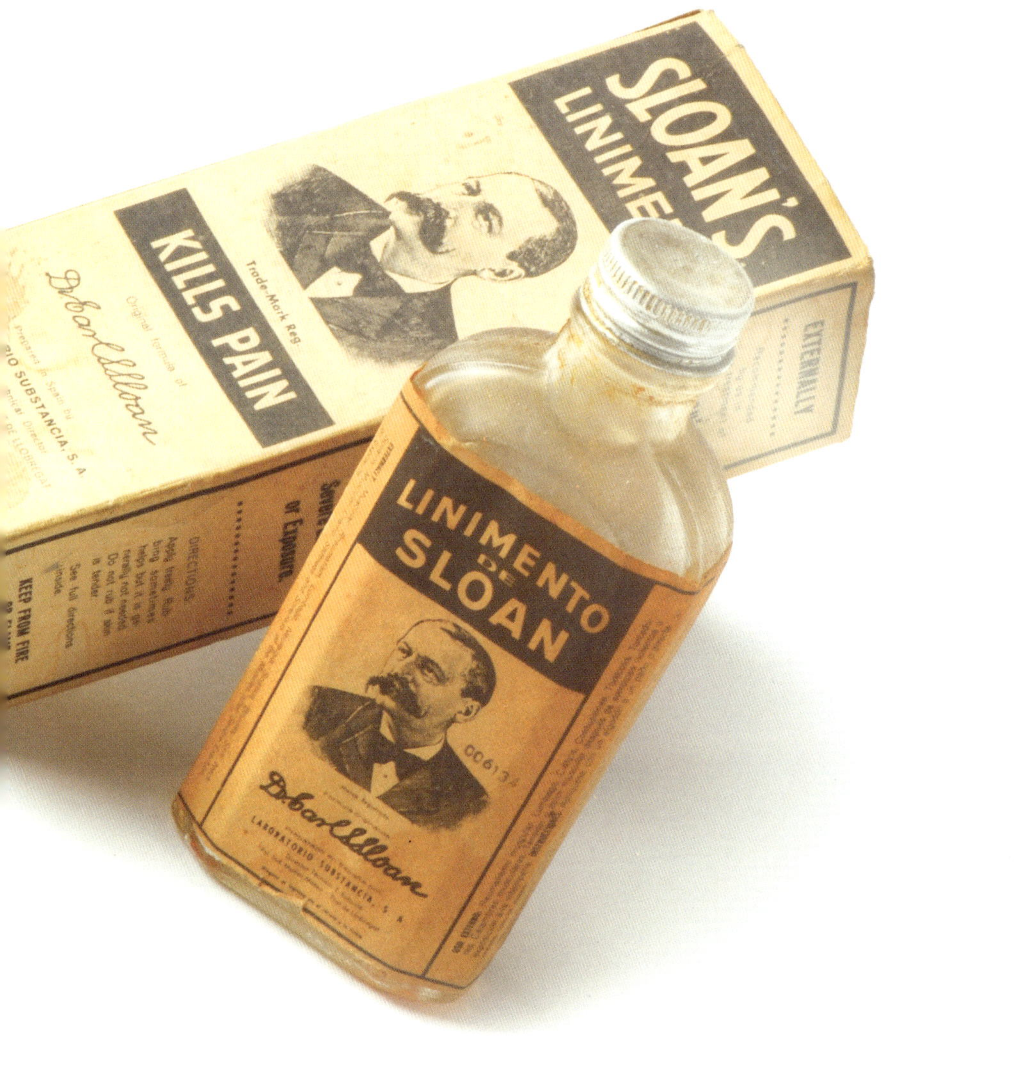
SLOAN'S
Trade-Mark Reg.
KILLS PAIN
DIRECTIONS
KEEP FROM FIRE
LINIMENTO DE SLOAN

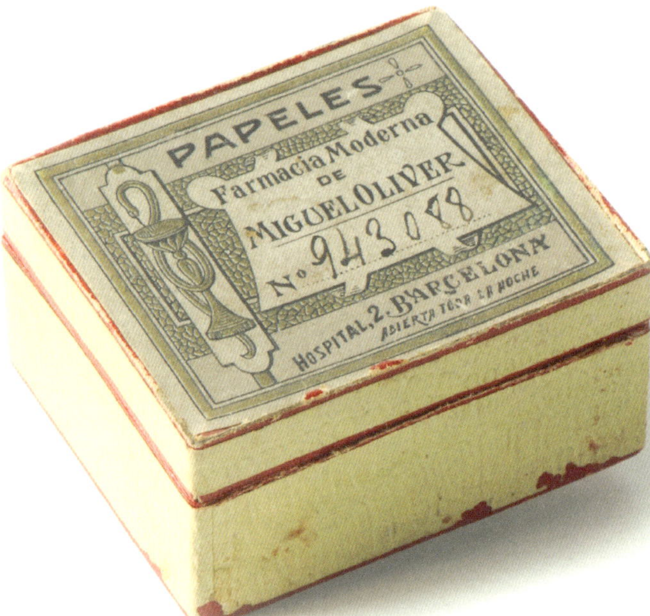
PAPELES
Farmacia Moderna
DE
MIGUEL OLIVER
No 943088
HOSPITAL, 2. BARCELONA
ABIERTA TODA LA NOCHE

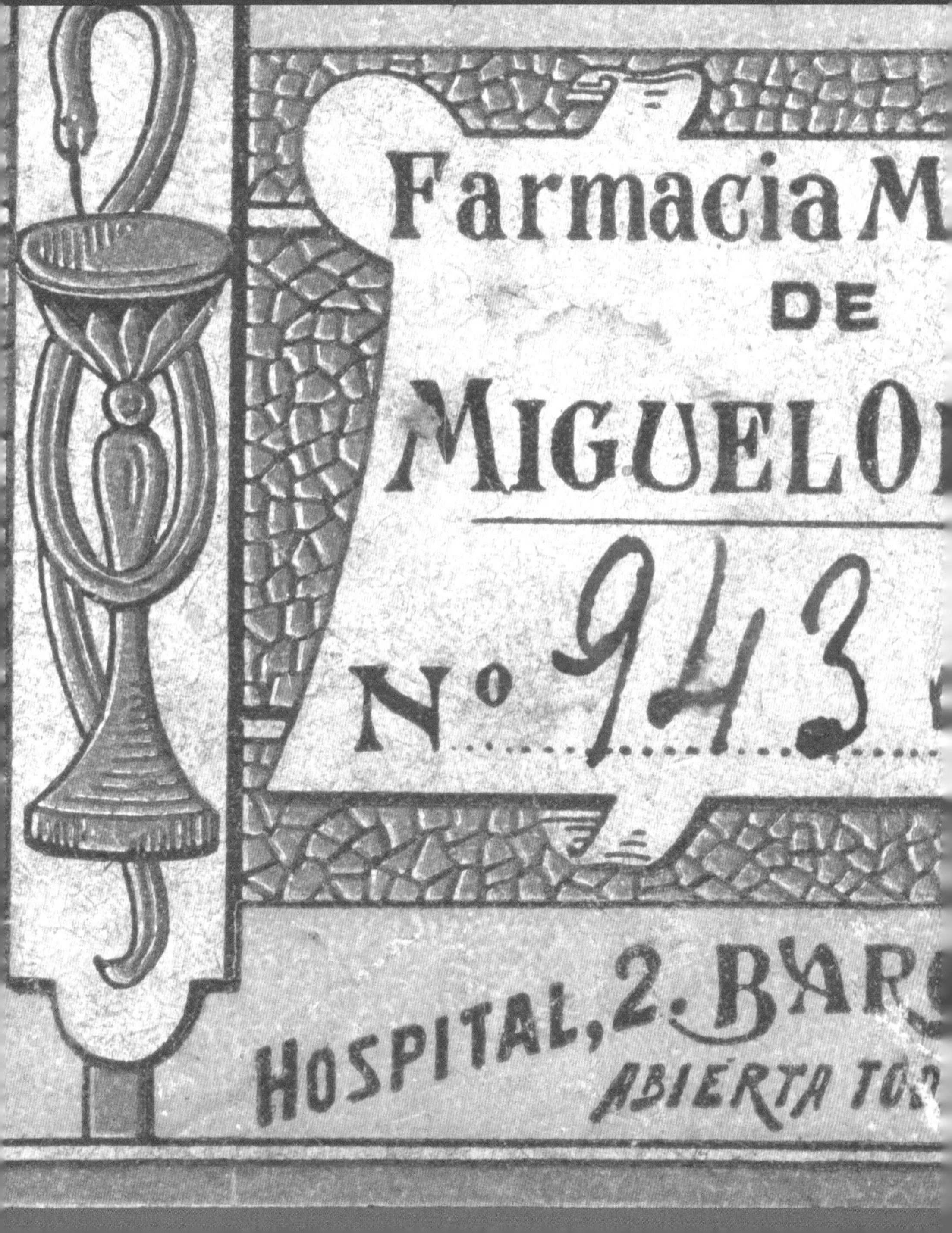
Farmacia M
DE
MIGUEL O
Nº 943
HOSPITAL, 2. BAR
ABIERTA TOD

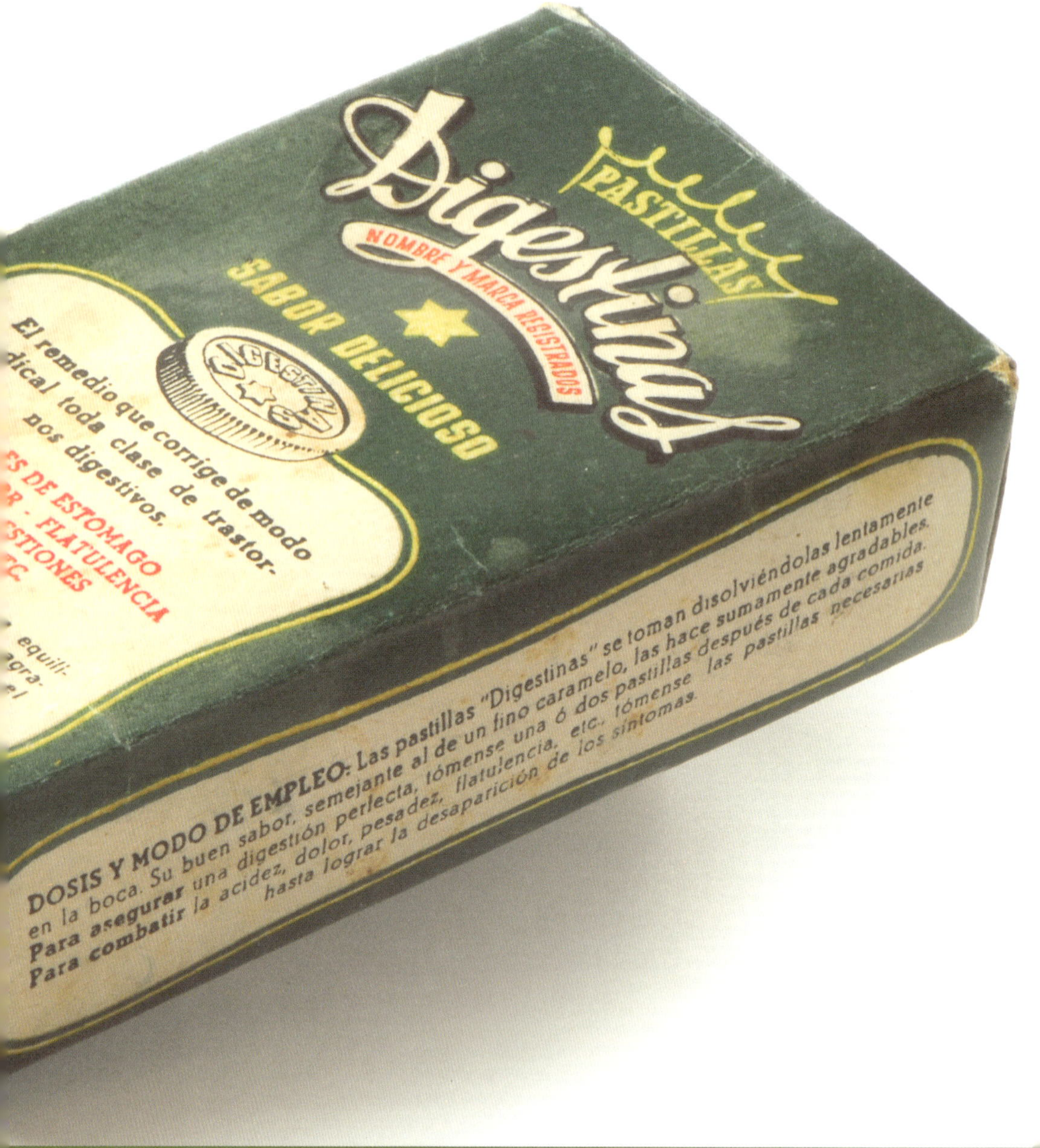
PASTILLAS
Digestinas
NOMBRE Y MARCA REGISTRADOS
SABOR DELICIOSO
El remedio que corrige de modo
toda clase de trastor-
nos digestivos.
DOSIS Y MODO DE EMPLEO: Las pastillas "Digestinas" se toman disolviéndolas lentamente
en la boca. Su buen sabor, semejante al de un fino caramelo, las hace sumamente agradables.
Para asegurar una digestión perfecta, tómense una ó dos pastillas después de cada comida.
Para combatir la acidez, dolor, pesadez, flatulencia, etc. tómense las pastillas necesarias
hasta lograr la desaparición de los síntomas.

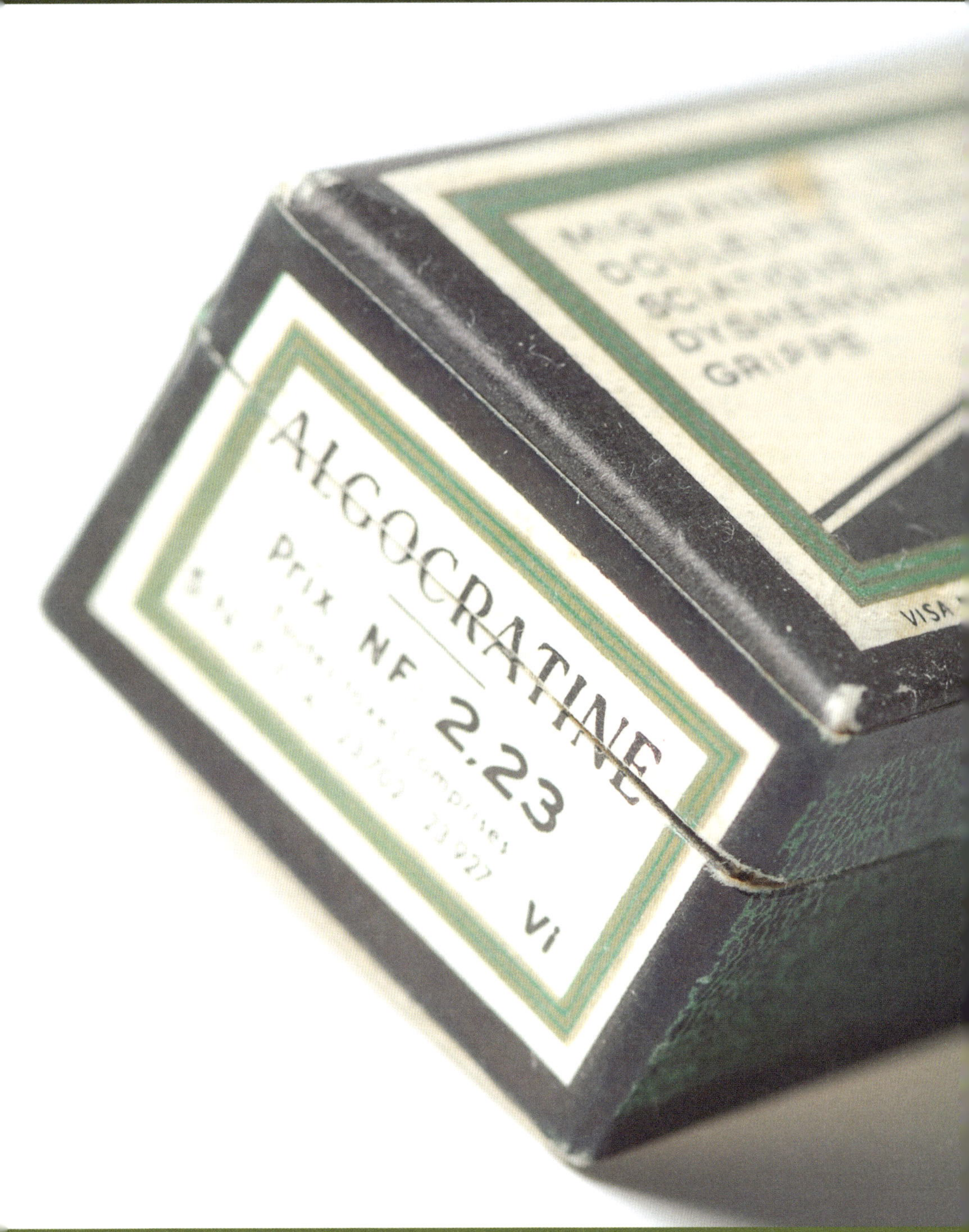
GRIPPE
VISA
ALGOCRATINE
Prix NF 2.23
VI

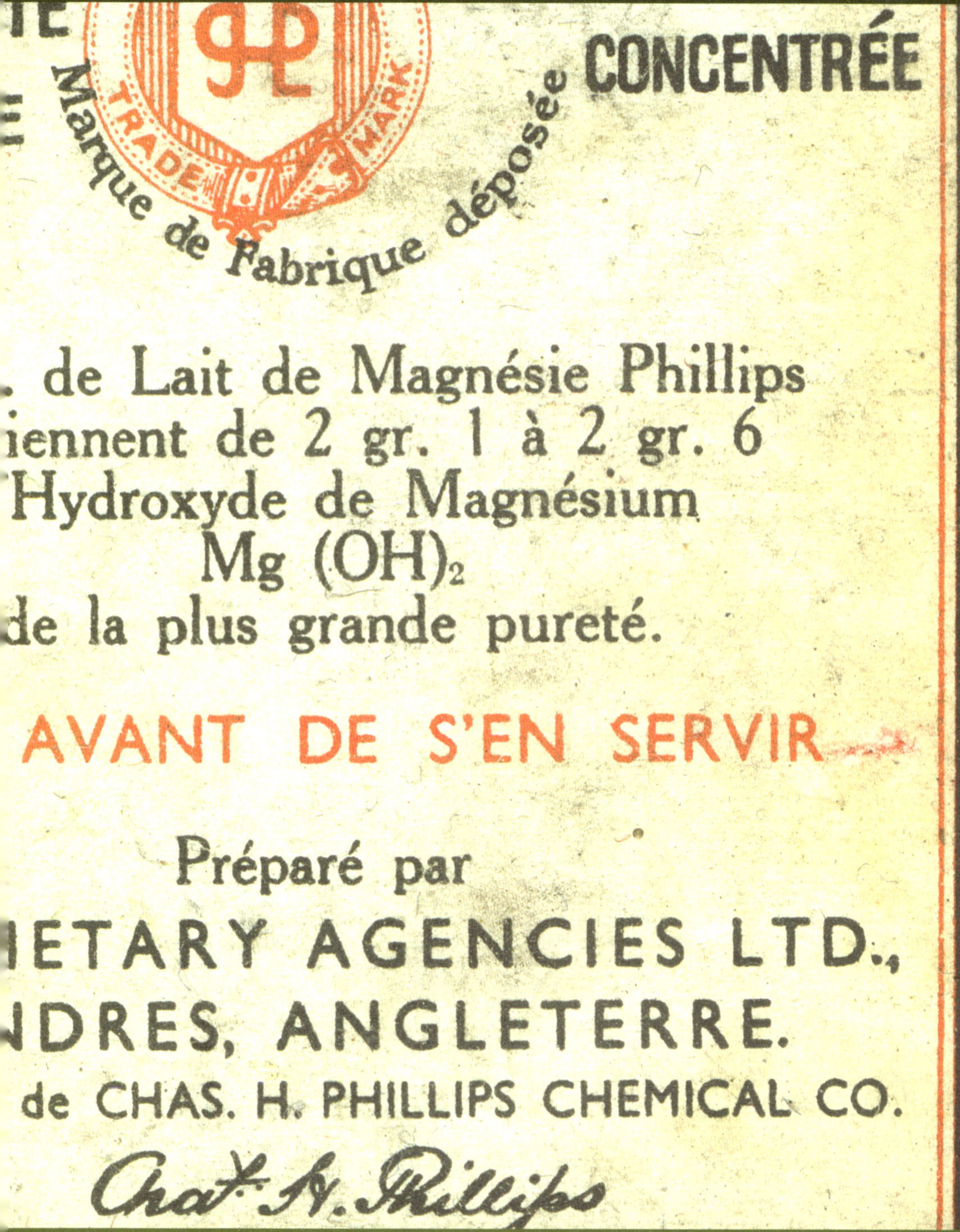
TRADE MARK
Marque de Fabrique déposée
CONCENTRÉE
. de Lait de Magnésie Phillips
iennent de 2 gr. 1 à 2 gr. 6
Hydroxyde de Magnésium
Mg (OH)2
de la plus grande pureté.
AVANT DE S'EN SERVIR
Préparé par
IETARY AGENCIES LTD.,
IDRES, ANGLETERRE.
de CHAS. H. PHILLIPS CHEMICAL CO.

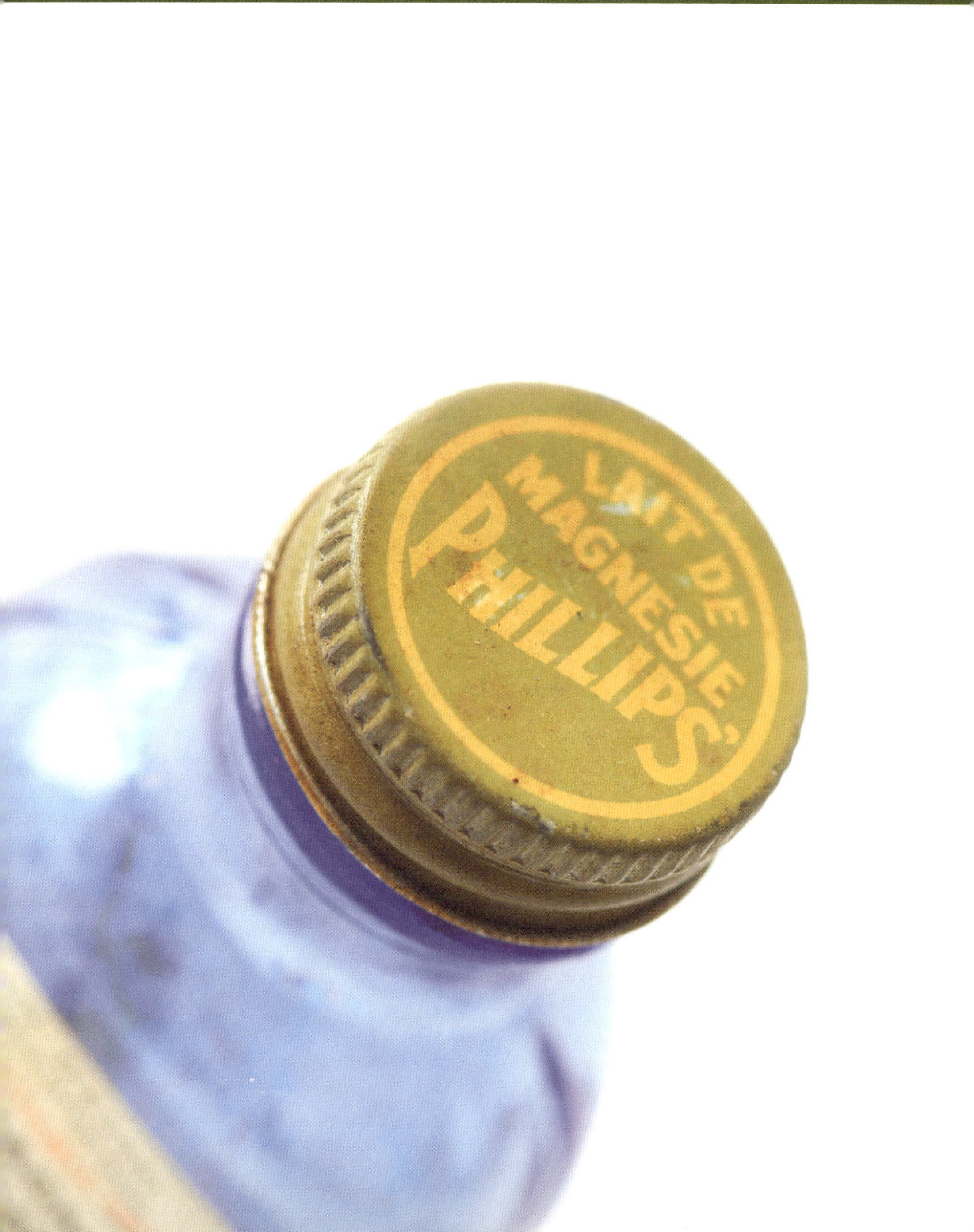
LAIT DE
MAGNESIE
PHILLIP'S

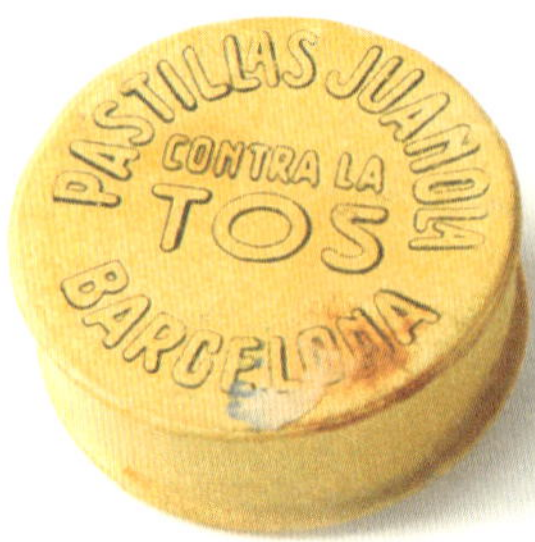
PASTILLAS JUANOLA
CONTRA LA
TOS
BARCELONA

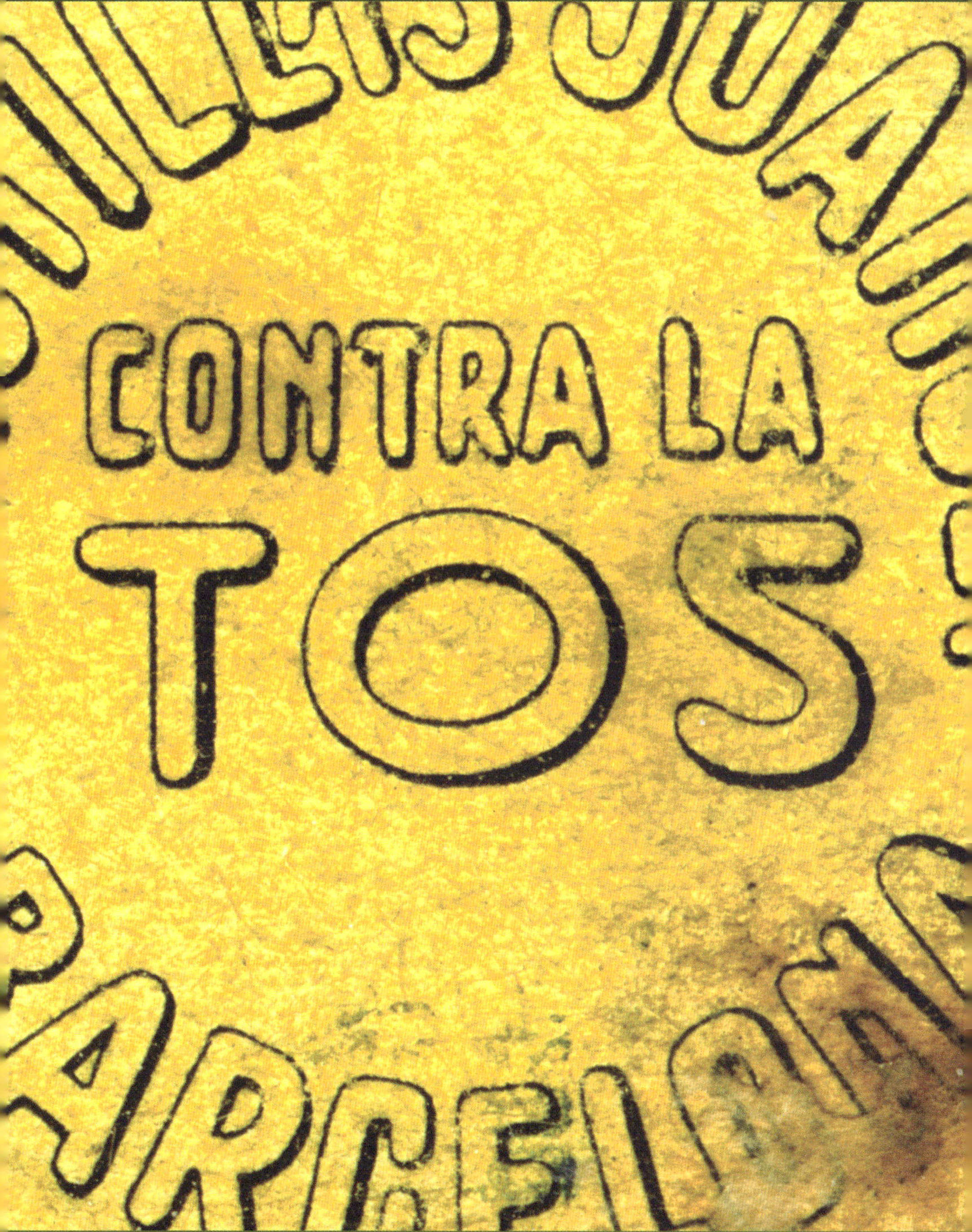
CONTRA LA
TOS

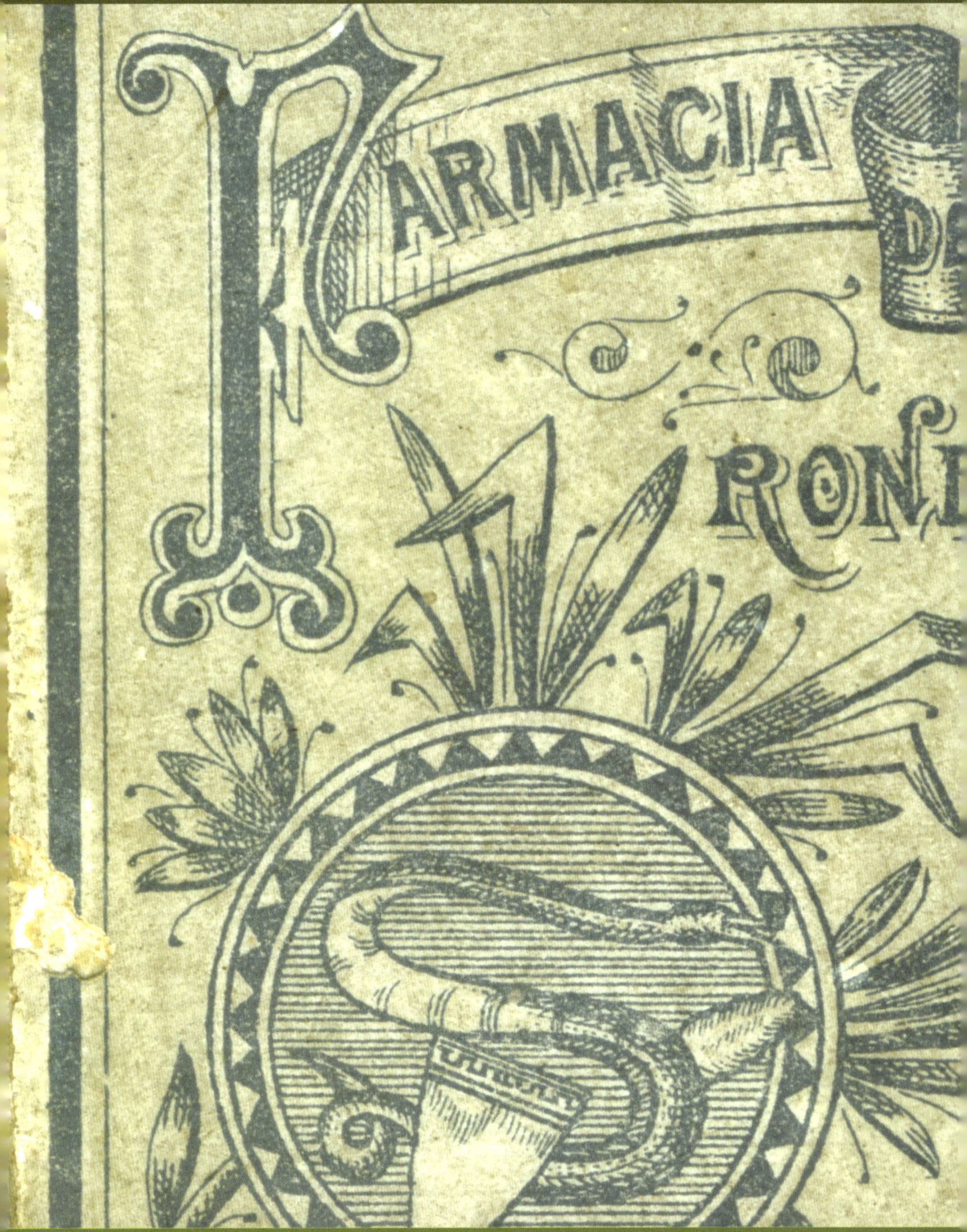
FARMACIA
DE
RONI

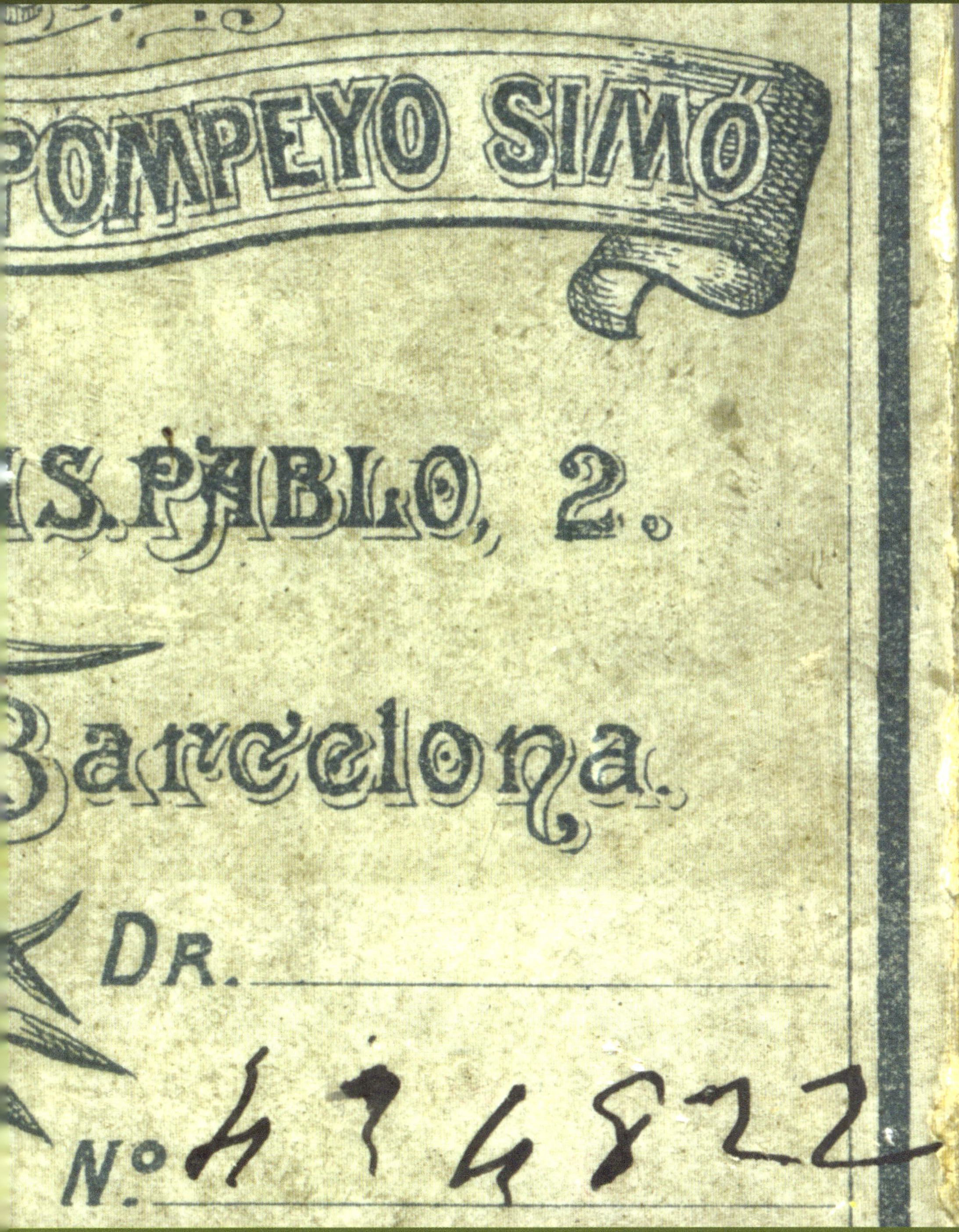
POMPEYO SIMÓ
S. PABLO, 2.
Barcelona.
DR.
N.° 414822

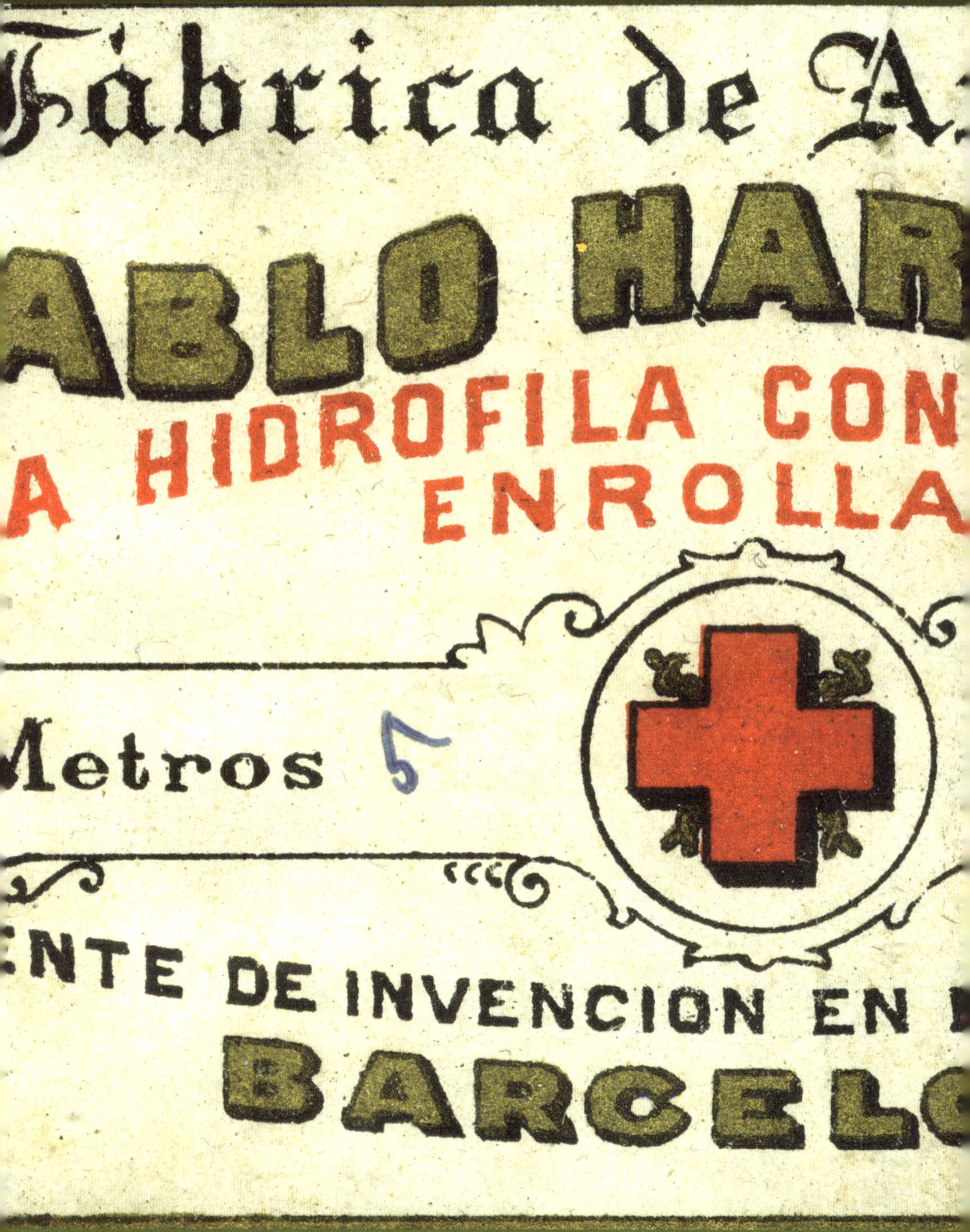
Fábrica de A
ABLO HAR
A HIDROFILA CON
ENROLLA
Metros 5
NTE DE INVENCION EN
BARCEL

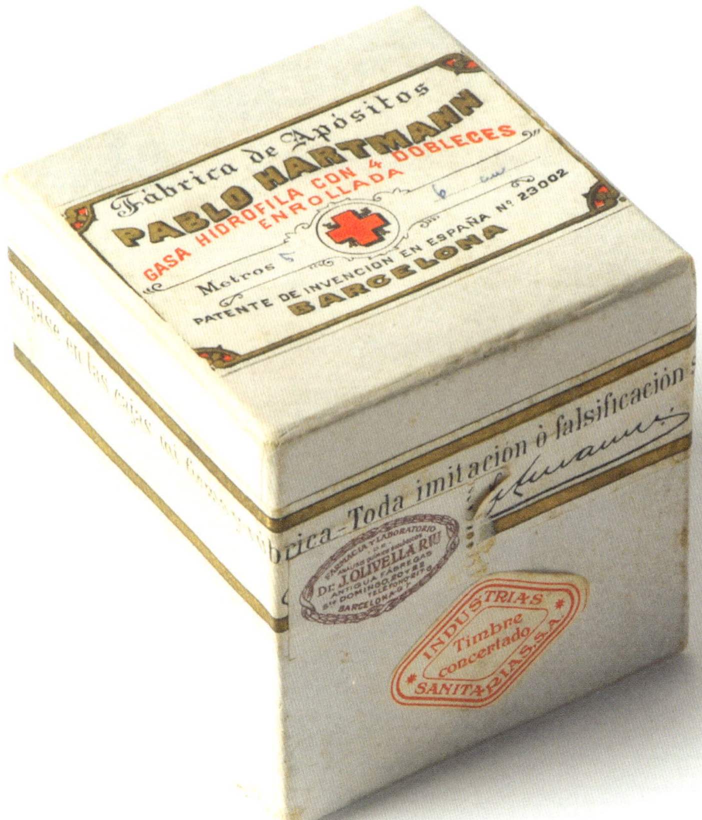
Fábrica de Apósitos
PABLO HARTMANN
GASA HIDROFILA CON 4 DOBLECES
ENROLLADA
Metros
PATENTE DE INVENCION EN ESPAÑA Nº 23002
BARCELONA
brica-Toda imitación ò falsificación
DR. J. OLIVELLA RIU
INDUSTRIAS
Timbre
concertado
SANITARIAS S.A.

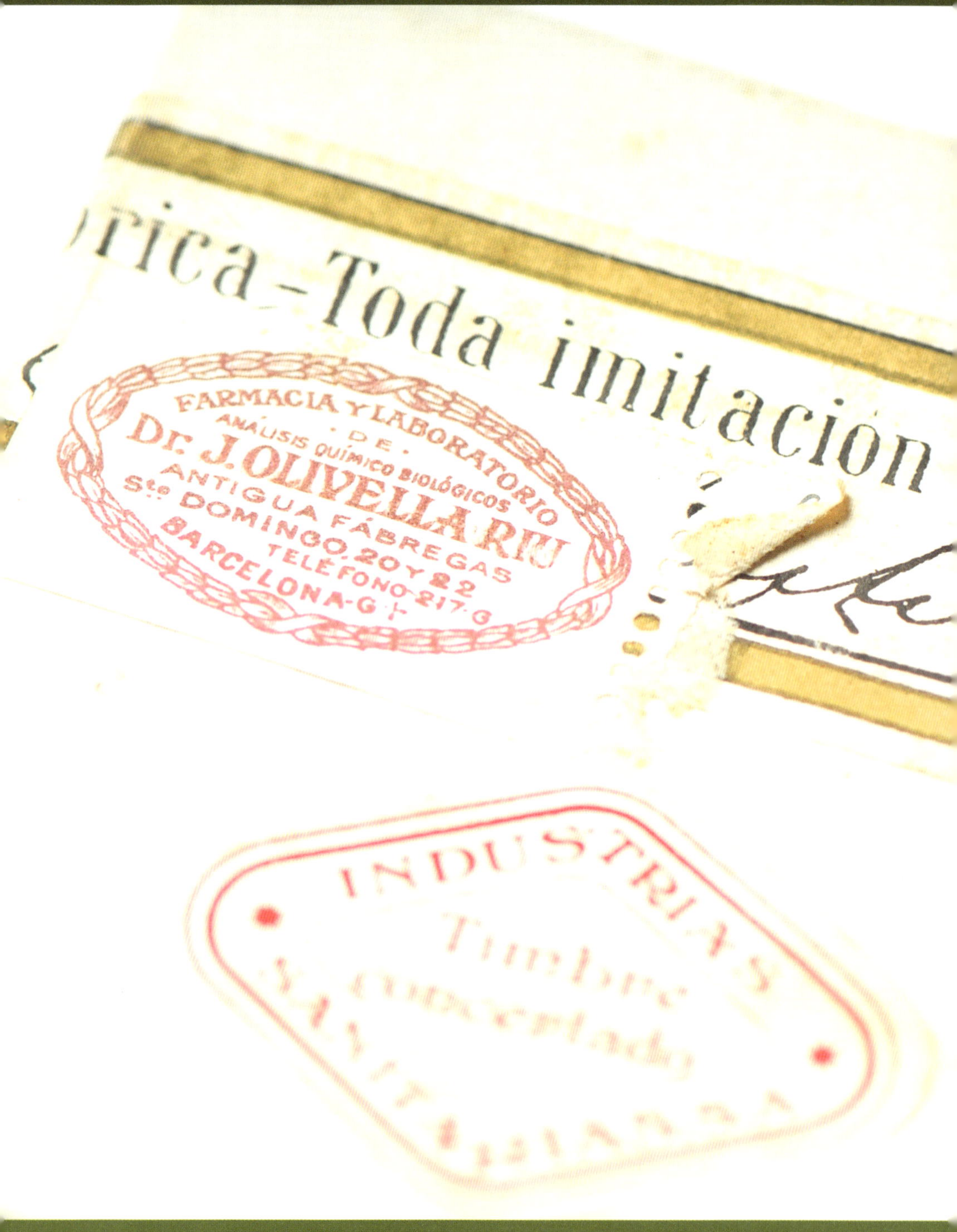
rica-Toda imitación
FARMACIA Y LABORATORIO
ANÁLISIS QUÍMICO BIOLÓGICOS
- DE -
Dr. J. OLIVELLA RIU
ANTIGUA FÁBREGAS
Sto DOMINGO, 20 Y 22
TELÉFONO 217 G
BARCELONA-G
INDUSTRIAS

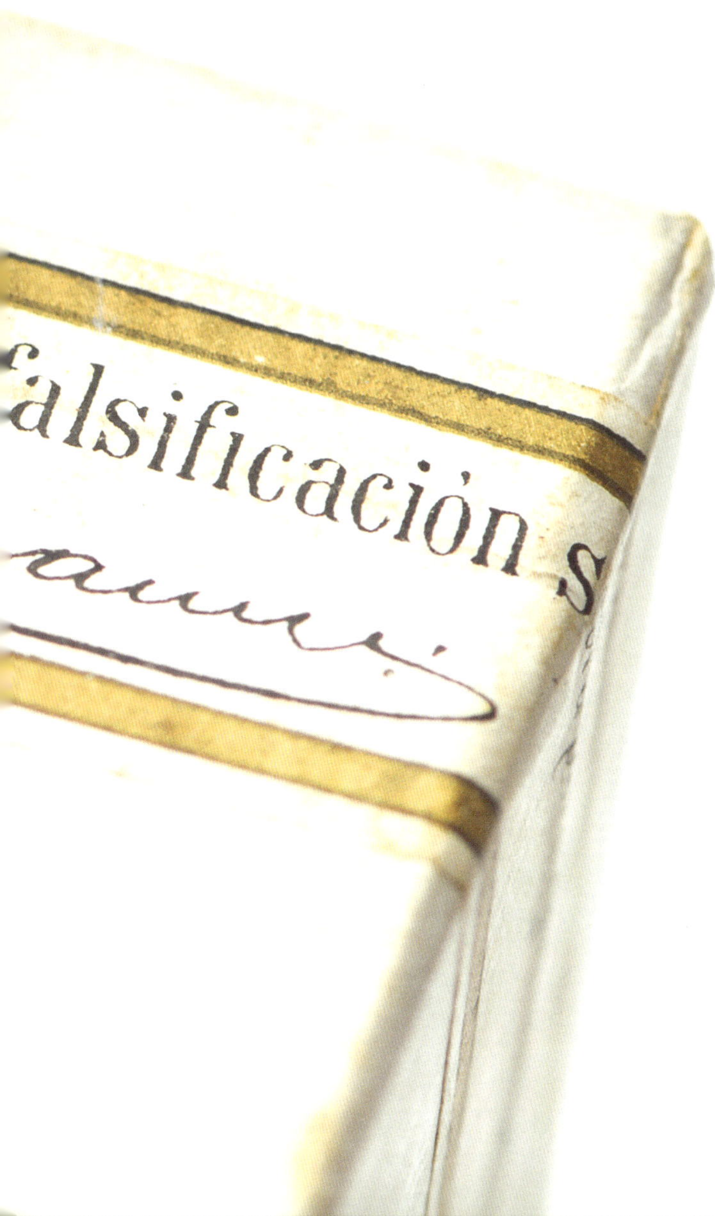
alsificación

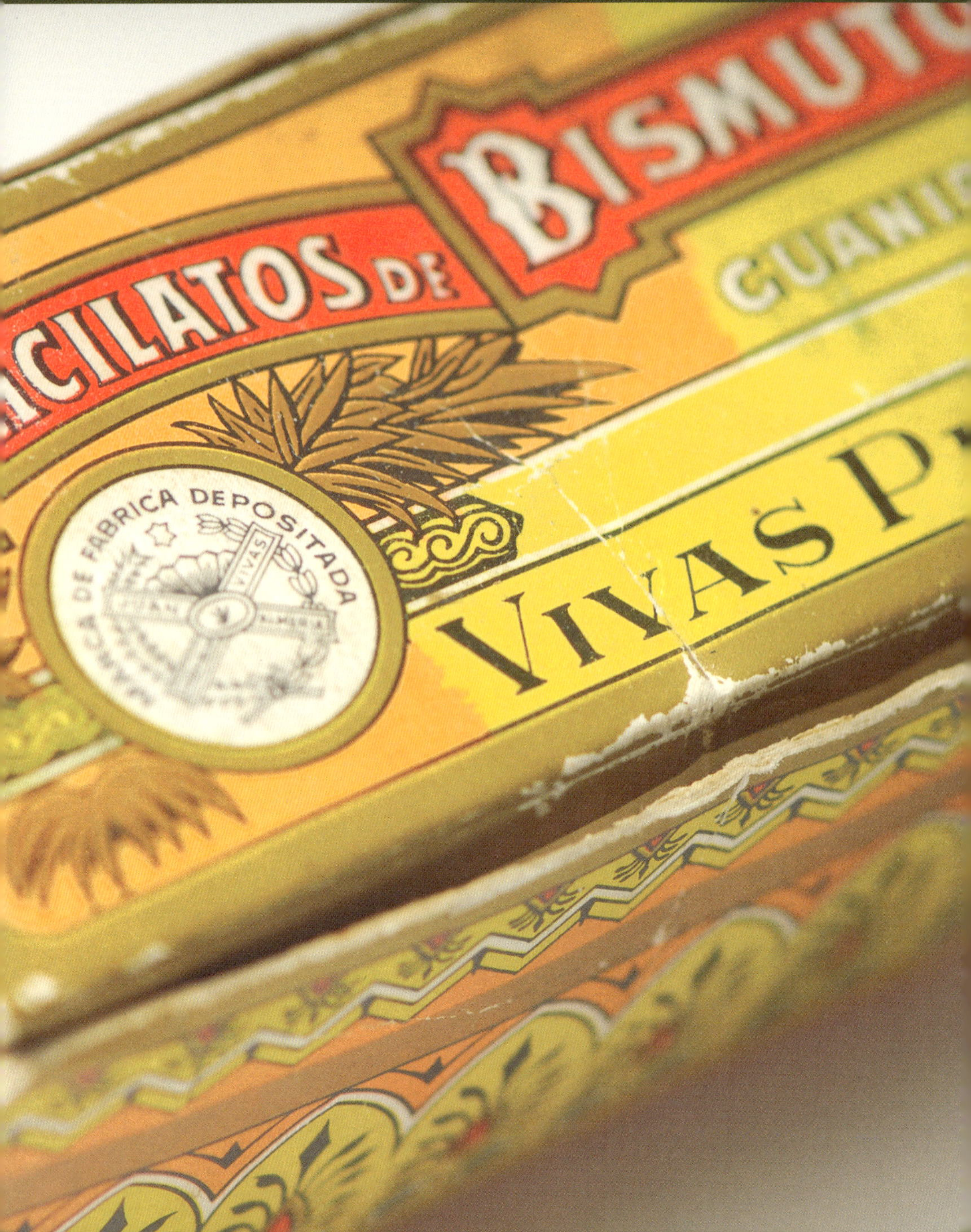
MARCA DE FABRICA DEPOSITADA
VIVAS

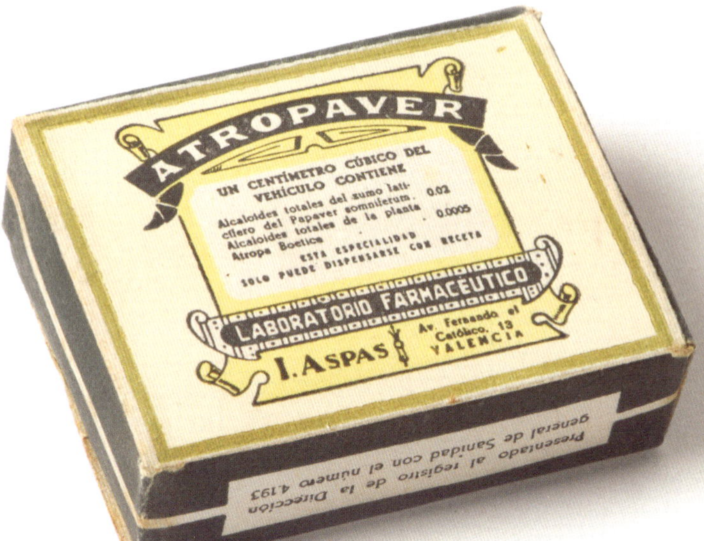
ATROPAVER
UN CENTÍMETRO CÚBICO DEL VEHÍCULO CONTIENE
Alcaloides totales del zumo lati-cífero del Papaver somniferum. 0.02
Alcaloides totales de la planta Atropa Boetica 0.0005
ESTA ESPECIALIDAD SOLO PUEDE DISPENSARSE CON RECETA
LABORATORIO FARMACEUTICO
I. ASPAS
Av. Fernando el Católico, 13 VALENCIA
Presentado al registro de la Dirección general de Sanidad con el número 4.193

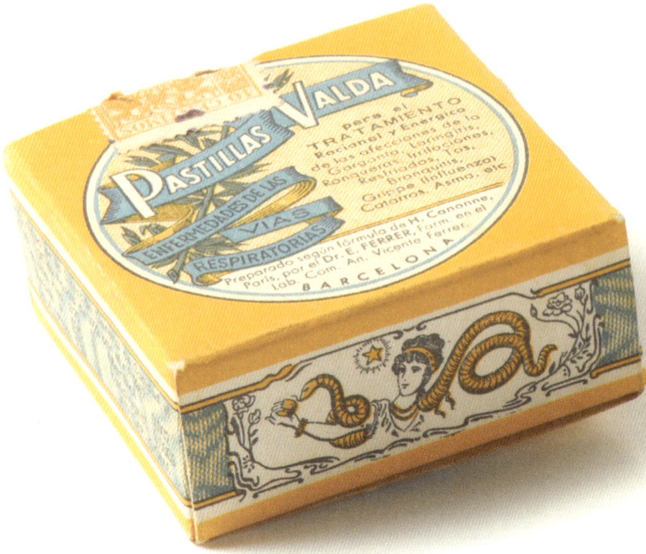
PASTILLAS VALDA
para el
TRATAMIENTO
Racional y Energico
de las afecciones de la
Garganta, Laringitis,
Ronqueras, Irritaciones,
Resfriados, Tos,
Bronquitis,
Grippe (Influenza)
Catarros, Asma, etc
ENFERMEDADES DE LAS
VIAS
RESPIRATORIAS
Preparado según fórmula de H. Canonne,
Paris, por el Dr. E. FERRER, Farm. en el
Lab. Com. An. Vicente Ferrer,
BARCELONA

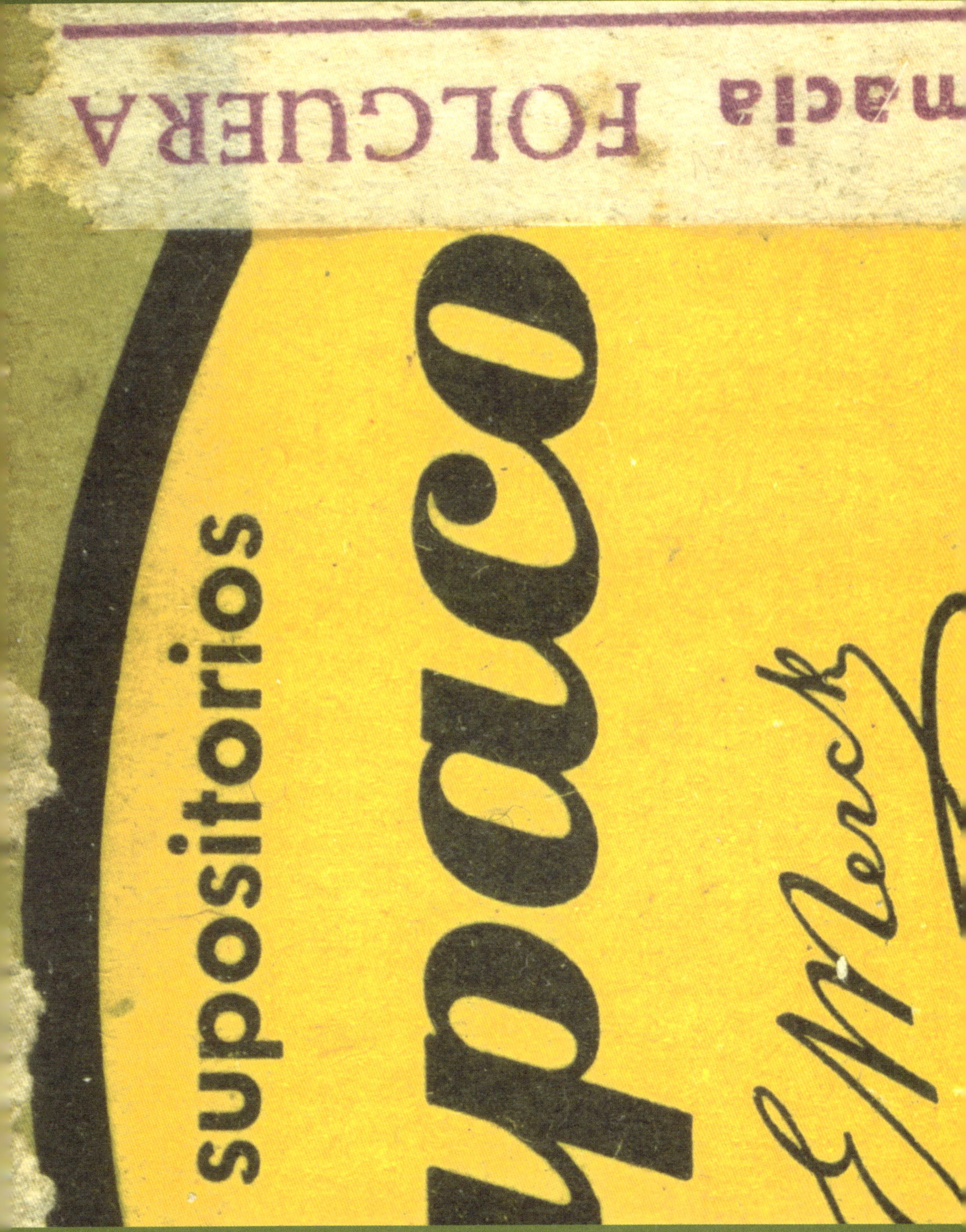
FOLGUERA
supositorios
Merck

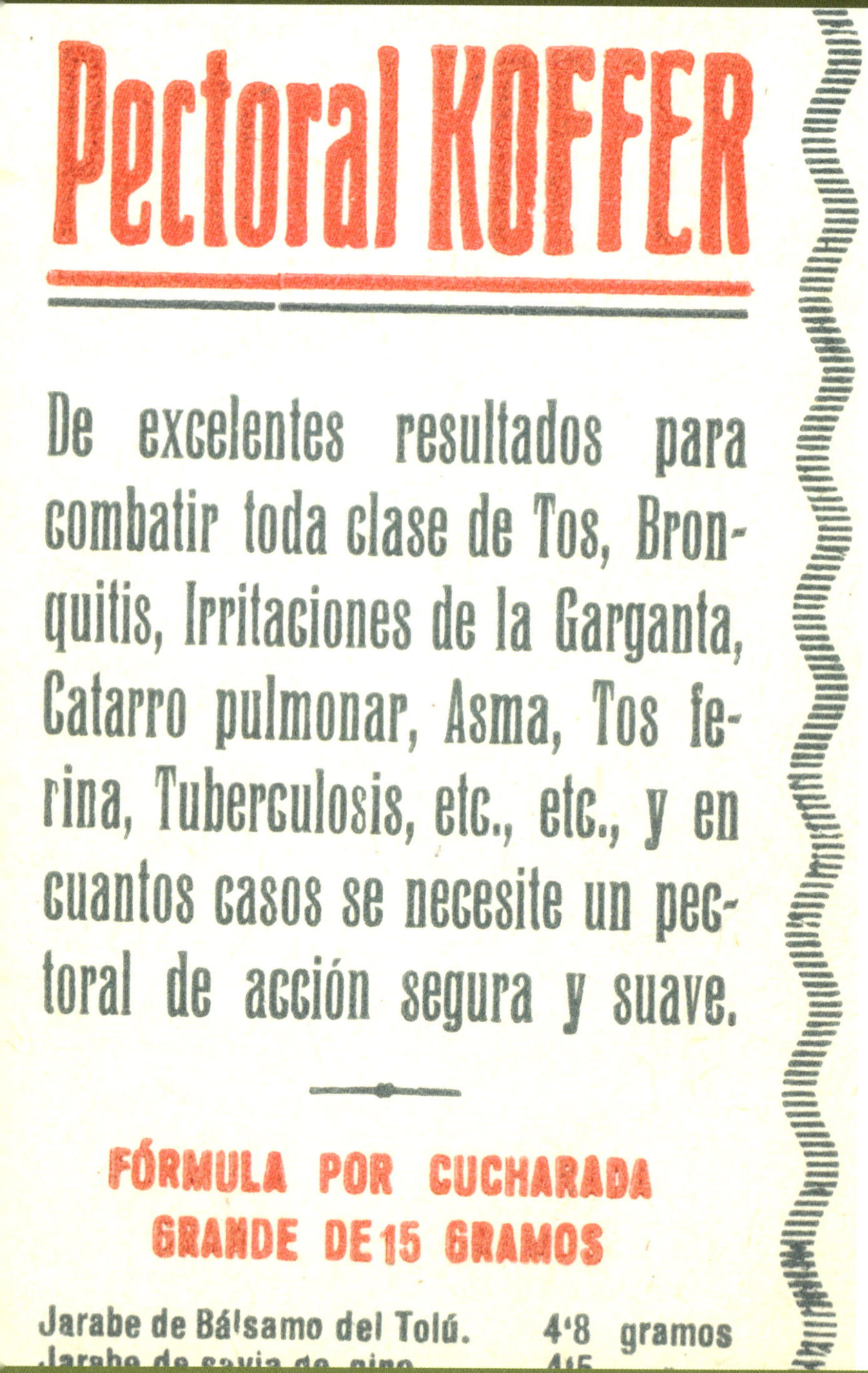

Pectoral KOFFER

De excelentes resultados para combatir toda clase de Tos, Bronquitis, Irritaciones de la Garganta, Catarro pulmonar, Asma, Tos ferina, Tuberculosis, etc., etc., y en cuantos casos se necesite un pectoral de acción segura y suave.

FÓRMULA POR CUCHARADA GRANDE DE 15 GRAMOS

Jarabe de Bálsamo del Tolú. 4'8 gramos

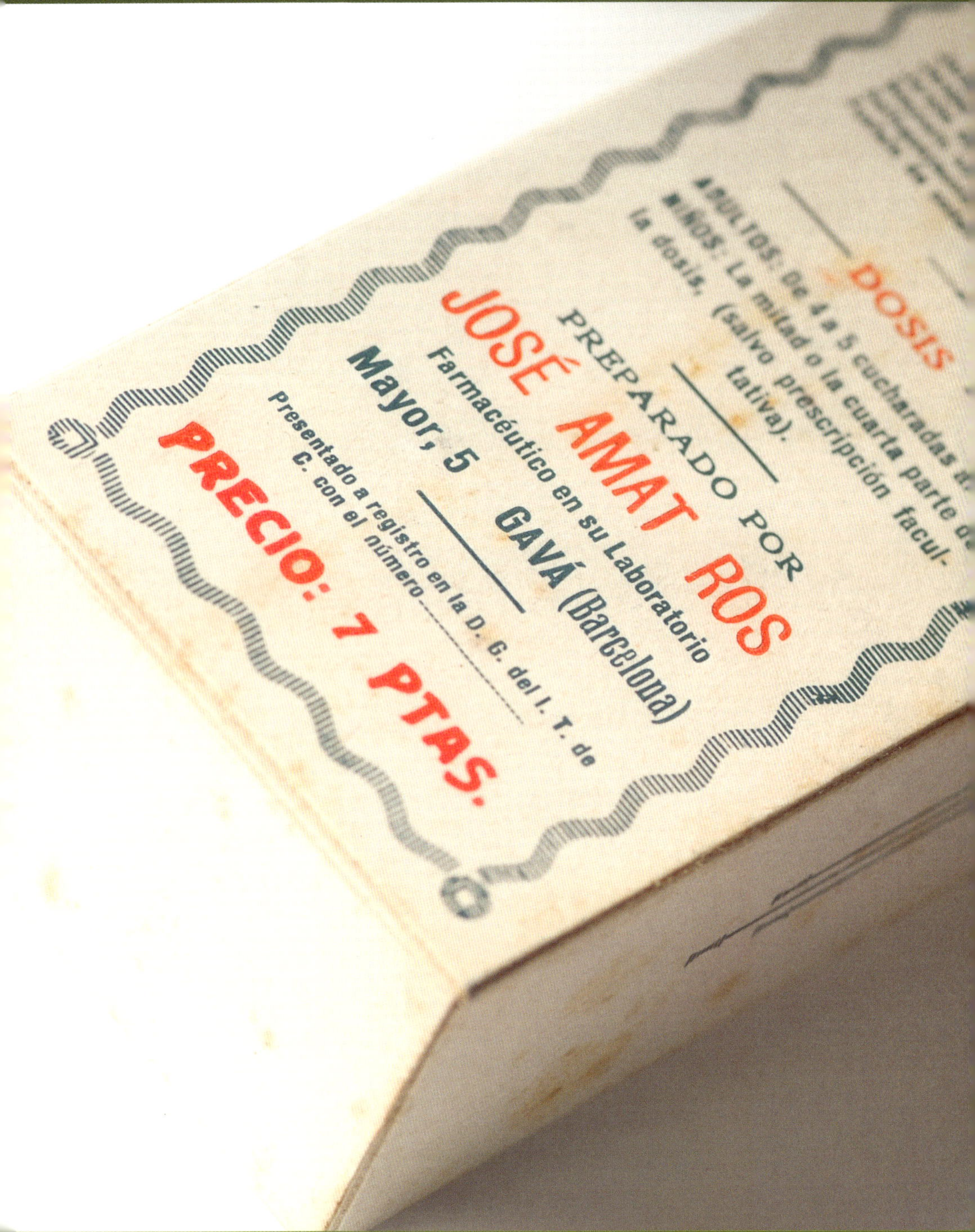
DOSIS
(salvo prescripción facul-
tativa).
PREPARADO POR
JOSÉ AMAT ROS
Farmacéutico en su Laboratorio
Mayor, 5 GAVÁ (Barcelona)
Presentado a registro en la D. G. del I. T. de
C. con el número
PRECIO: 7 PTAS.

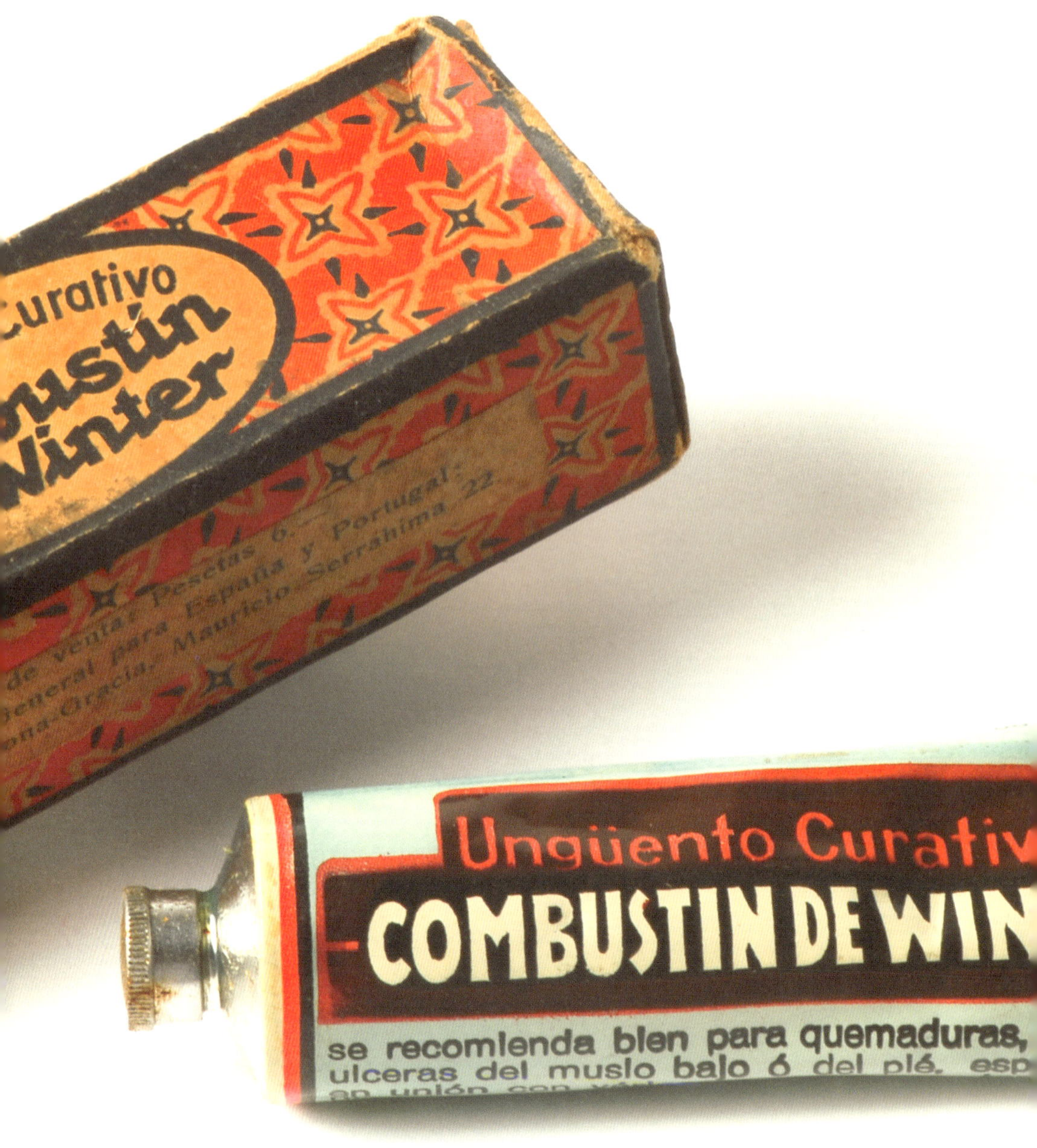
Curativo
ustin
Winter
de venta: Pesetas 6.
General para España y Portugal:
Mauricio Serrahima 22
Ungüento Curativ
COMBUSTIN DE WIN
se recomienda bien para quemaduras,
ulceras del muslo bajo ó del pié. esp

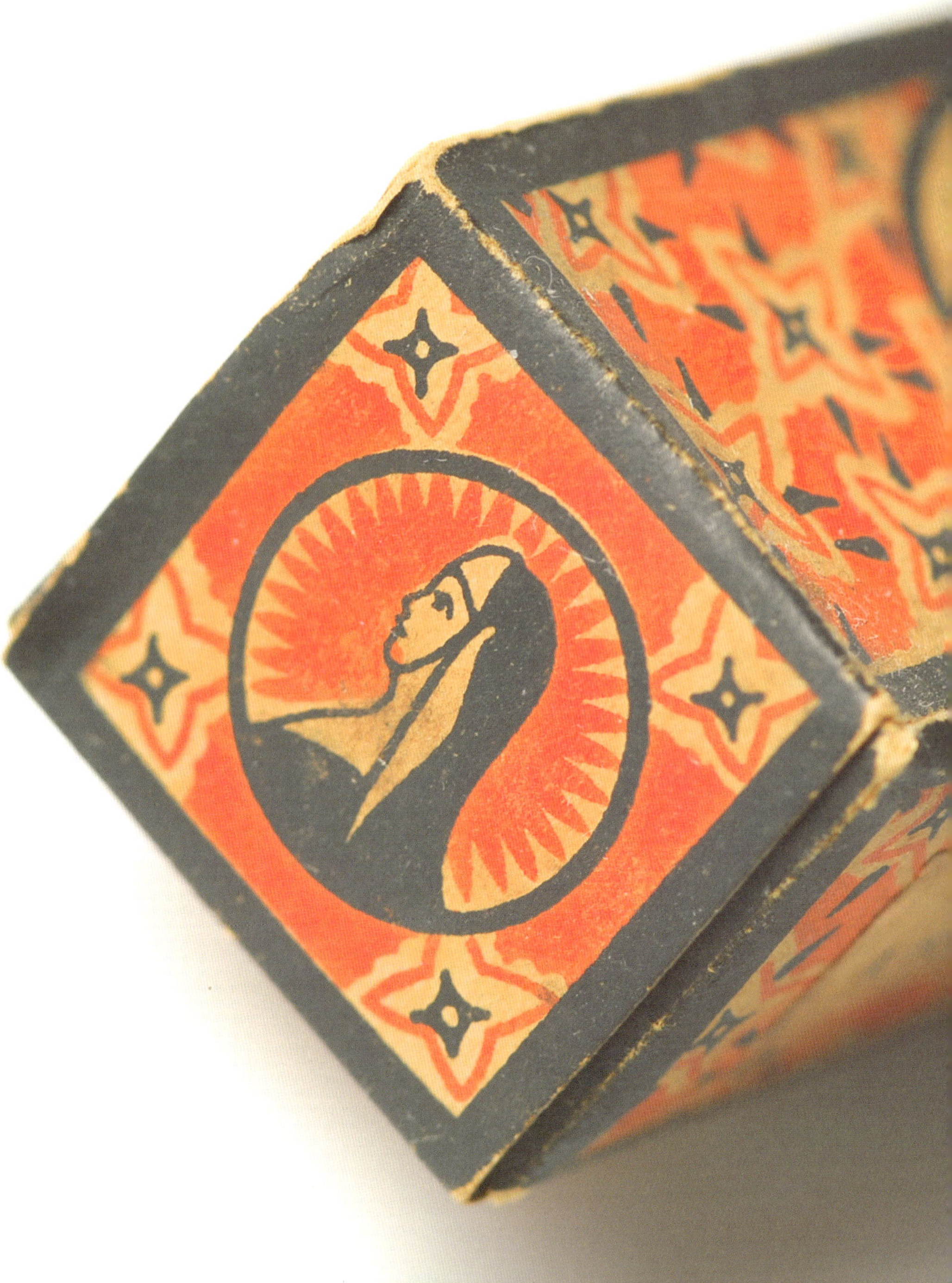

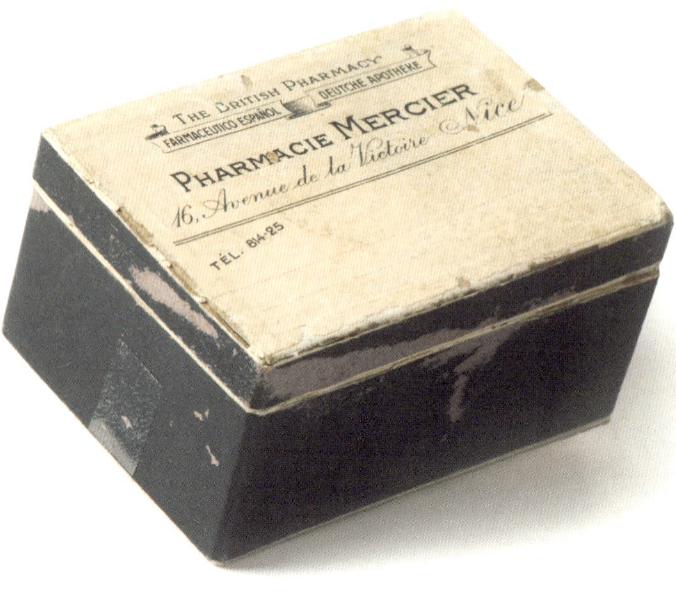
THE BRITISH PHARMACY
FARMACEUTICO ESPAÑOL
DEUTCHE APOTHEKE
PHARMACIE MERCIER
16, Avenue de la Victoire Nice
TÉL. 814-25

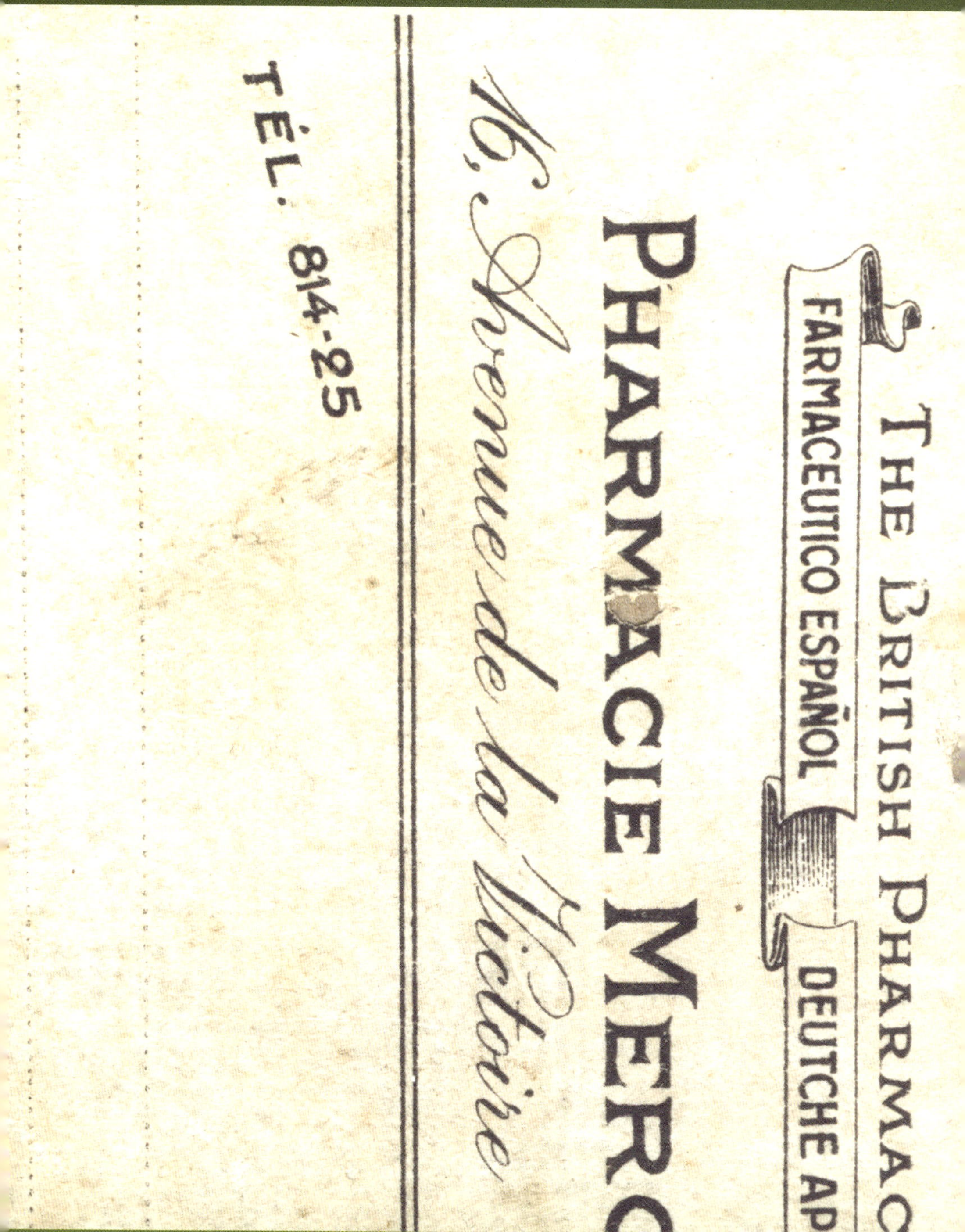
THE BRITISH PHARMAC
FARMACEUTICO ESPAÑOL
DEUTCHE AP
PHARMACIE MERC
16, Avenue de la Victoire
TÉL. 814-25

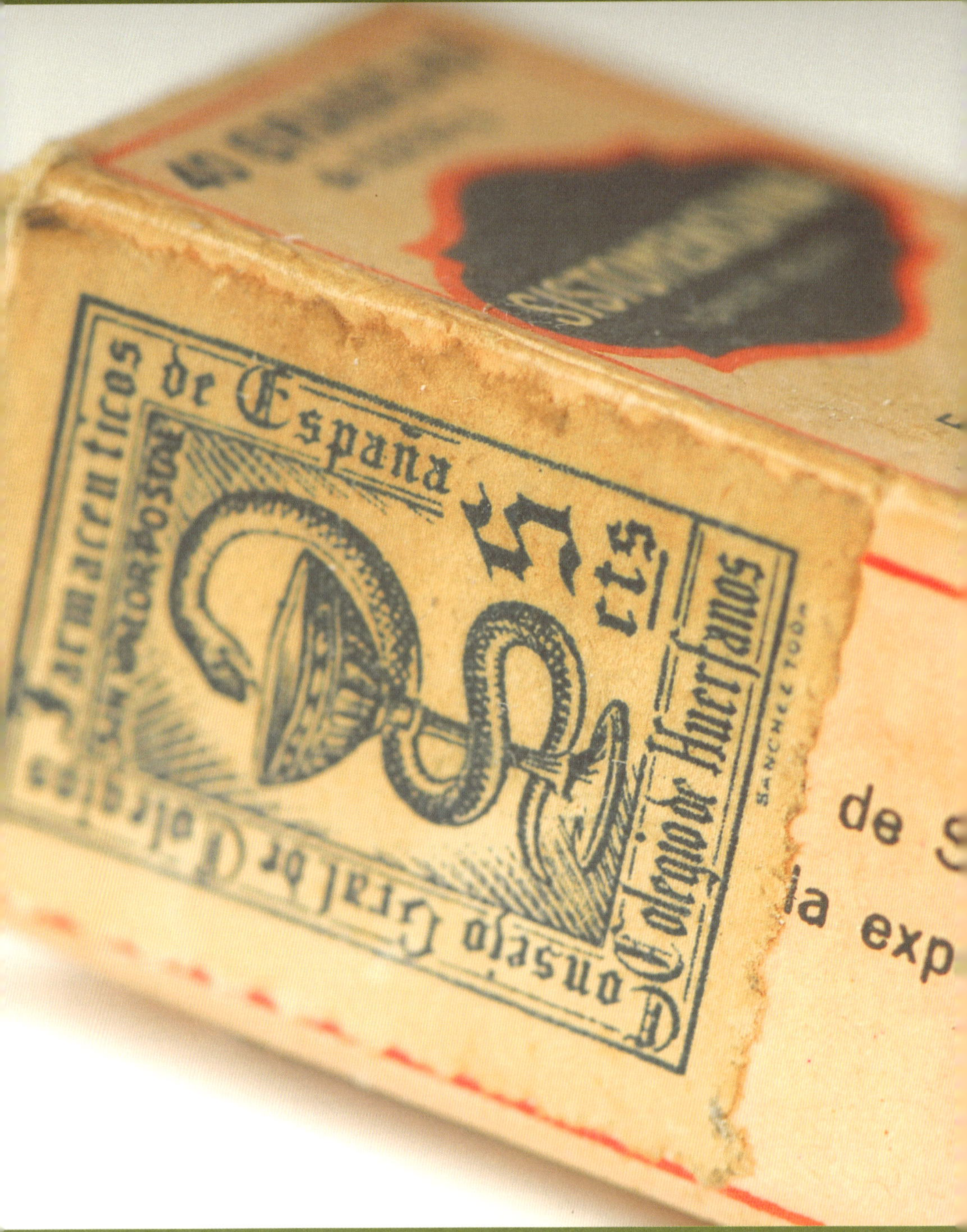
Farmacéuticos de España
5 cts
Colegio de Huérfanos
Consejo Gral de

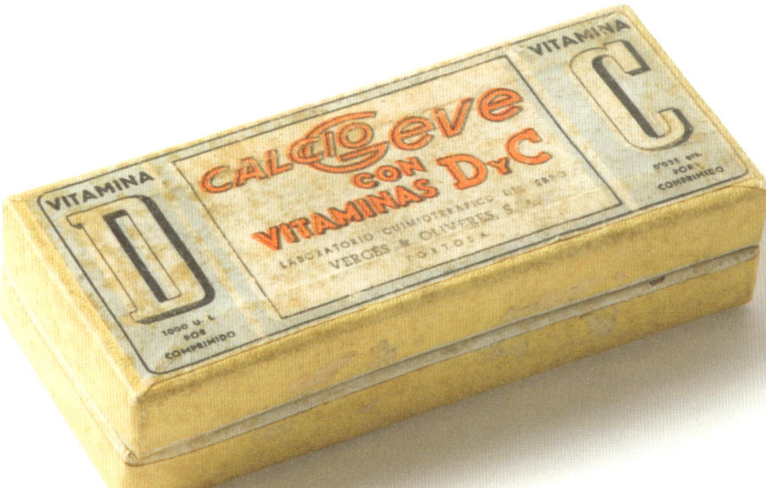
VITAMINA
D
1000 U. I.
POR
COMPRIMIDO
CALCIOCEVE
CON
VITAMINAS DyC
VERGÉS
TORTOSA
VITAMINA
C
POR
COMPRIMIDO

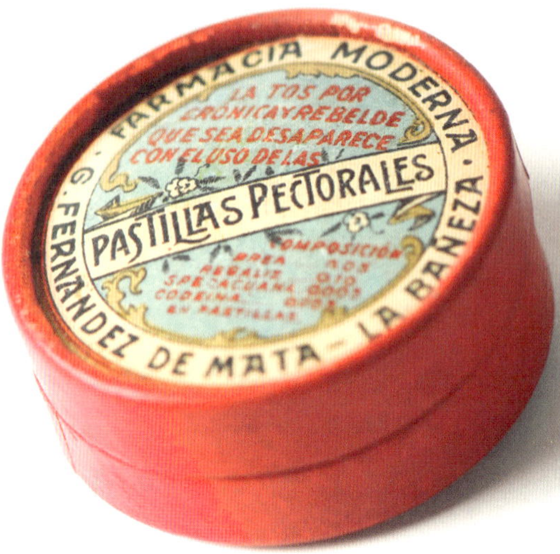
FARMACIA MODERNA
LA TOS POR
CRÓNICA Y REBELDE
QUE SEA DESAPARECE
CON EL USO DE LAS
PASTILLAS PECTORALES
COMPOSICIÓN
G. FERNANDEZ DE MATA - LA BAÑEZA

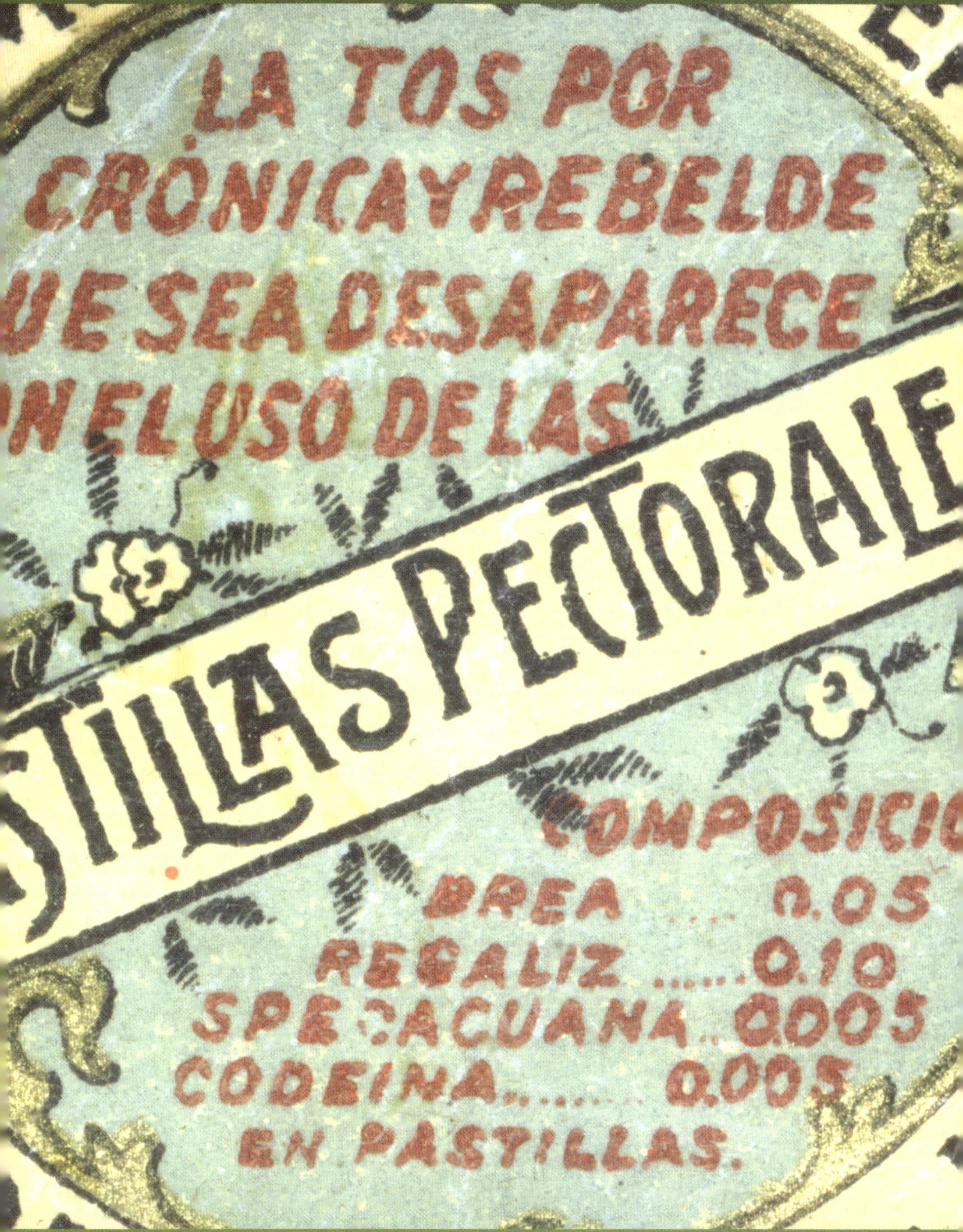
LA TOS POR
CRÓNICA Y REBELDE
UE SEA DESAPARECE
N EL USO DE LAS
STILLAS PECTORALE
COMPOSICIO
BREA 0.05
REGALIZ 0.10
SPECACUANA .. 0.005
CODEINA 0.005
EN PASTILLAS.

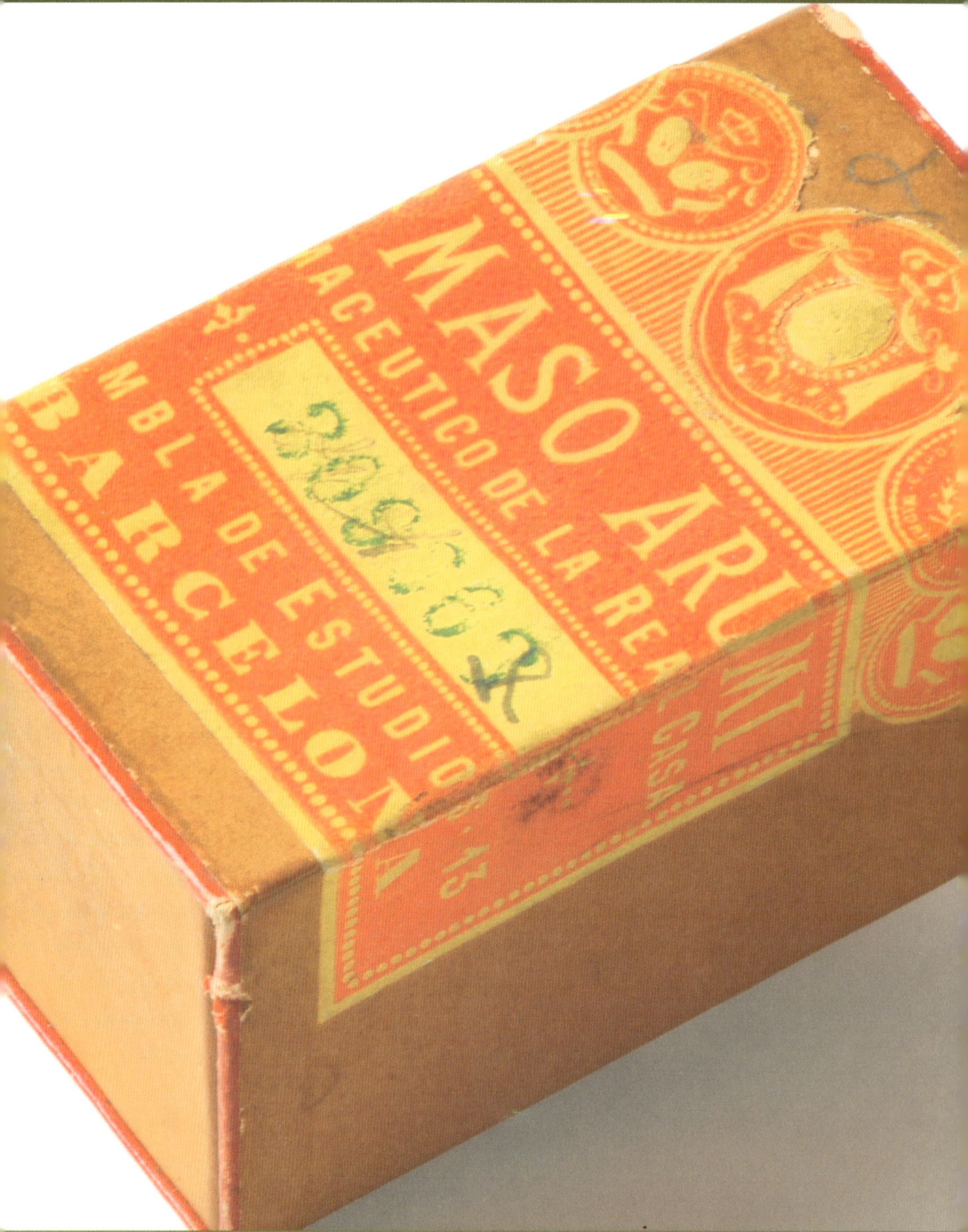

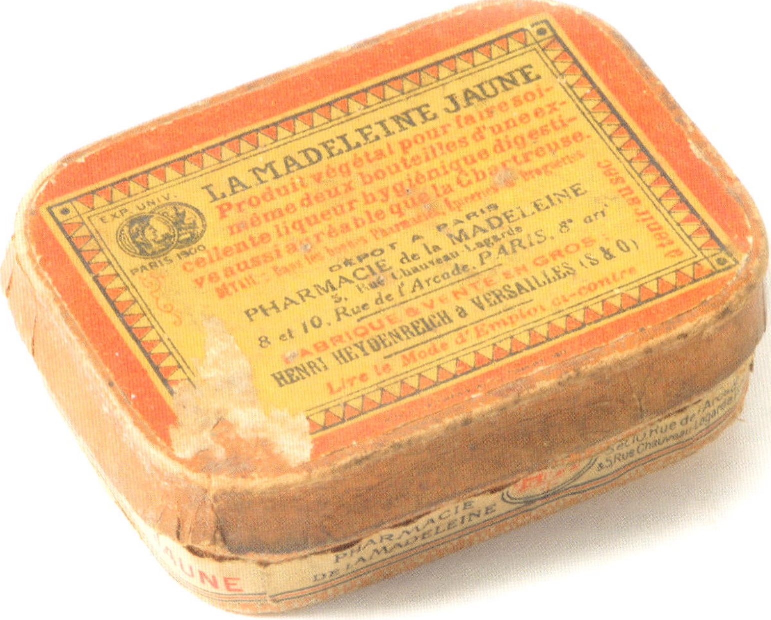
EXP. UNIV.
PARIS 1900
LA MADELEINE JAUNE
Produit végétal pour faire soi-
même deux bouteilles d'une ex-
cellente liqueur hygiénique digesti-
ve aussi agréable que la Chartreuse.
DÉPOT A PARIS
PHARMACIE de la MADELEINE
3, Rue Chauveau-Lagarde
8 et 10, Rue de l'Arcade, PARIS, 8e arr.
FABRIQUE & VENTE EN GROS
HENRI HEYDENREICH à VERSAILLES (S & O)
Lire le Mode d'Emploi ci-contre
PHARMACIE
DE LA MADELEINE
8 et 10, Rue de l'Arcade
& 3 Rue Chauveau-Lagarde
JAUNE

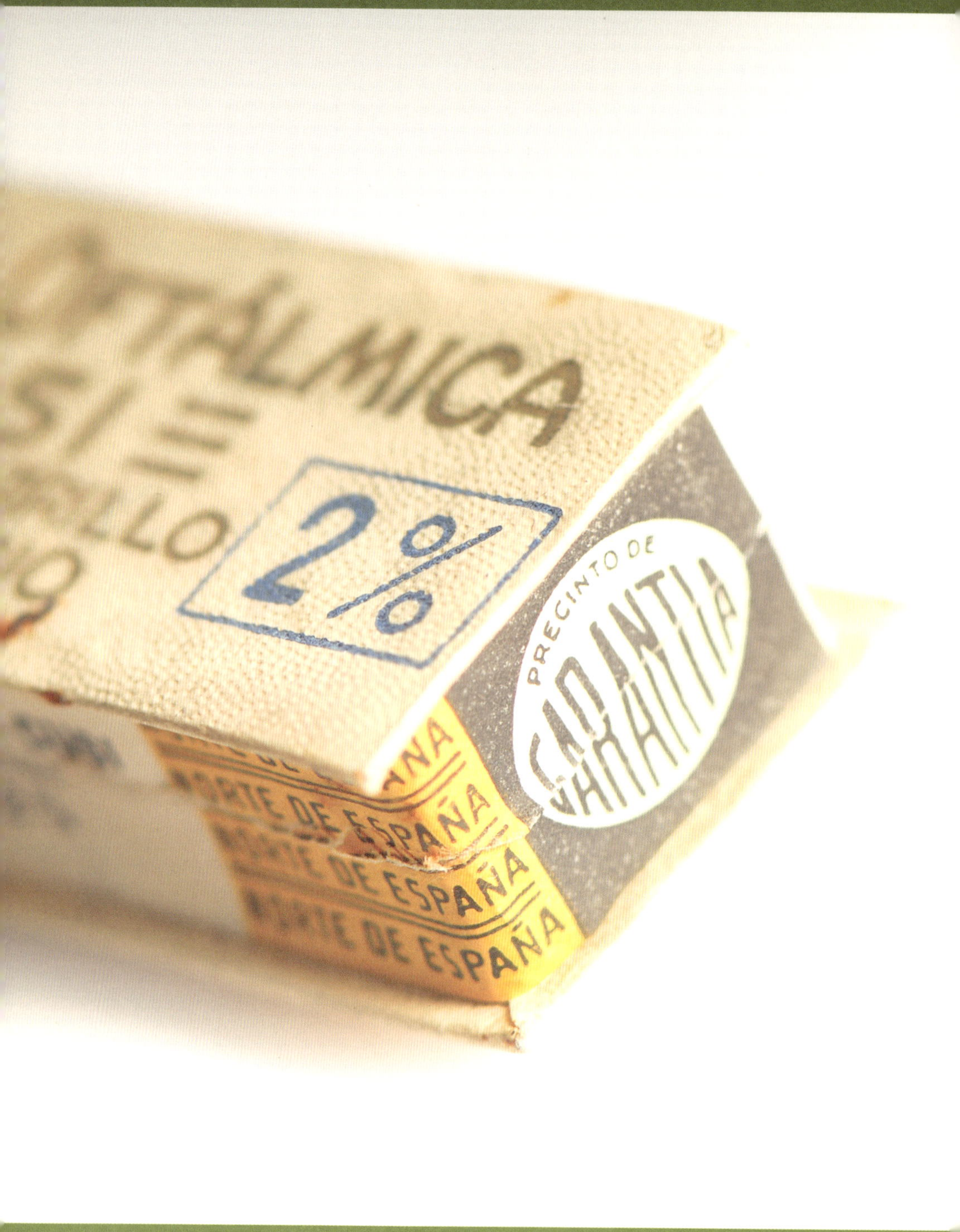
2%
PRECINTO DE

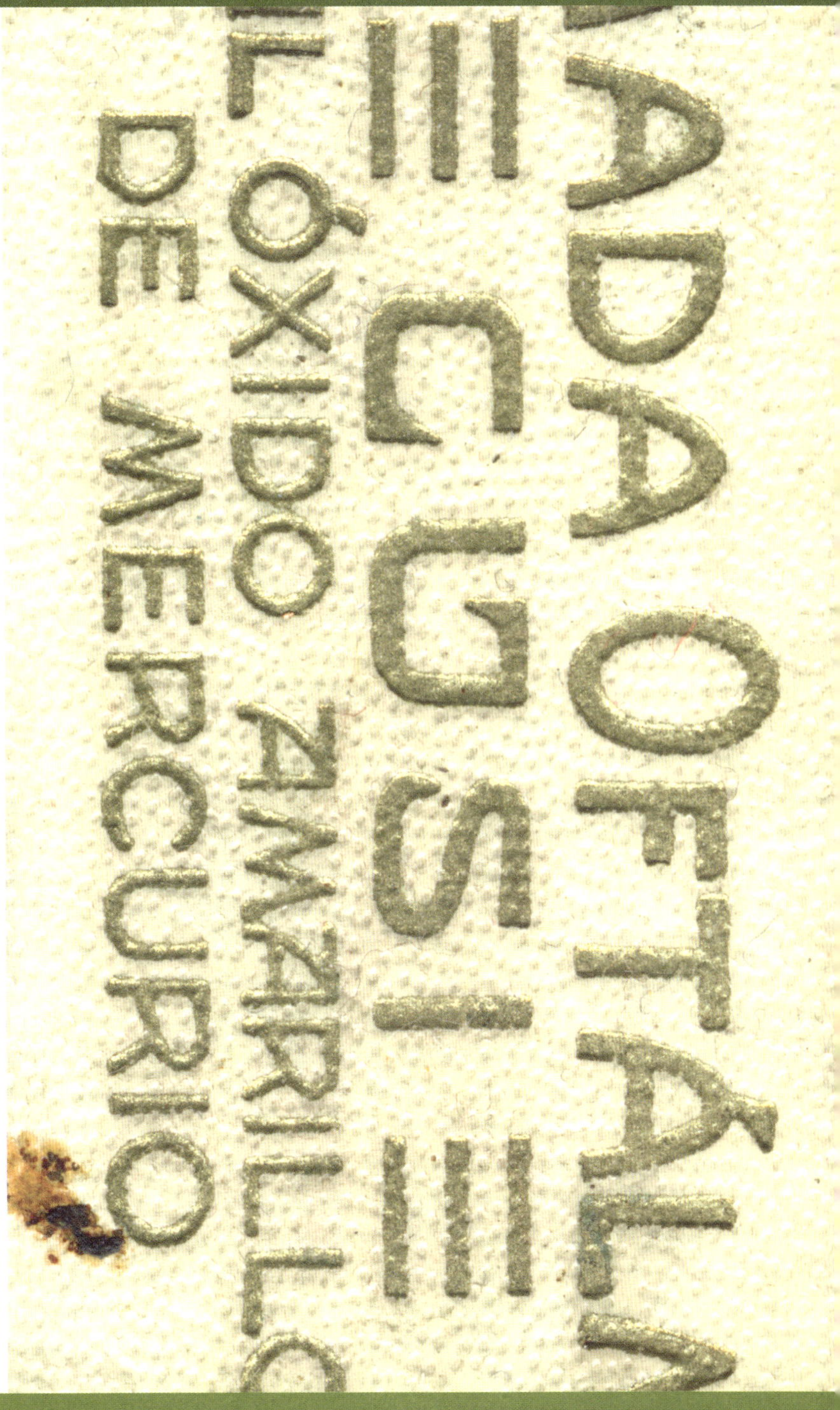
ÓXIDO
DE MERCURIO

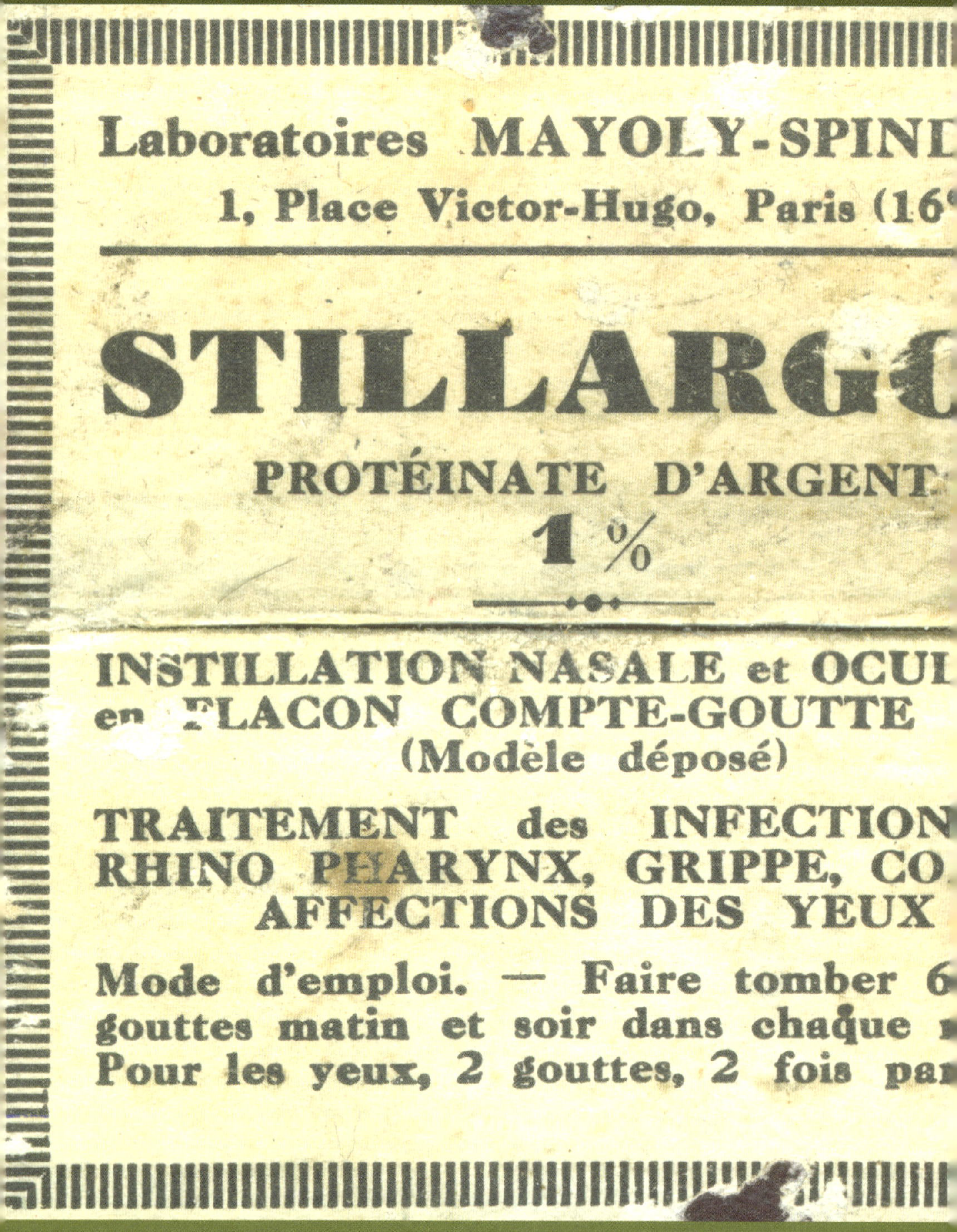
Laboratoires MAYOLY-SPIND
1, Place Victor-Hugo, Paris (16
STILLARGO
PROTÉINATE D'ARGENT
1 %
INSTILLATION NASALE et OCUL
en FLACON COMPTE-GOUTTE
(Modèle déposé)
TRAITEMENT des INFECTION
RHINO PHARYNX, GRIPPE, CO
AFFECTIONS DES YEUX
Mode d'emploi. — Faire tomber 6
gouttes matin et soir dans chaque
Pour les yeux, 2 gouttes, 2 fois par

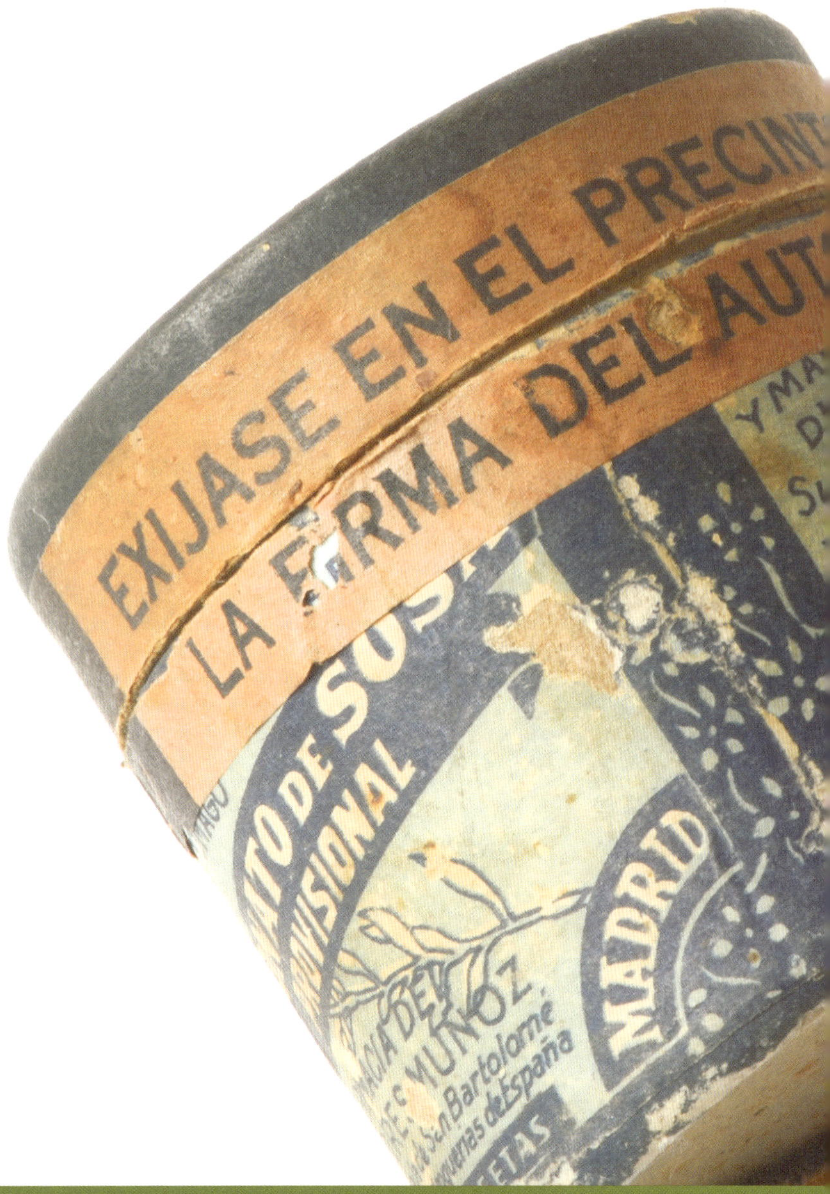
EXIJASE EN EL
DEL
MADRID

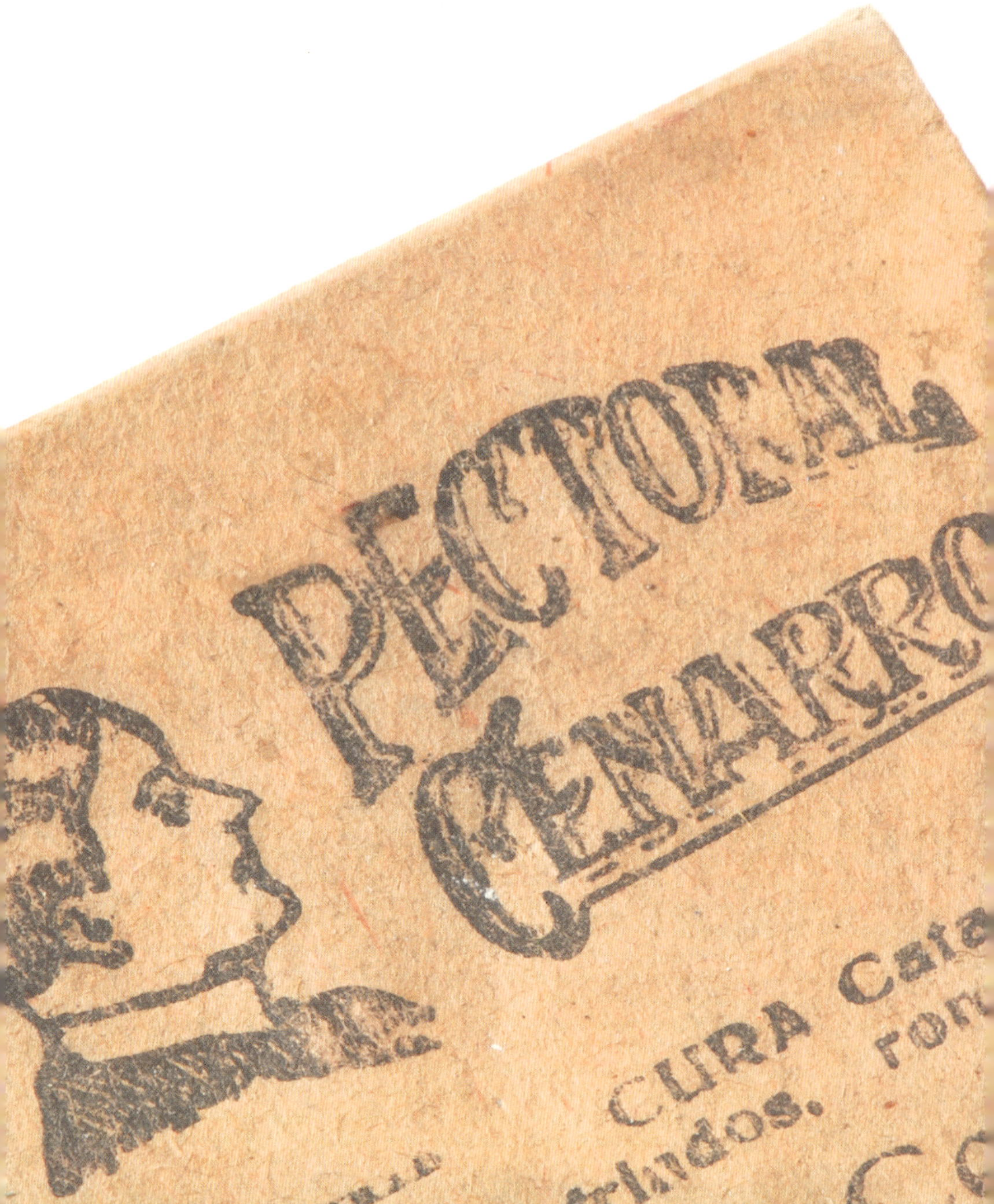
PECTORAL
CENARRO
CURA Cata
rom
rtidos.

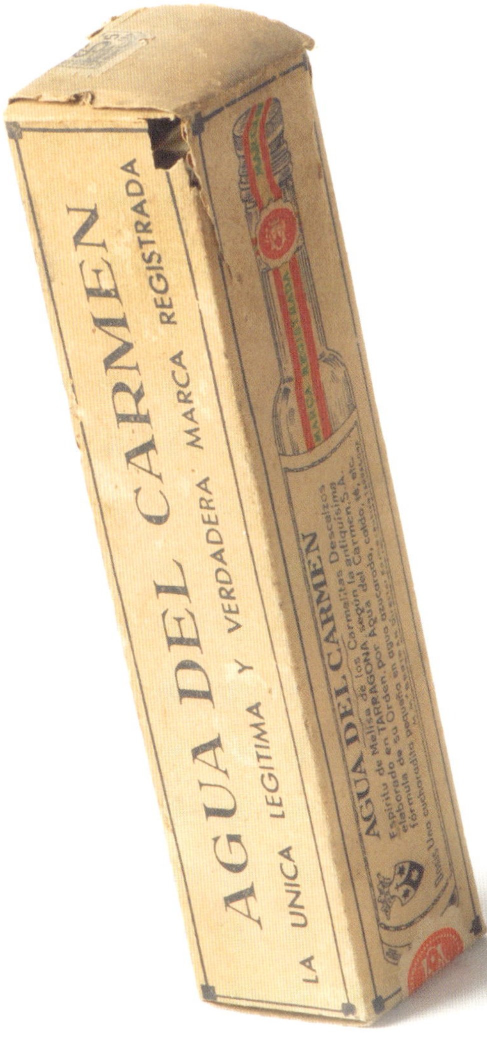
AGUA DEL CARMEN
LA UNICA LEGITIMA Y VERDADERA MARCA REGISTRADA
AGUA DEL CARMEN

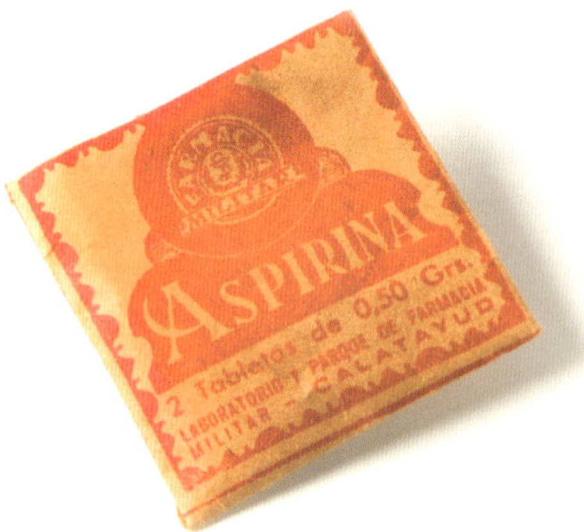
ASPIRINA
2 Tabletas de 0,50 Grs.
LABORATORIO Y PARQUE DE FARMACIA
MILITAR - CALATAYUD

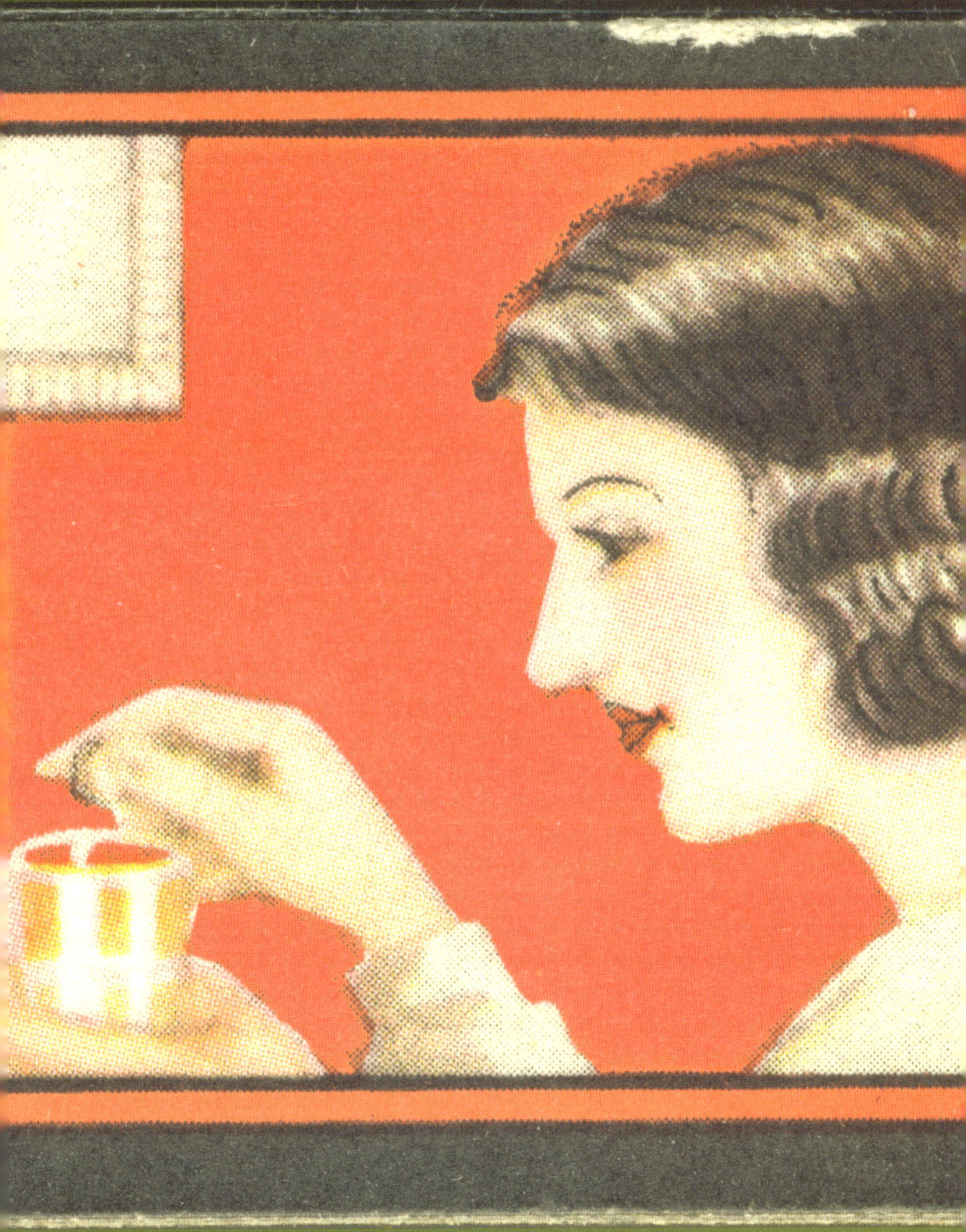

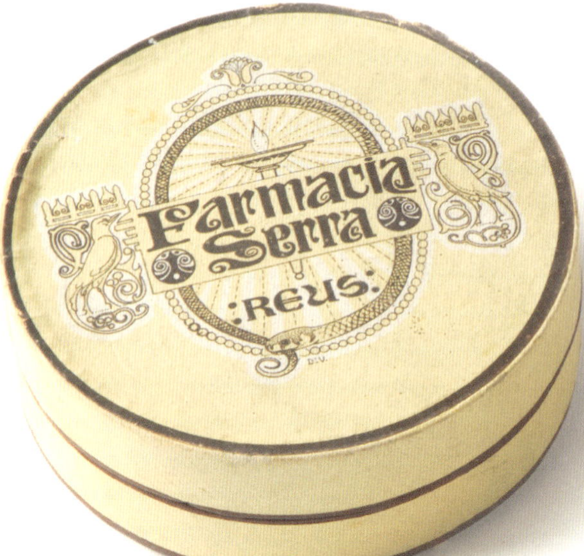
Farmacia
Serra
REUS

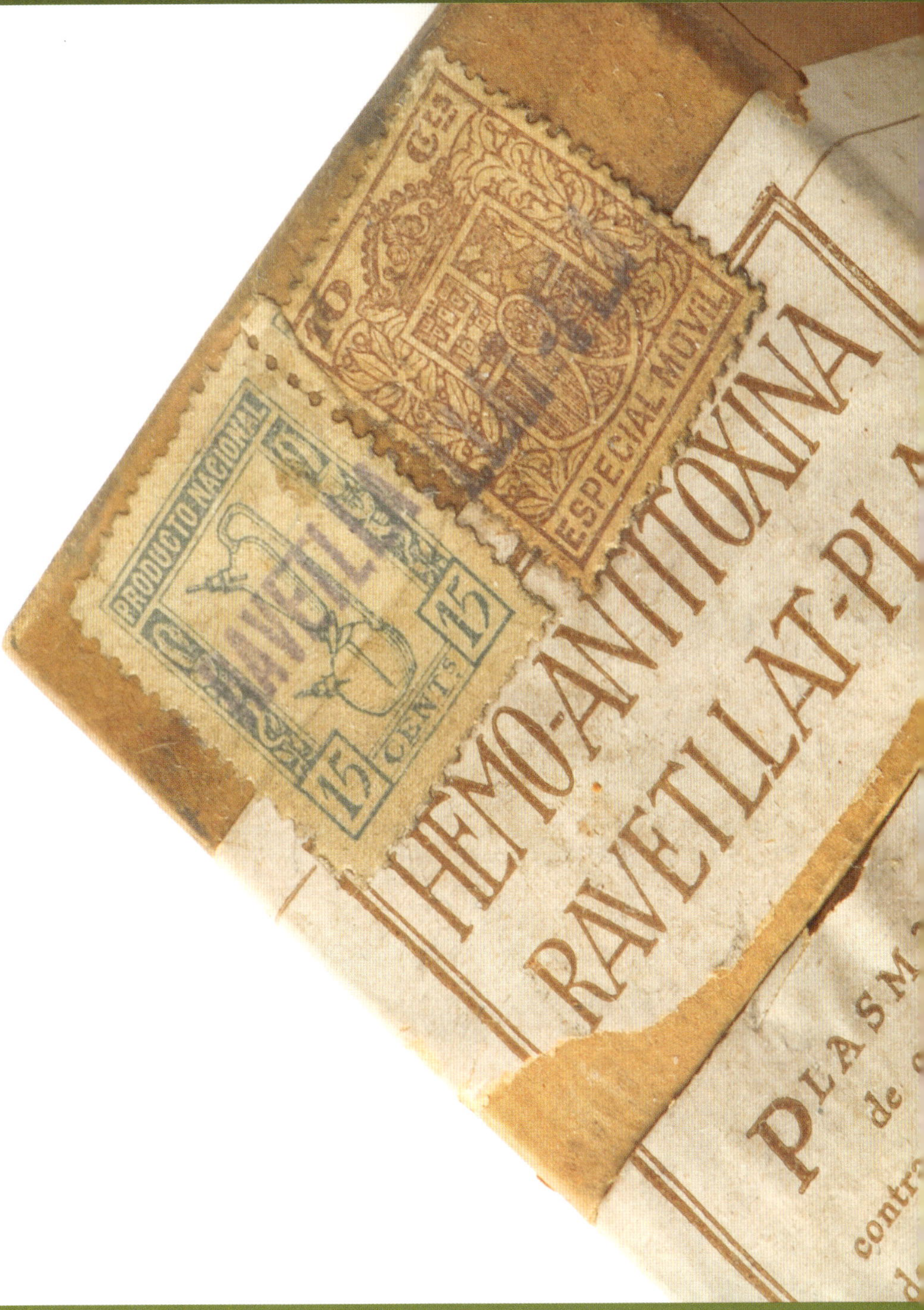
HEMO-ANTITOXINA
RAVETLLAT-PLA
PRODUCTO NACIONAL
15 CENTS 15
ESPECIAL MOVIL

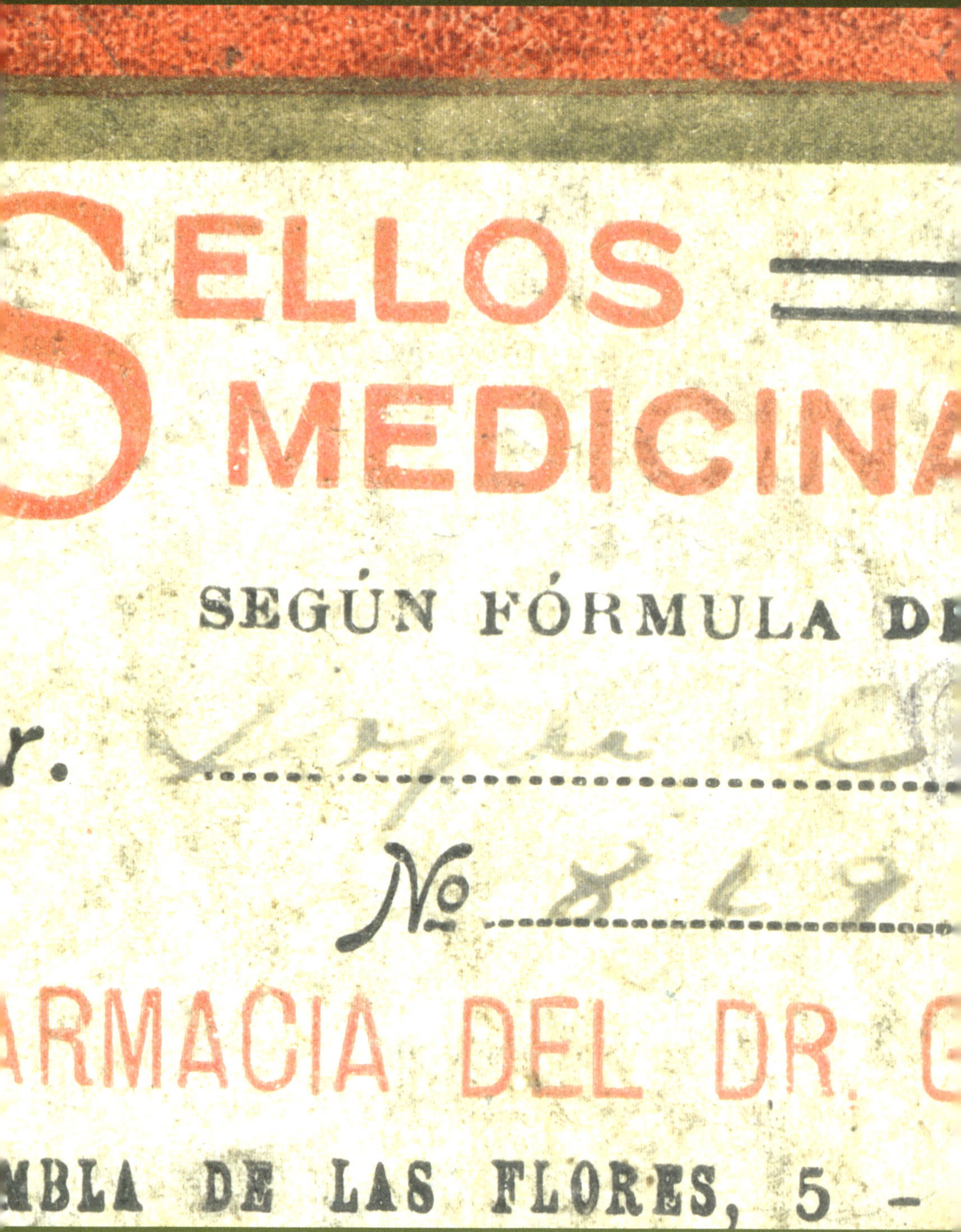
SELLOS
MEDICINA
SEGÚN FÓRMULA
№
ARMACIA DEL DR.
MBLA DE LAS FLORES, 5 -

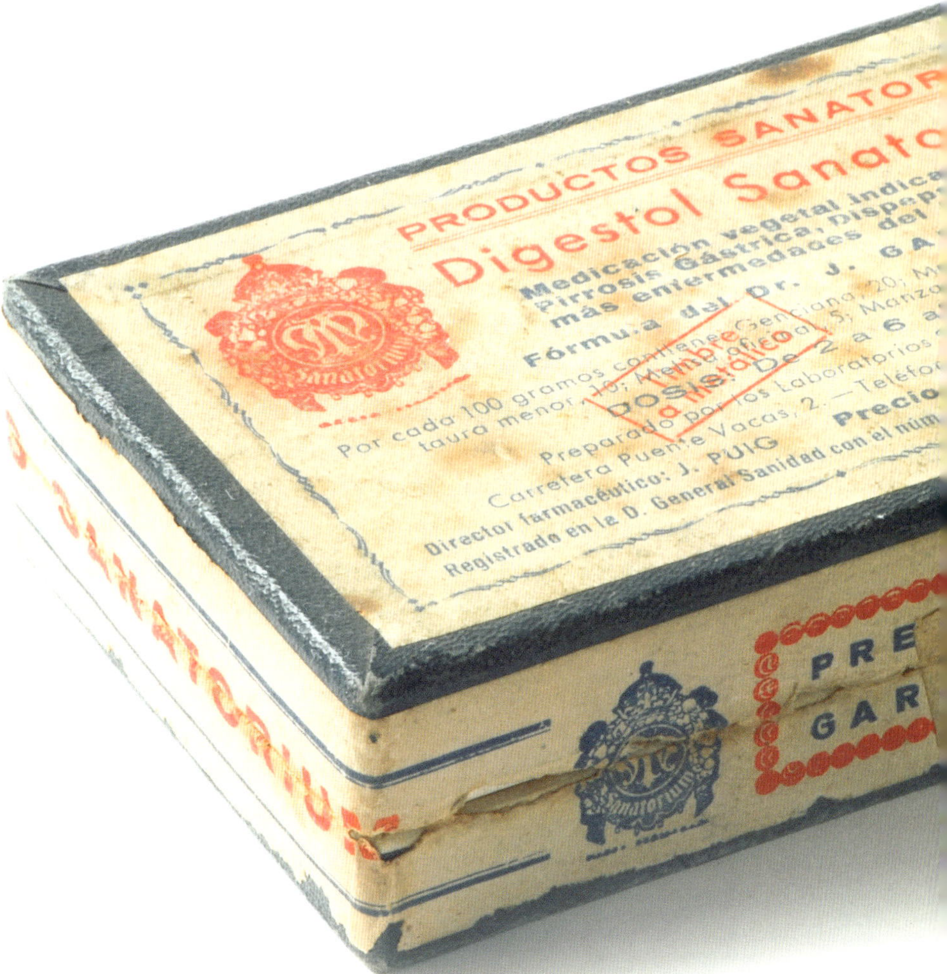
PRODUCTOS SANATOR
Digestol Sanato
Medicación vegetal indica
Pirrosis Gástrica, Dispep
más enfermedades del
Fórmula del Dr. J. GA
Timbre
a metálico
DOSIS: De 2 a 6 a
Por cada 100 gramos contiene Genciana 20; M
Preparado por los Laboratorios
Carretera Puente Vacas, 2.—Teléfo
Director farmacéutico: J. PUIG
Precio
Registrado en la D. General Sanidad con el núm
Sanatorium
SANATORIU
PRE
GAR

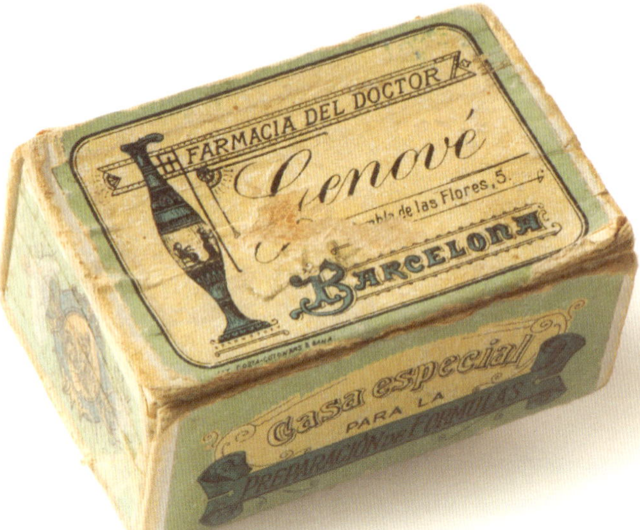
FARMACIA DEL DOCTOR
Genové
de las Flores, 5.
BARCELONA
Casa especial
PARA LA
PREPARACIÓN DE FÓRMULAS

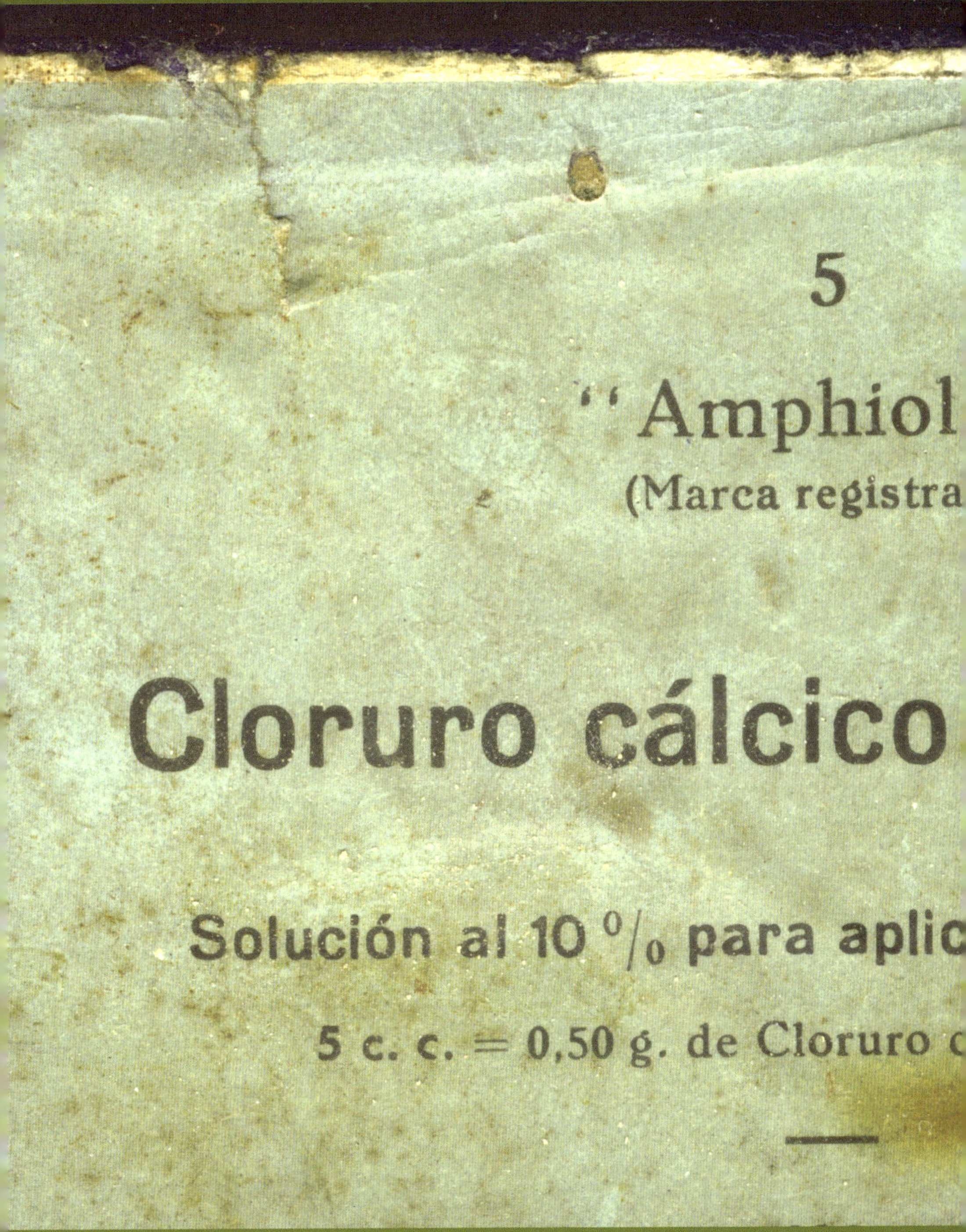
5
"Amphiol
(Marca registra
Cloruro cálcico
Solución al 10 % para aplic
5 c. c. = 0,50 g. de Cloruro c

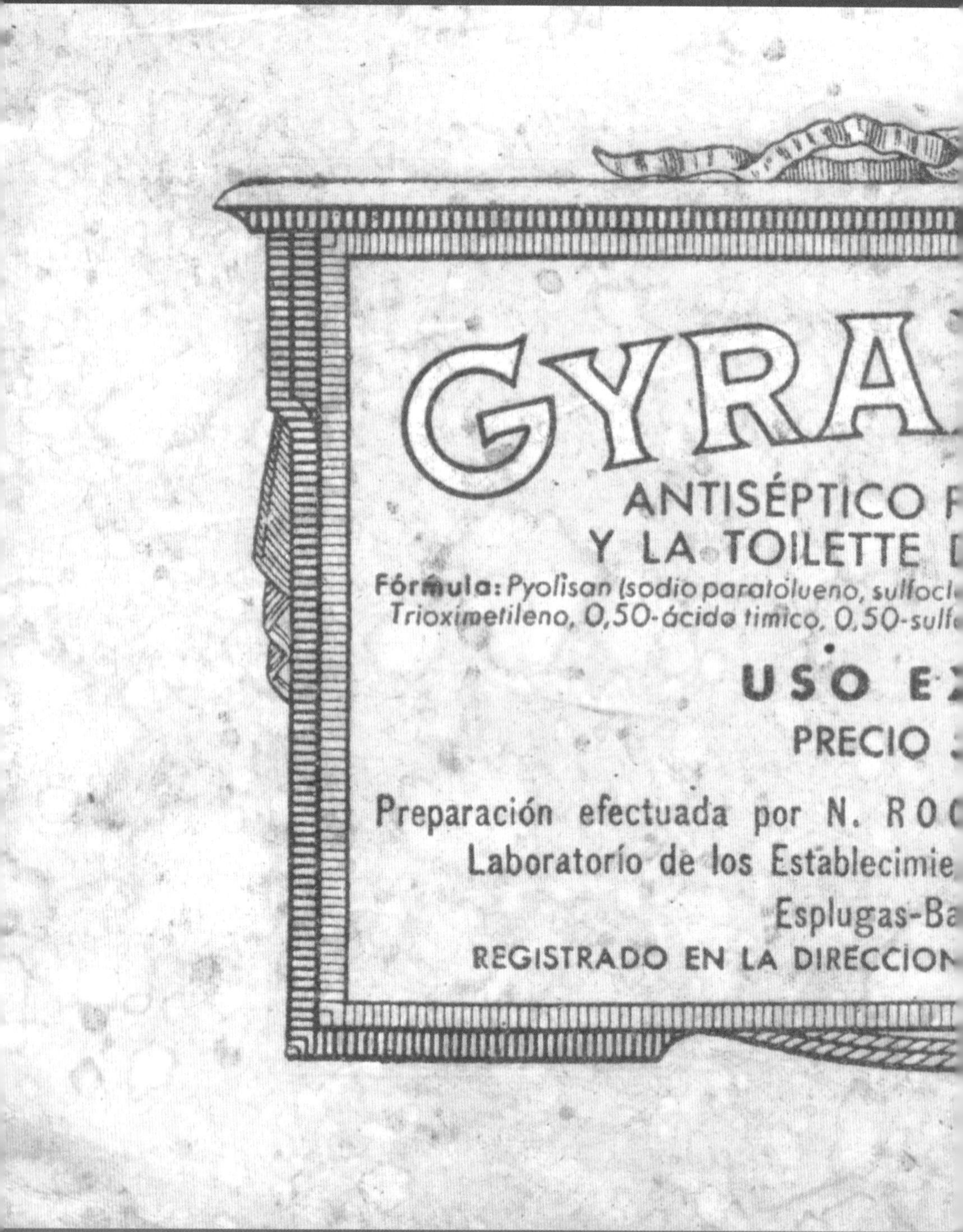
GYRA
ANTISÉPTICO
Y LA TOILETTE
Fórmula: Pyolisan (sodio paratolueno, sulfocl
Trioximetileno, 0,50-ácido timico, 0,50-sulf
USO E
PRECIO
Preparación efectuada por N. RO
Laboratorio de los Establecimie
Esplugas-Ba
REGISTRADO EN LA DIRECCION

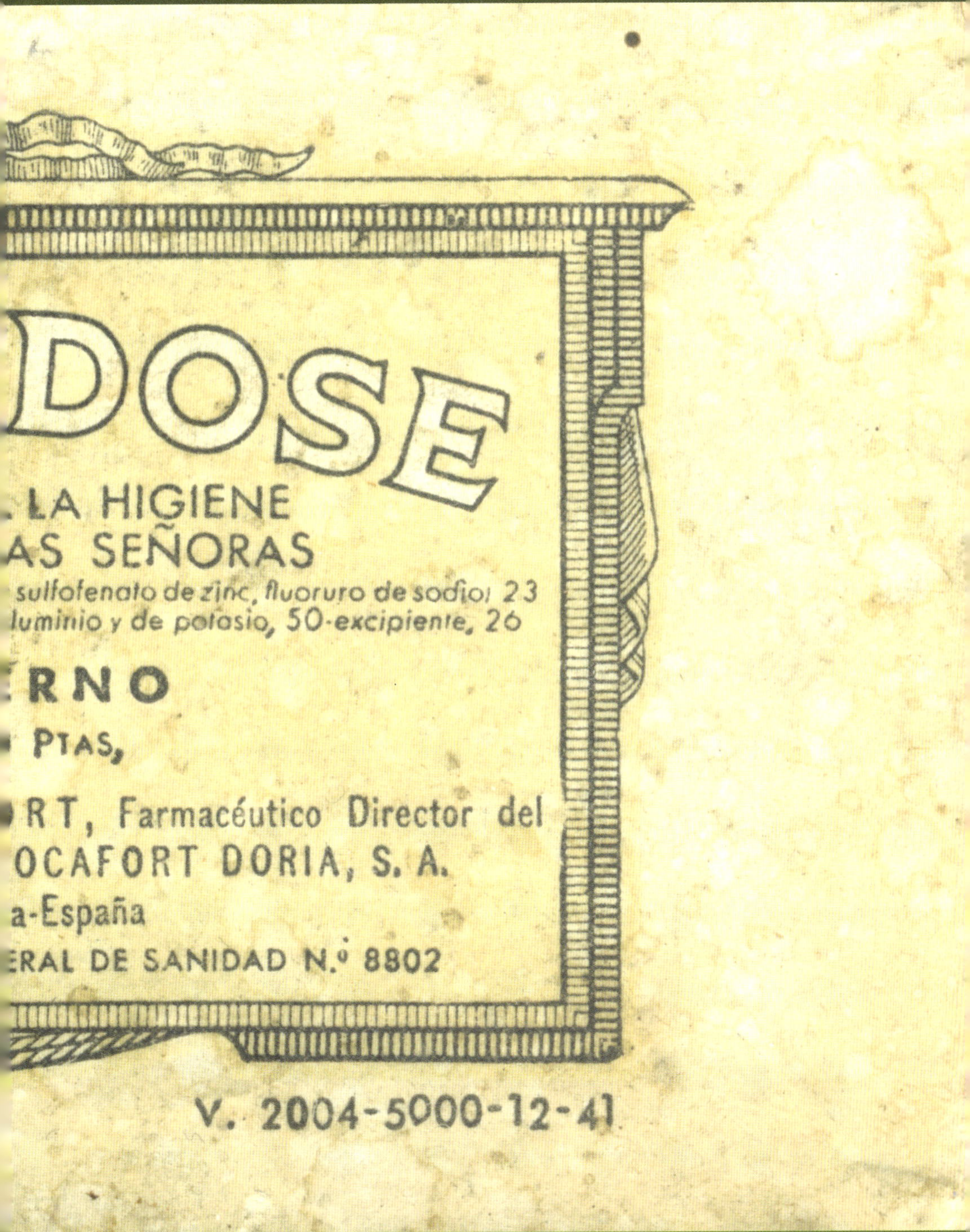
DOSE
LA HIGIENE
AS SEÑORAS
sulfofenato de zinc, fluoruro de sodio; 23
luminio y de potasio, 50-excipiente, 26
RNO
PTAS,
RT, Farmacéutico Director del
OCAFORT DORIA, S. A.
a-España
ERAL DE SANIDAD N.º 8802
V. 2004-5000-12-41

UDIOVIS

UDIOVISUALAUDI

AUDIOVISUAL

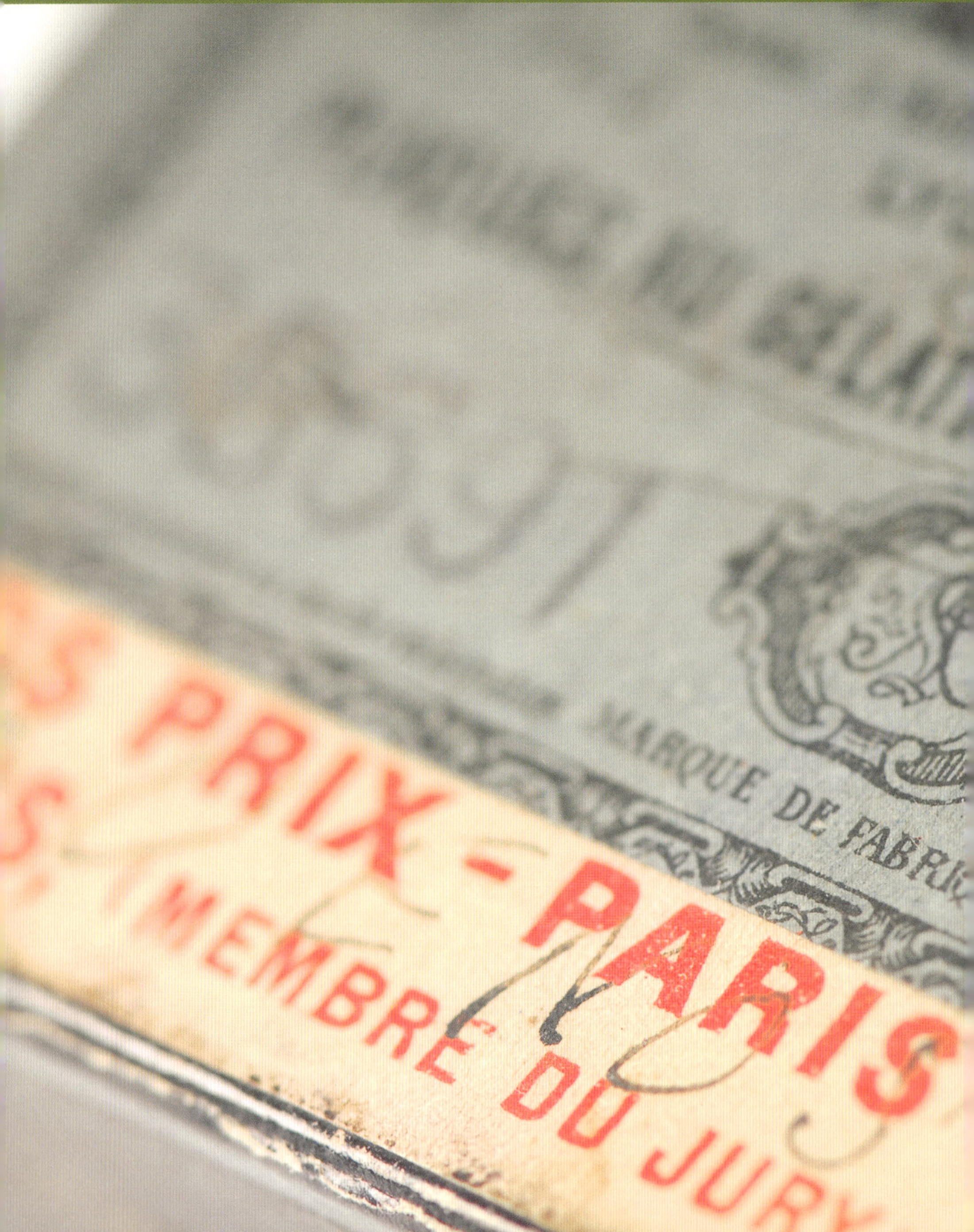
PRIX - PARIS
MEMBRE DU
MARQUE DE

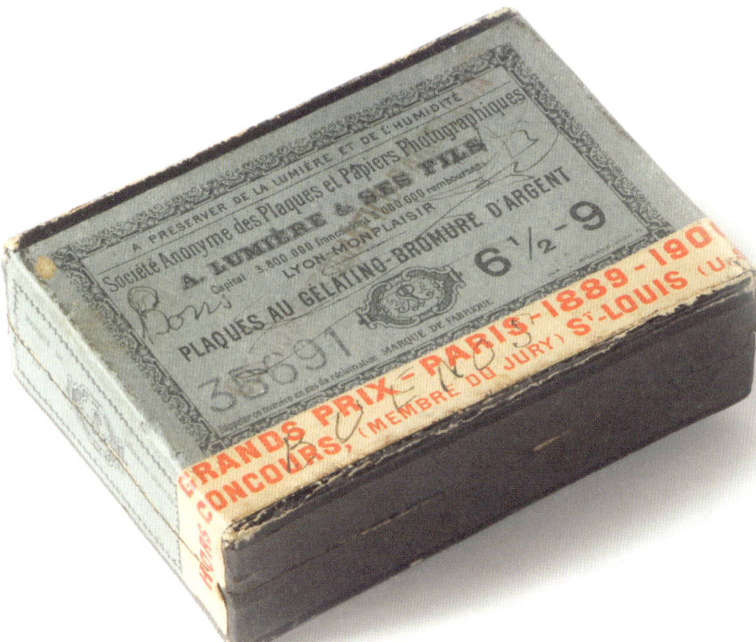
A PRESERVER DE LA LUMIÈRE ET DE L'HUMIDITÉ
Société Anonyme des Plaques et Papiers Photographiques
A. LUMIÈRE & SES FILS
LYON-MONPLAISIR
PLAQUES AU GÉLATINO-BROMURE D'ARGENT
6 1/2 - 9
MARQUE DE FABRIQUE
GRANDS PRIX - PARIS - 1889 - 190
CONCOURS, (MEMBRE DU JURY) St-LOUIS

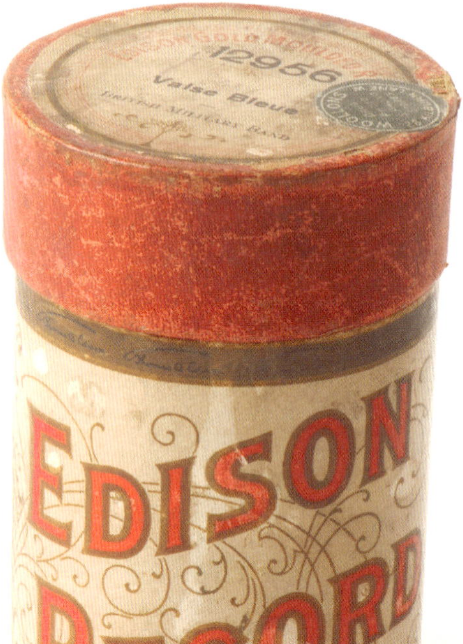
12956
Valse Bleue
EDISON
RECORD

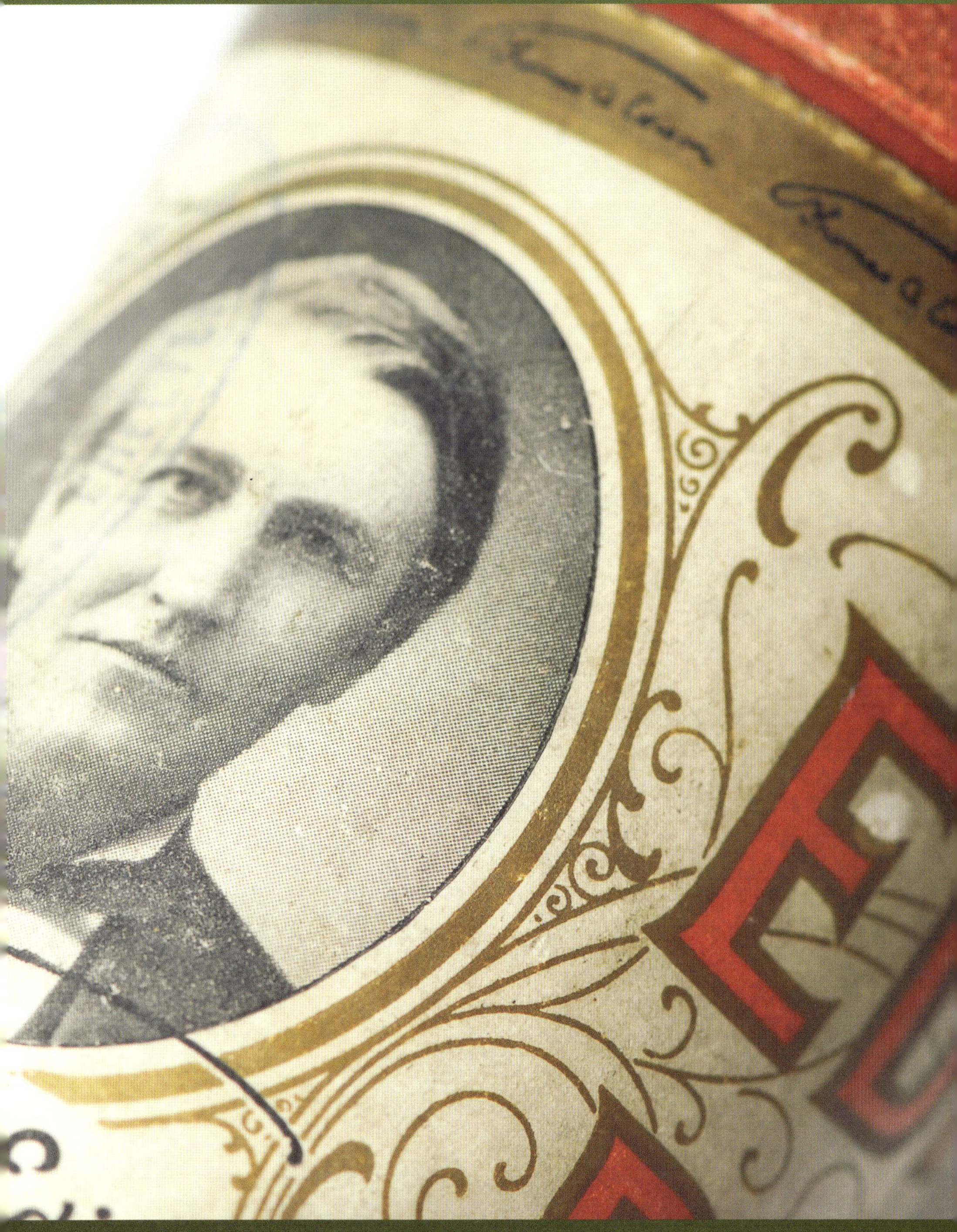

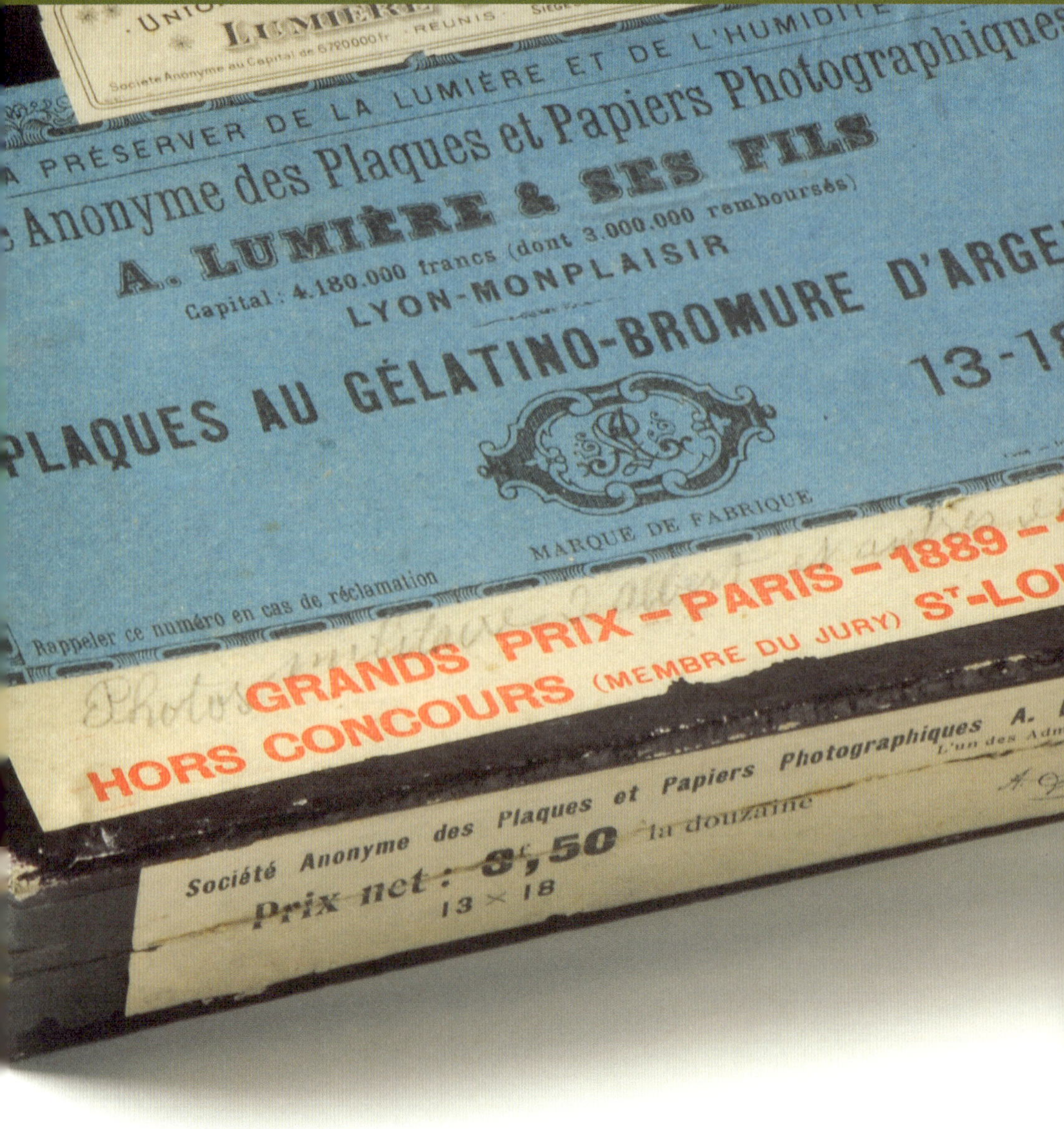
LUMIÈRE
A PRÉSERVER DE LA LUMIÈRE ET DE L'HUMIDITÉ
Anonyme des Plaques et Papiers Photographiques
A. LUMIÈRE & SES FILS
Capital: 4.180.000 francs (dont 3.000.000 remboursés)
LYON-MONPLAISIR
PLAQUES AU GÉLATINO-BROMURE D'ARGE
13-18
MARQUE DE FABRIQUE
Rappeler ce numéro en cas de réclamation
GRANDS PRIX – PARIS – 1889 –
HORS CONCOURS (MEMBRE DU JURY) ST-LO
Société Anonyme des Plaques et Papiers Photographiques A.
Prix net : 8f,50 la douzaine
13 × 18

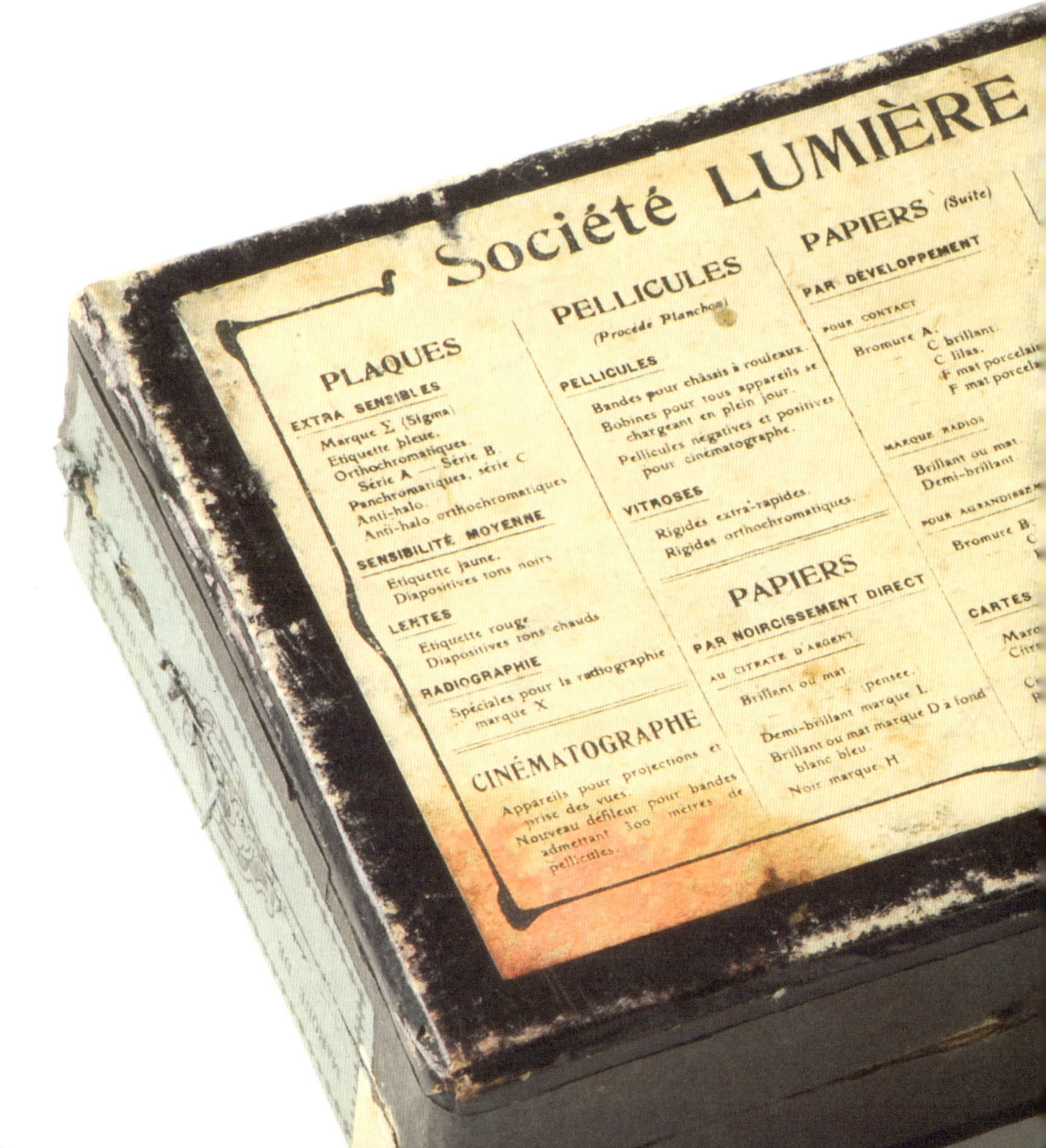
Société LUMIÈRE
PLAQUES
EXTRA SENSIBLES
Marque Σ (Sigma)
Etiquette bleue.
Orthochromatiques.
Série A — Série B
Panchromatiques, série C
Anti-halo.
Anti-halo orthochromatiques
SENSIBILITÉ MOYENNE
Etiquette jaune.
Dispositives tons noirs
LENTES
Etiquette rouge
Dispositives tons chauds
RADIOGRAPHIE
Spéciales pour la radiographie
marque X
CINÉMATOGRAPHE
Appareils pour projections et
prise des vues.
Nouveau défileur pour bandes
admettant 300 mètres de
pellicules.
PELLICULES
(Procédé Planchon)
PELLICULES
Bandes pour châssis à rouleaux.
Bobines pour tous appareils se
chargeant en plein jour.
Pellicules négatives et positives
pour cinématographe.
VITROSES
Rigides extra-rapides.
Rigides orthochromatiques.
PAPIERS
PAR NOIRCISSEMENT DIRECT
AU CITRATE D'ARGENT
Brillant ou mat.
Demi-brillant marque L
Brillant ou mat marque D à fond
blanc bleu.
Noir marque H
PAPIERS (Suite)
PAR DÉVELOPPEMENT
POUR CONTACT
Bromure A.
C brillant.
C lilas.
MARQUE RADIOS
Brillant ou mat.
Demi-brillant.
Bromure B.
CARTES

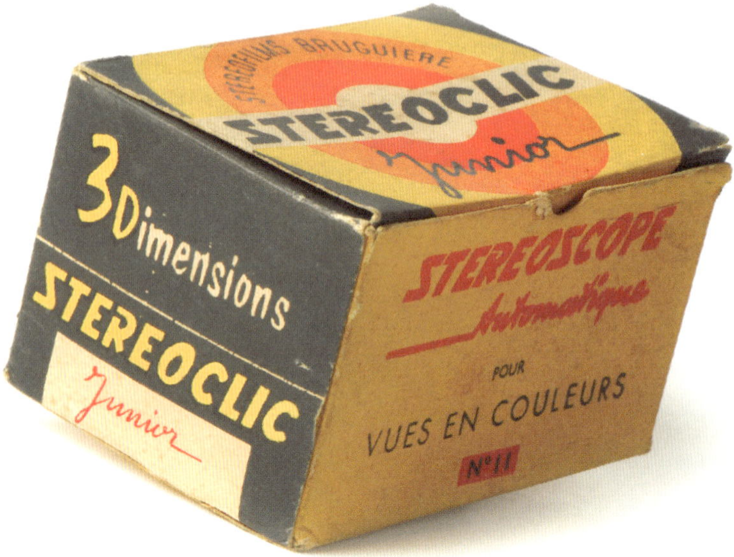
STEREOFILMS BRUGUIERE
STEREOCLIC
Junior
3 Dimensions
STEREOCLIC
Junior
STEREOSCOPE
Automatique
POUR
VUES EN COULEURS
N°11

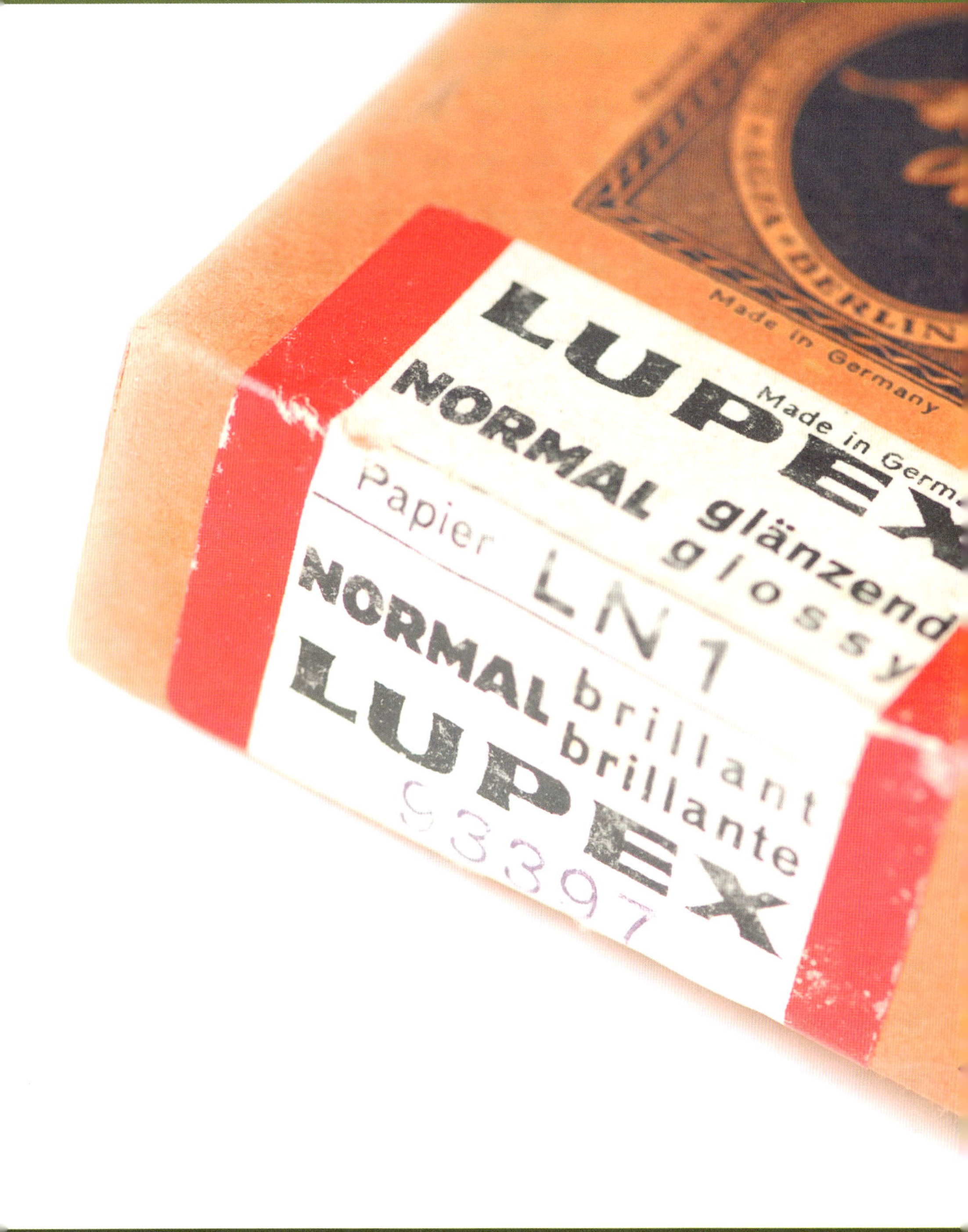
BERLIN
Made in Germany
LUPEX
Made in Germany
NORMAL glänzend
glossy
Papier LN1
NORMAL brillant
brillante
LUPEX
93397

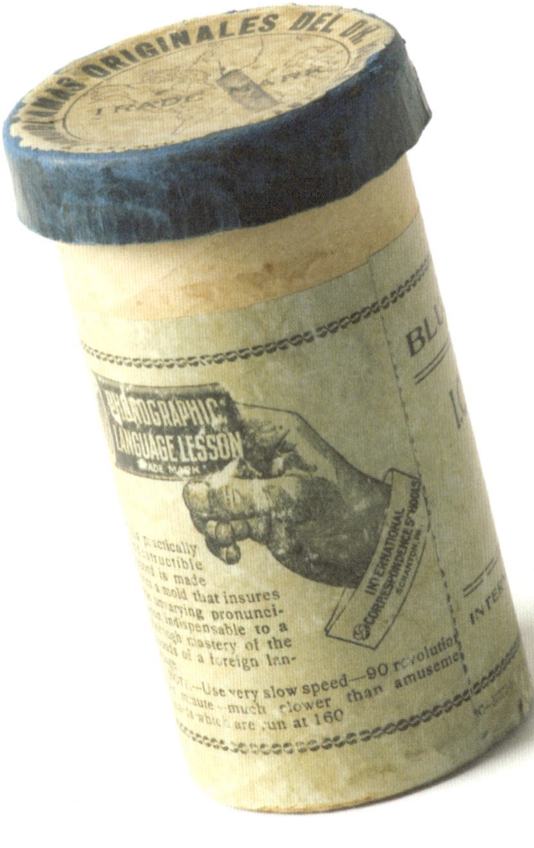
ORIGINALES DEL DR.
LANGUAGE LESSON
INTERNATIONAL CORRESPONDENCE SCHOOLS
Use very slow speed—90

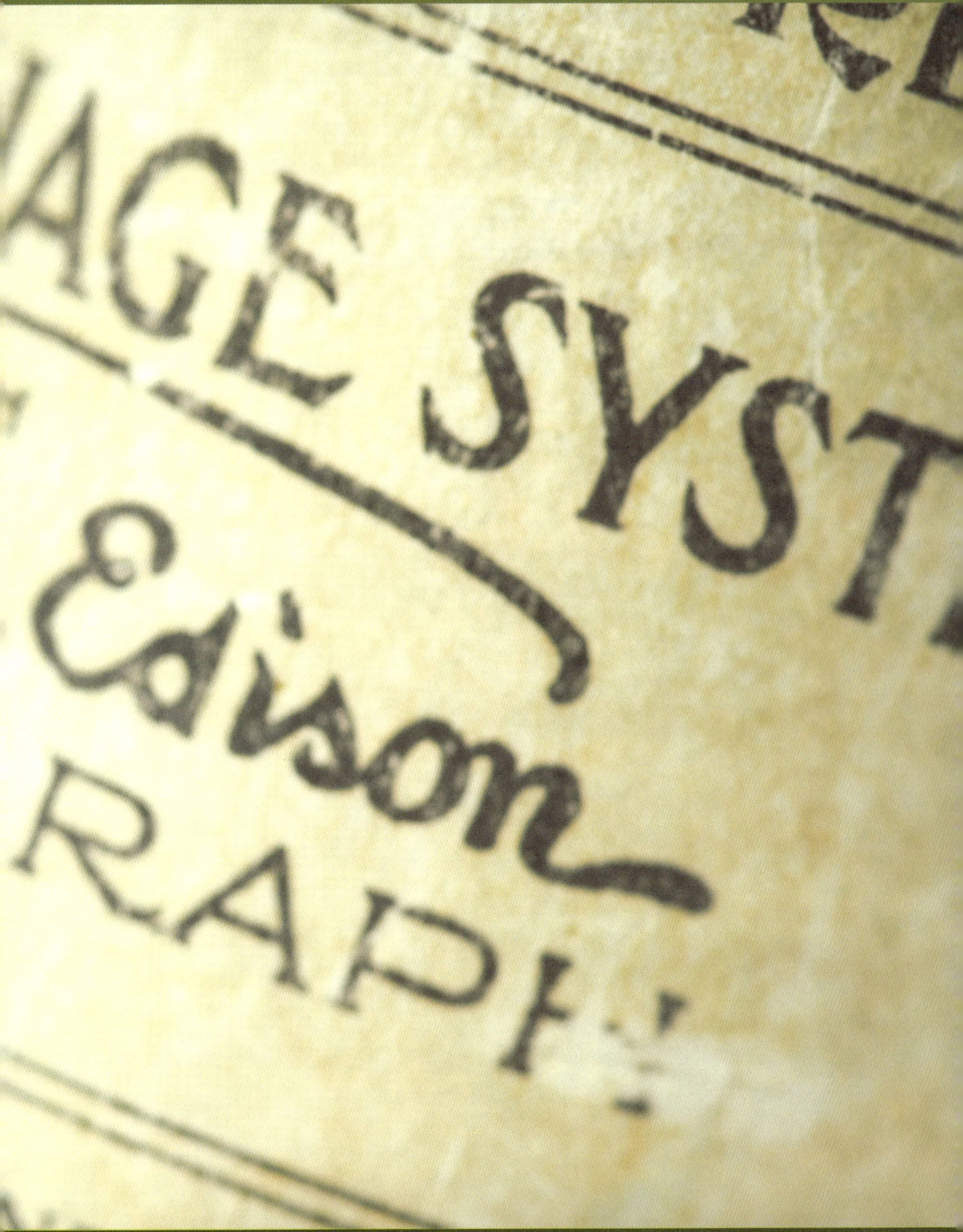
Edison

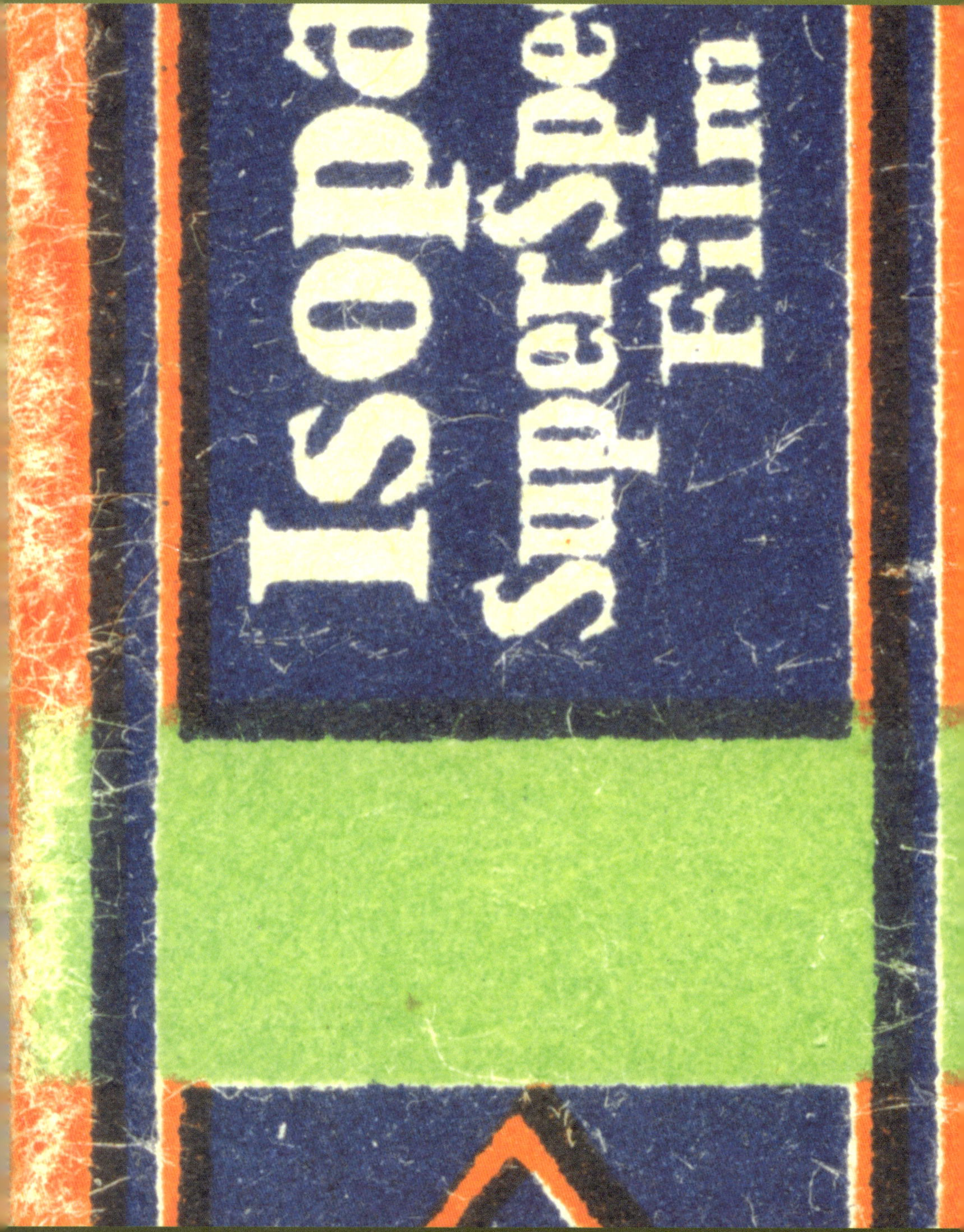

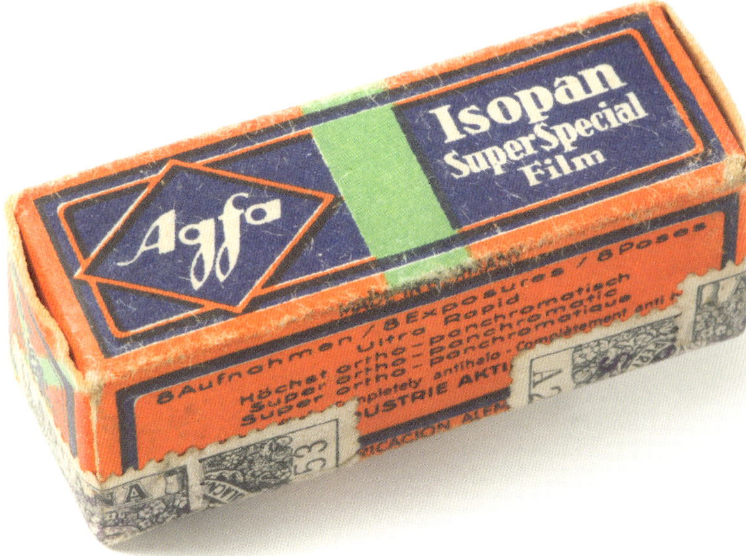
Agfa
Isopan
SuperSpecial
Film
8 Aufnahmen / 8 Exposures / 8 Poses
Ultra Rapid
Höchst ortho-panchromatisch
Super ortho-panchromatic
Super ortho-panchromatique

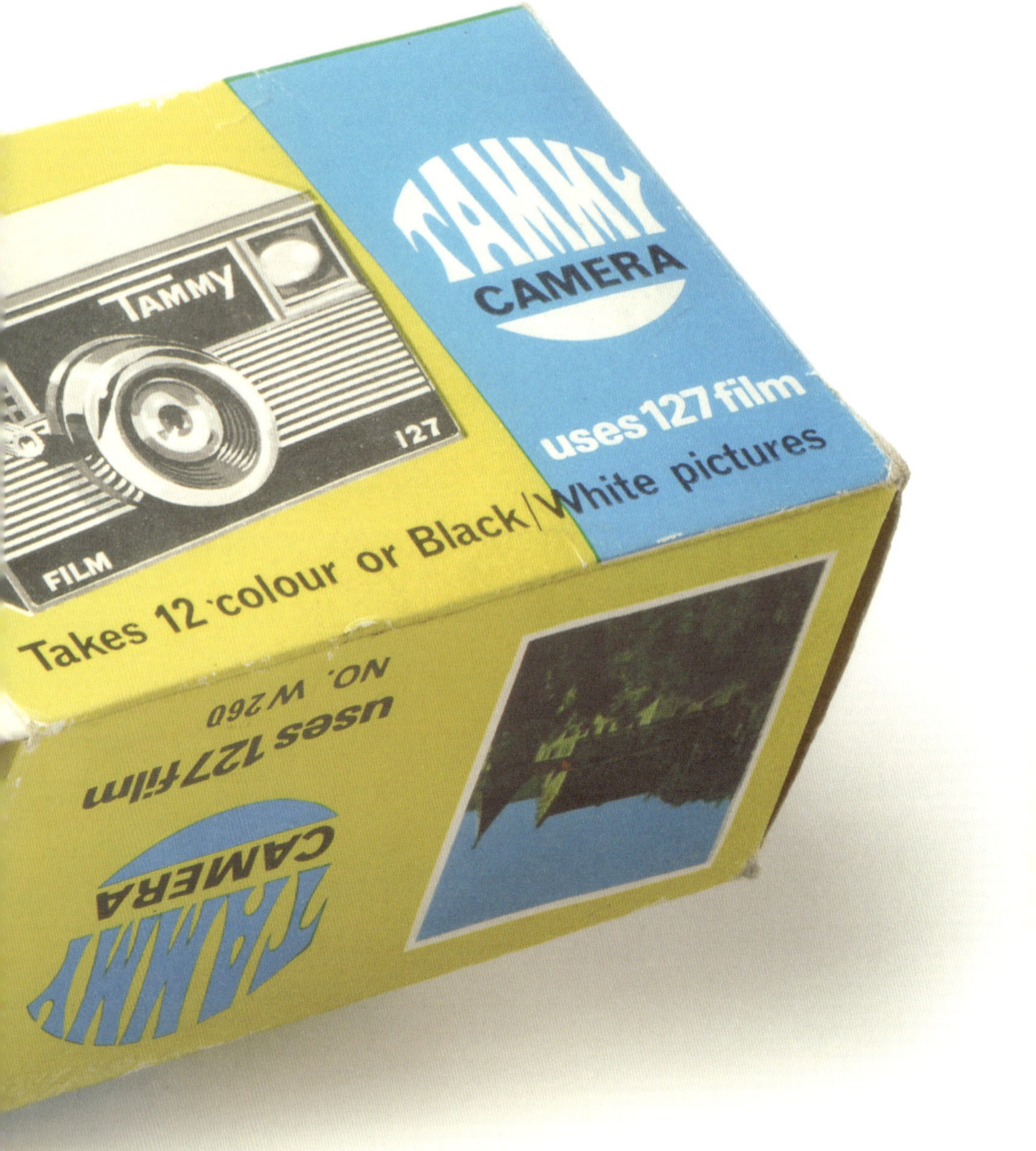
TAMMY
CAMERA
uses 127 film
Takes 12 colour or Black/White pictures
TAMMY
127
FILM
uses 127 film
NO. W260
TAMMY
CAMERA

83

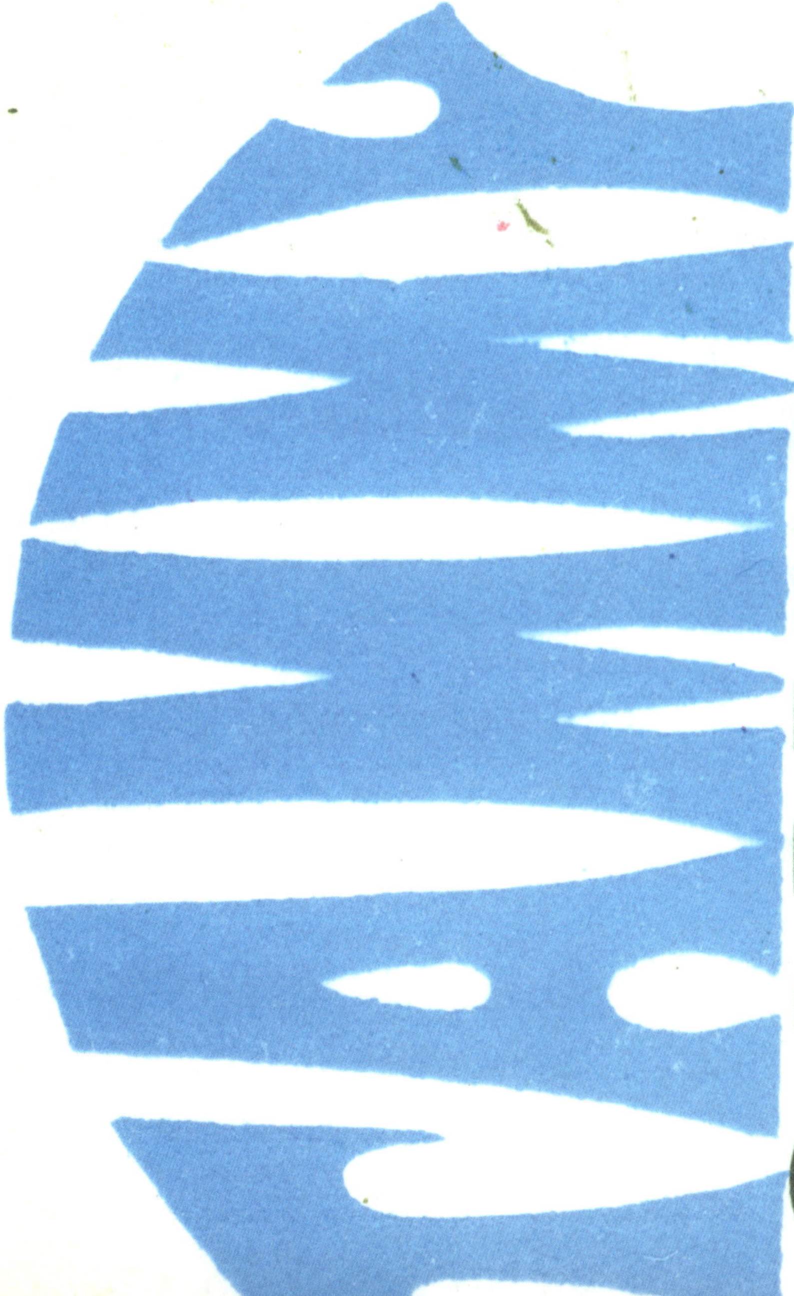

CAMERA

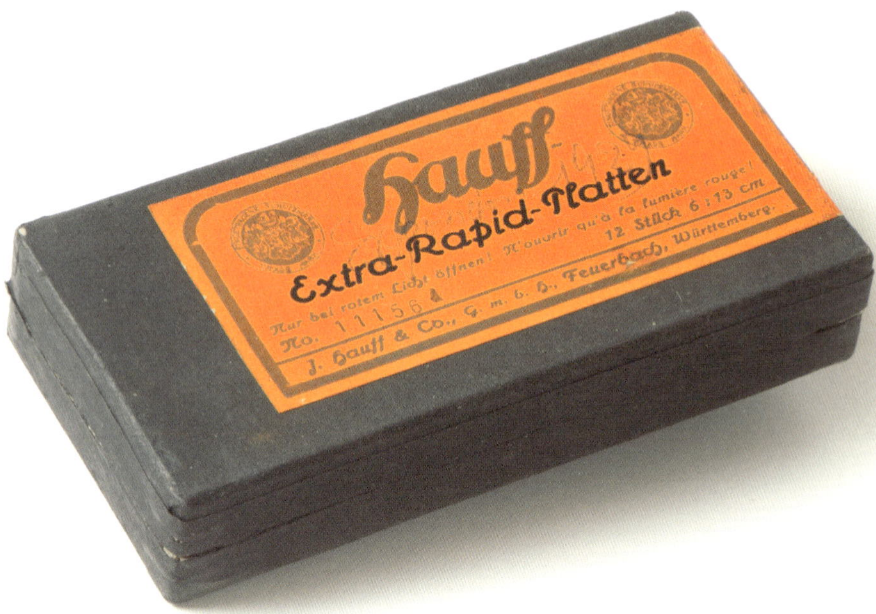
Hauff
Extra-Rapid-Platten
Nur bei rotem Licht öffnen! N'ouvrir qu'à la lumière rouge!
12 Stück 6:13 cm
No. 11156
J. Hauff & Co., G. m. b. H., Feuerbach, Württemberg.

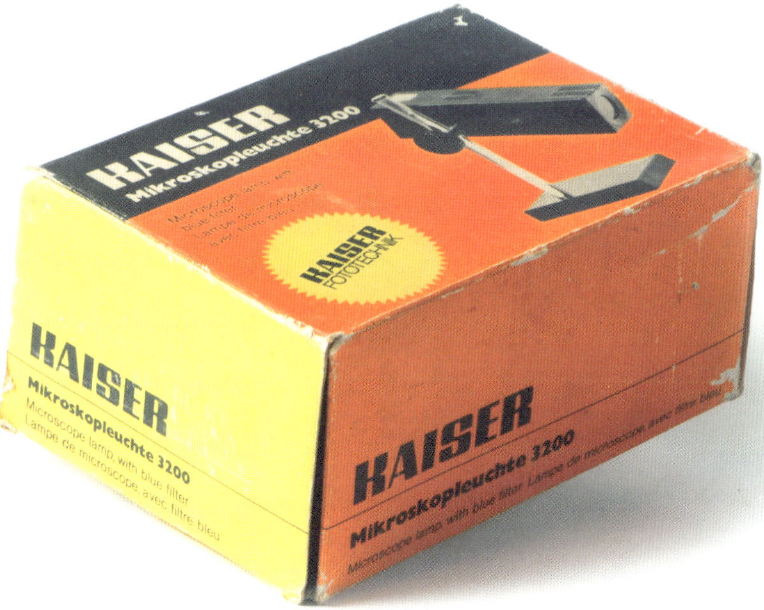
KAISER
Mikroskopleuchte 3200
Microscope lamp with blue filter
Lampe de microscope avec filtre bleu
KAISER FOTOTECHNIK
KAISER
Mikroskopleuchte 3200
Microscope lamp, with blue filter
Lampe de microscope, avec filtre bleu
KAISER
Mikroskopleuchte 3200
Microscope lamp, with blue filter Lampe de microscope, avec filtre bleu

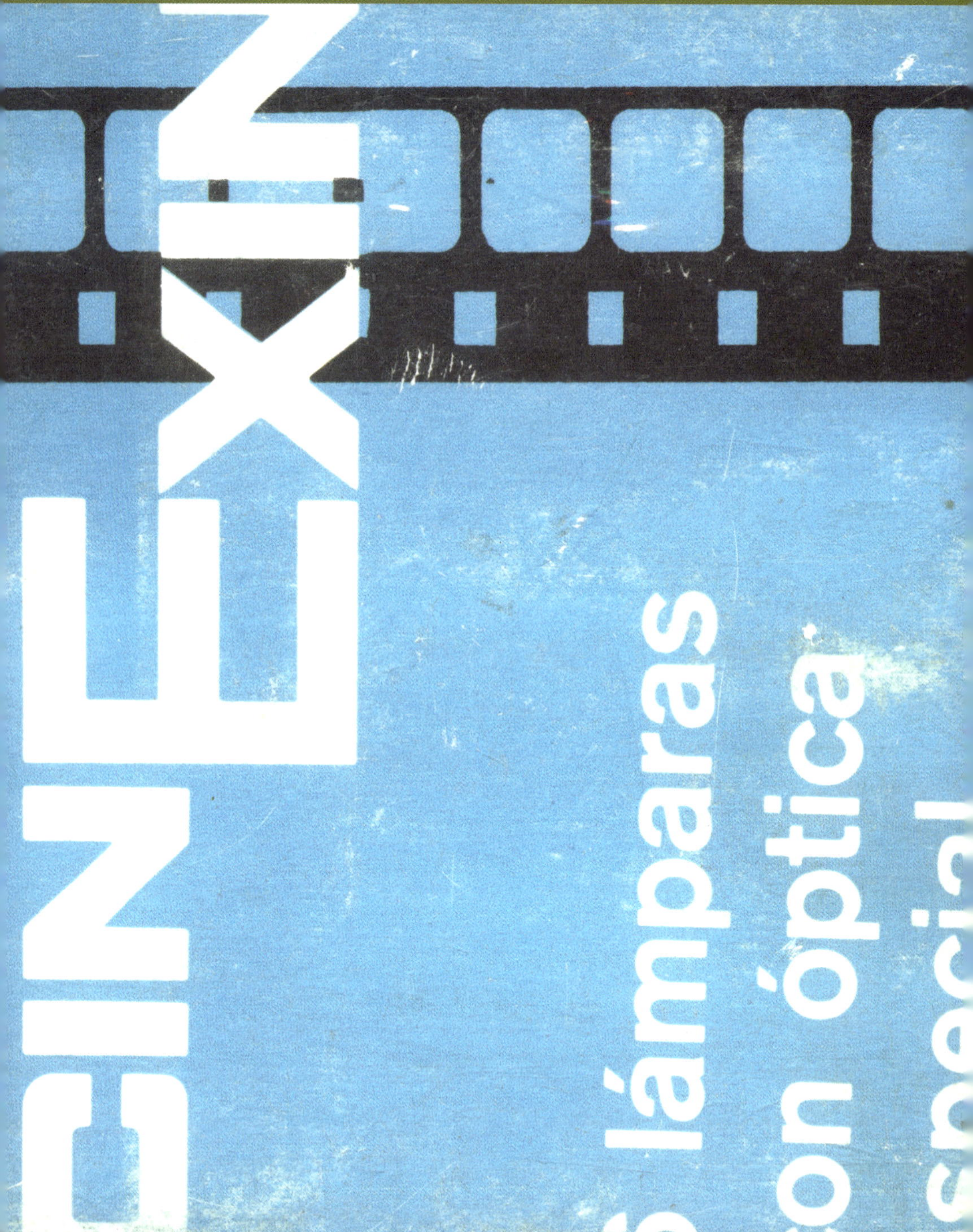
CINEXIN
lámparas
óptica

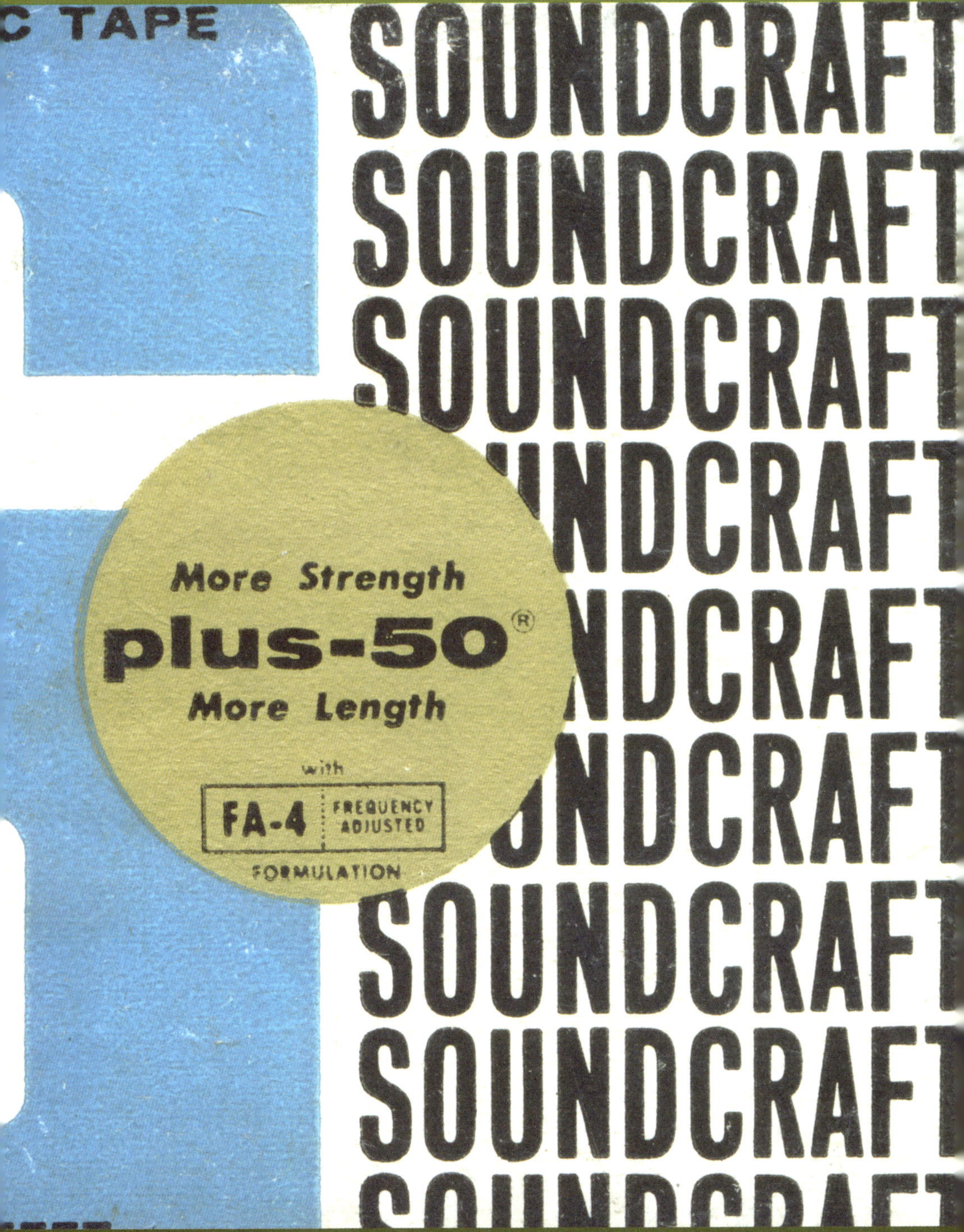
C TAPE
SOUNDCRAFT
SOUNDCRAFT
SOUNDCRAFT
More Strength
plus-50®
More Length
with
FA-4
FREQUENCY ADJUSTED
FORMULATION
SOUNDCRAFT
SOUNDCRAFT

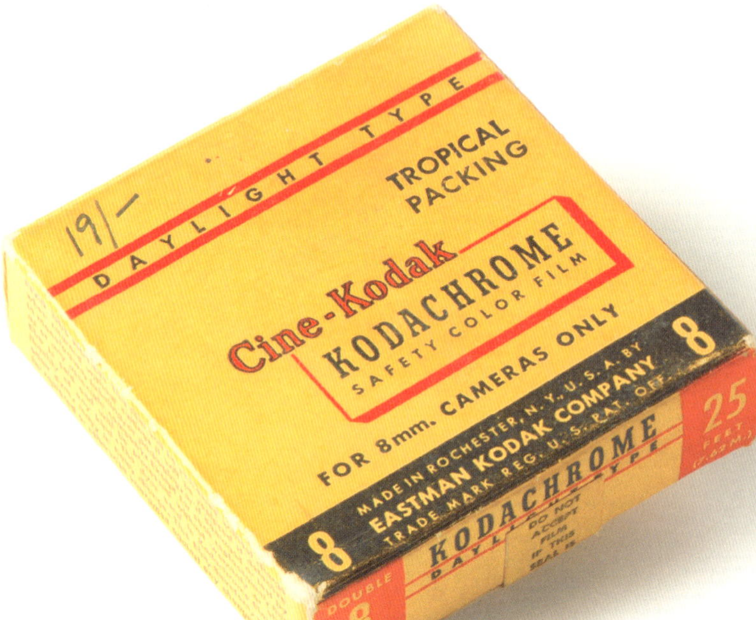
DAYLIGHT TYPE
TROPICAL PACKING
Cine-Kodak
KODACHROME
SAFETY COLOR FILM
FOR 8mm. CAMERAS ONLY
8
MADE IN ROCHESTER, N.Y., U.S.A. BY
EASTMAN KODAK COMPANY
TRADE MARK REG. U.S. PAT. OFF.
8
KODACHROME
25
DOUBLE
8

BRITAIN
Kodak

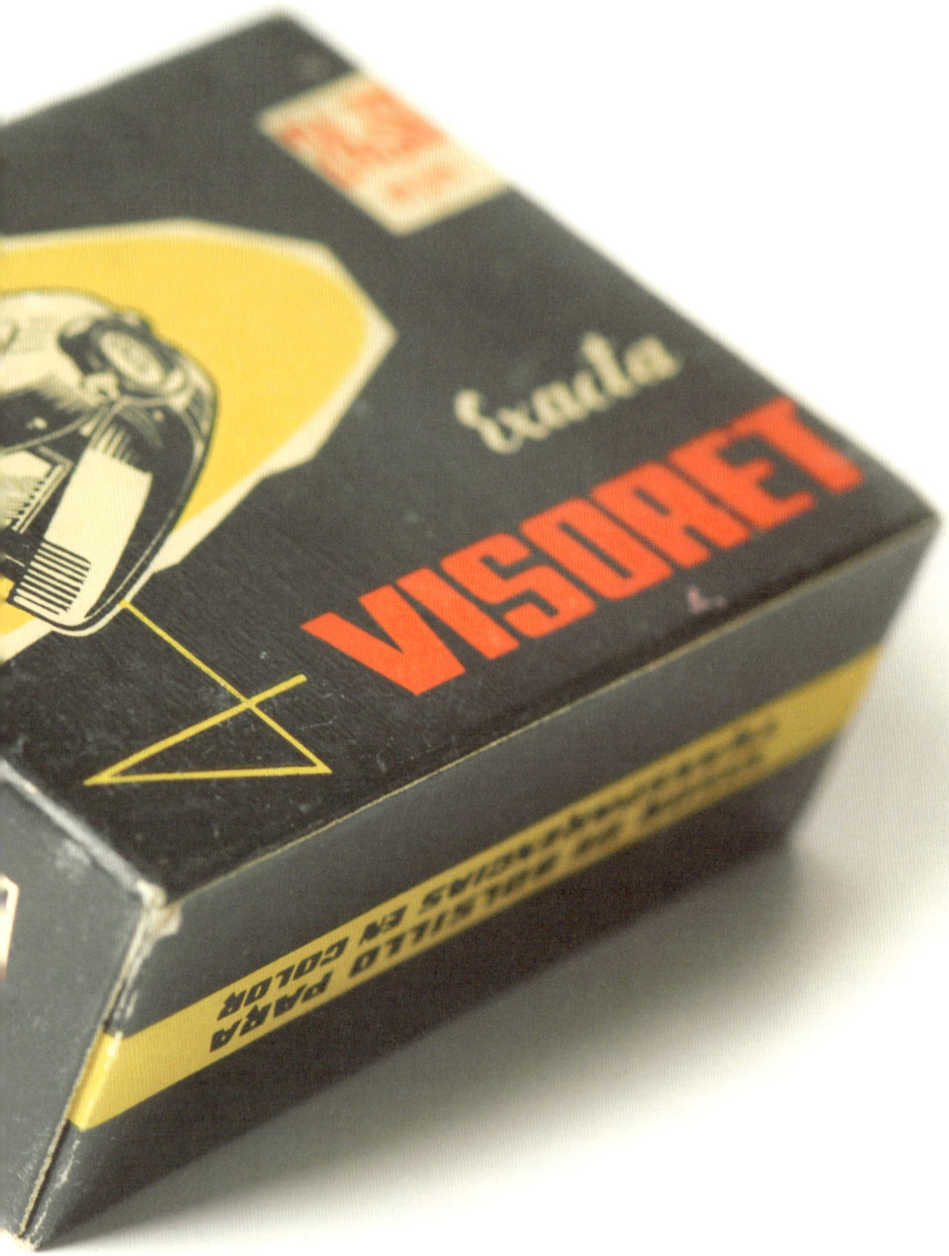
Exacta
VISORET

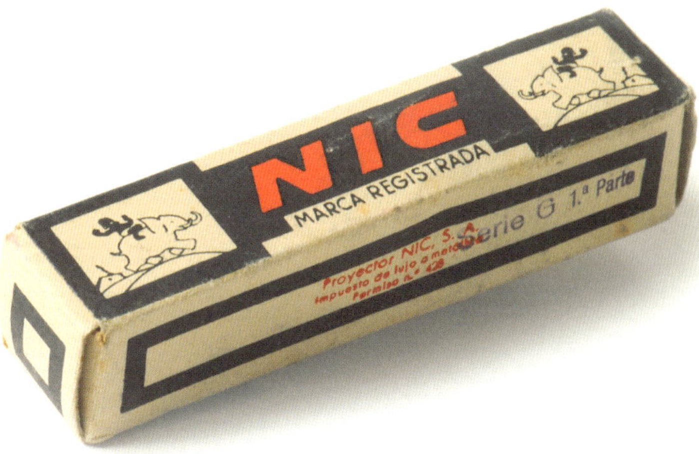
NIC
MARCA REGISTRADA
Serie G 1.ª Parte
Proyector NIC, S. A.

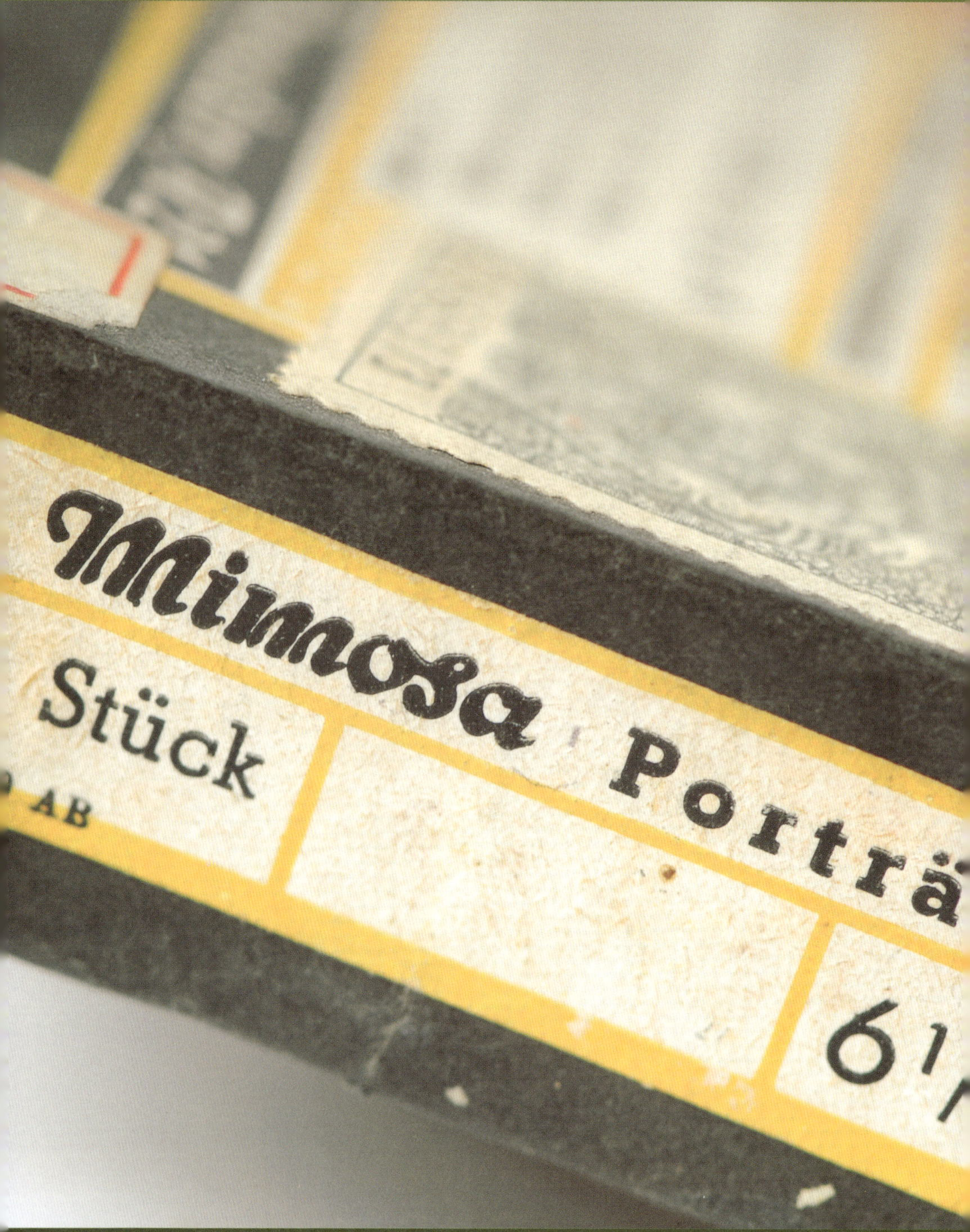
Mimosa Porträ
Stück
AB
61

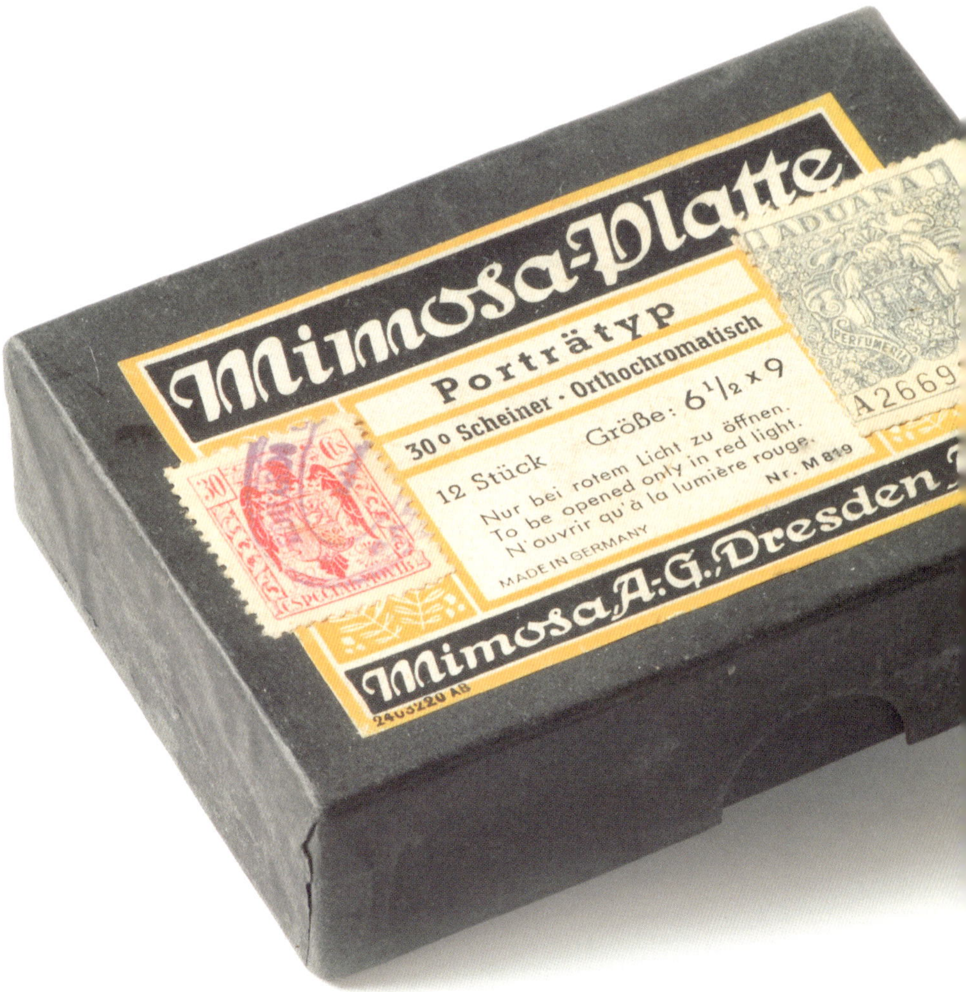
Mimosa-Platte
Porträtyp
30° Scheiner · Orthochromatisch
Größe: 6½ x 9
12 Stück
Nur bei rotem Licht zu öffnen.
To be opened only in red light.
N'ouvrir qu'à la lumière rouge.
Nr. M 819
MADE IN GERMANY
Mimosa A.-G., Dresden
ADUANA
PERFUMERIA

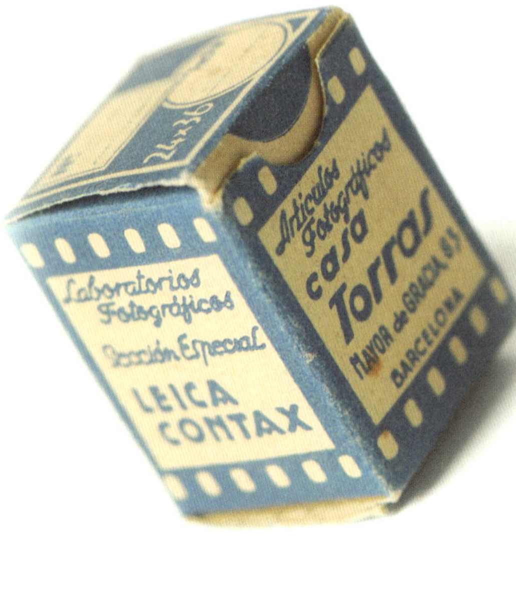
Laboratorios
Fotográficos
Sección Especial
LEICA
CONTAX
Artículos
Fotográficos
casa
Torras
MAYOR de GRACIA
BARCELONA

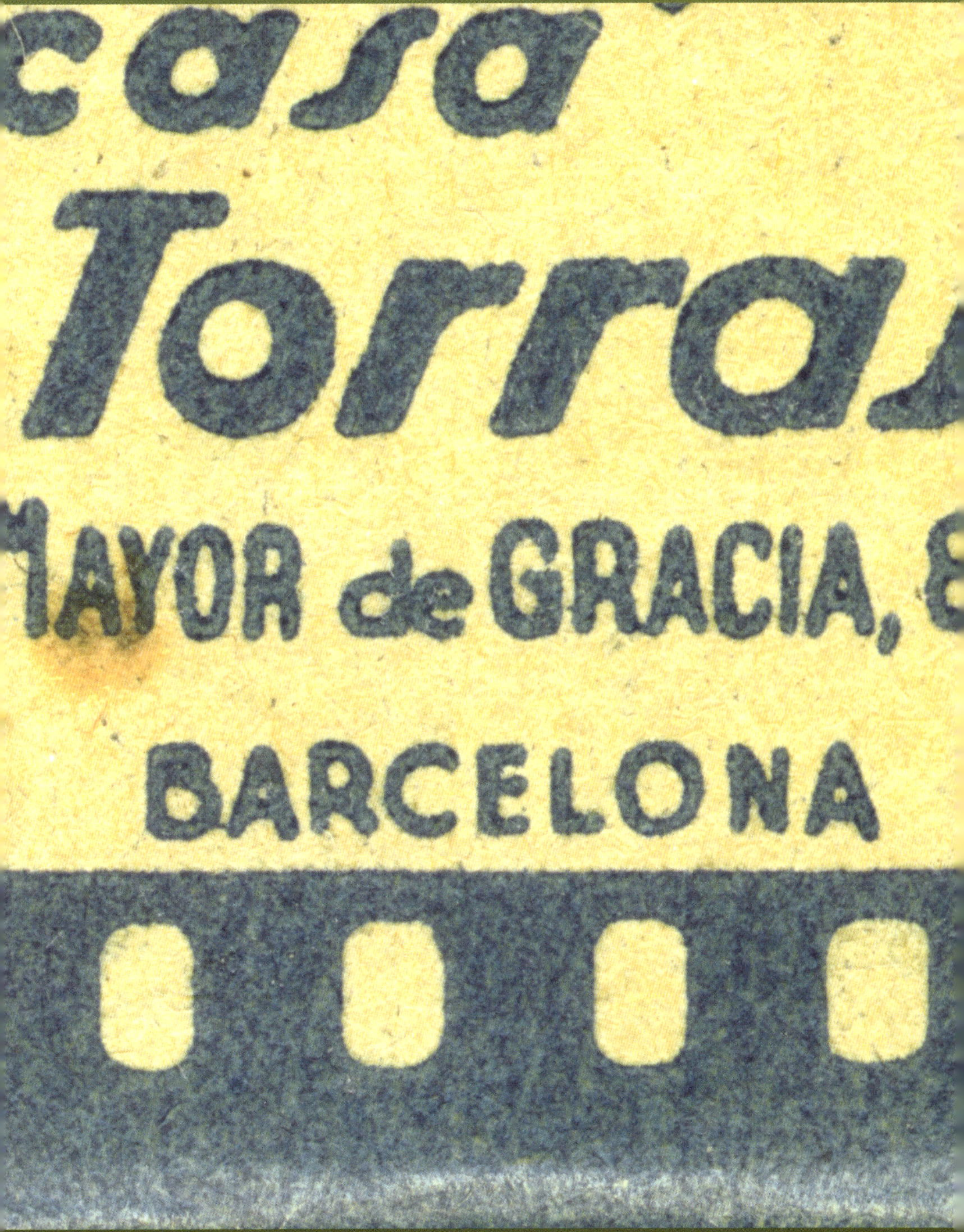
casa
Torra
AYOR de GRACIA,
BARCELONA

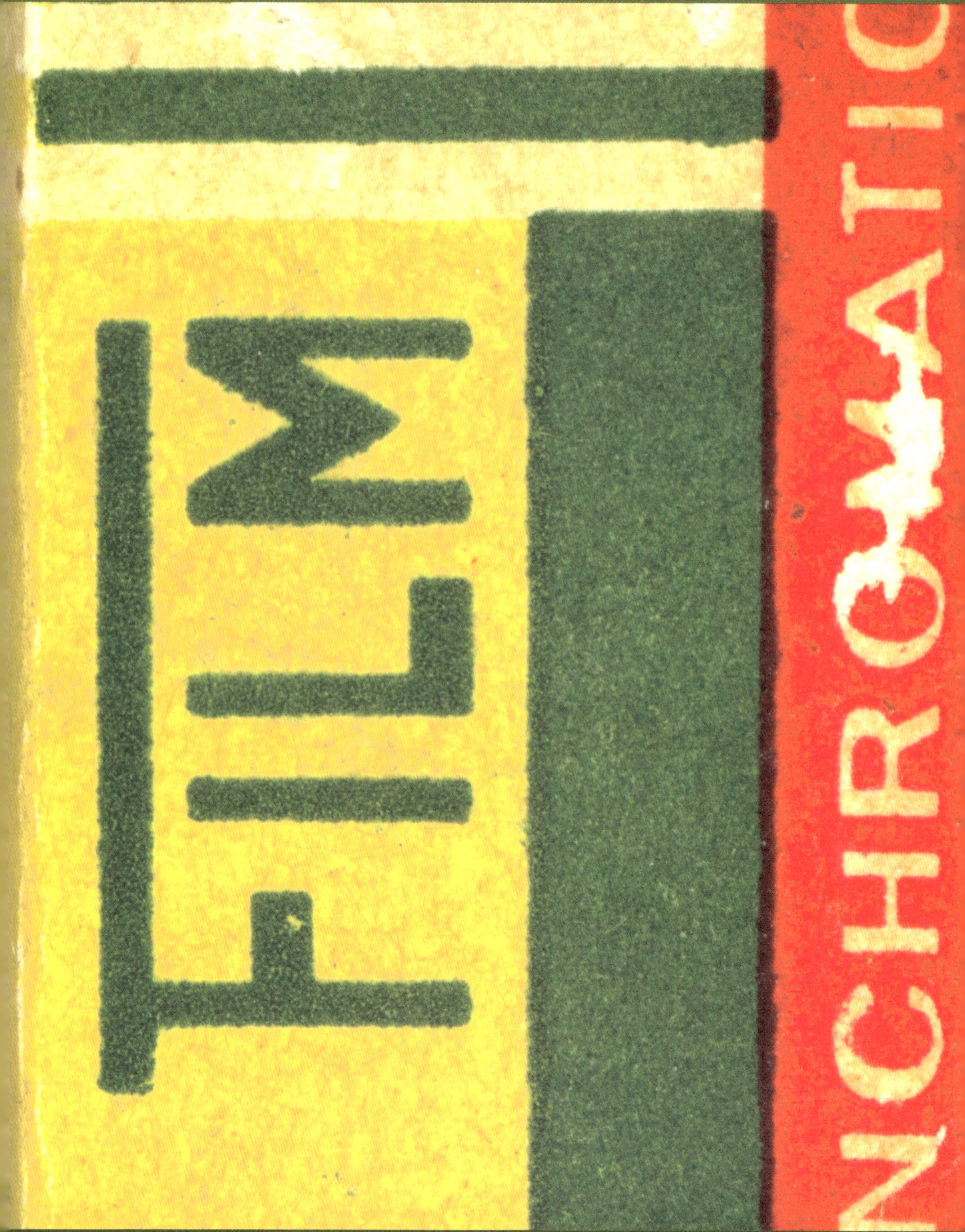
FILM

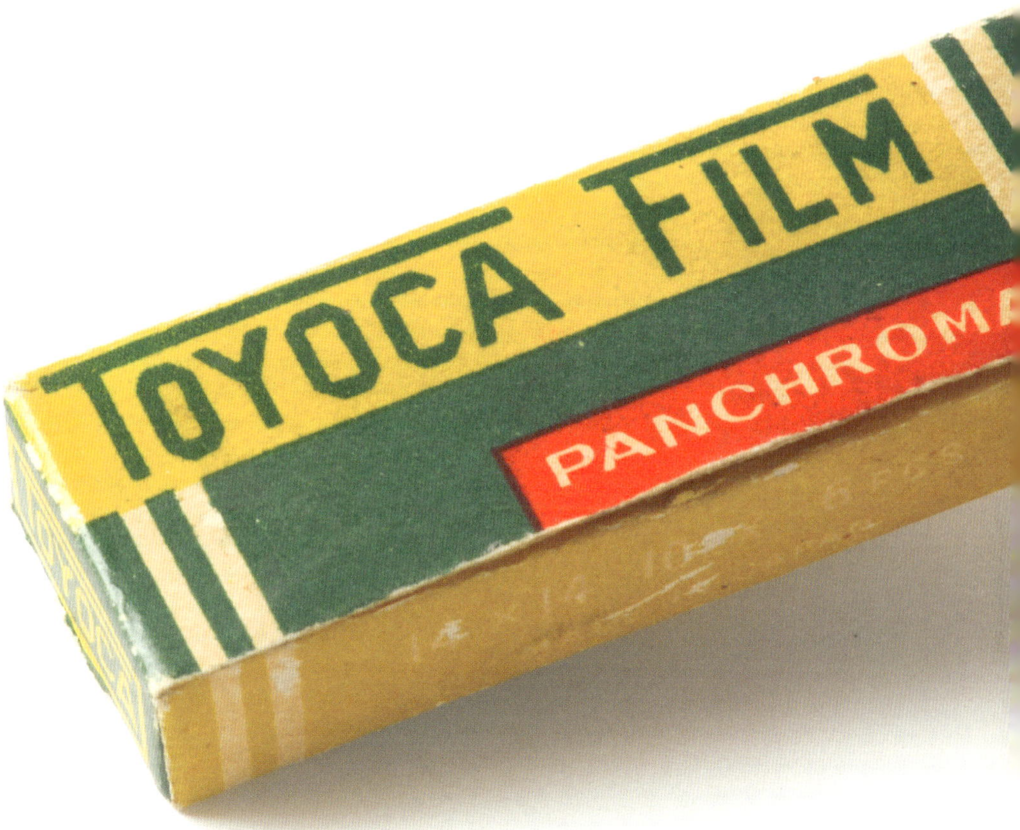
TOYOCA FILM
PANCHROMA
TOYOCA

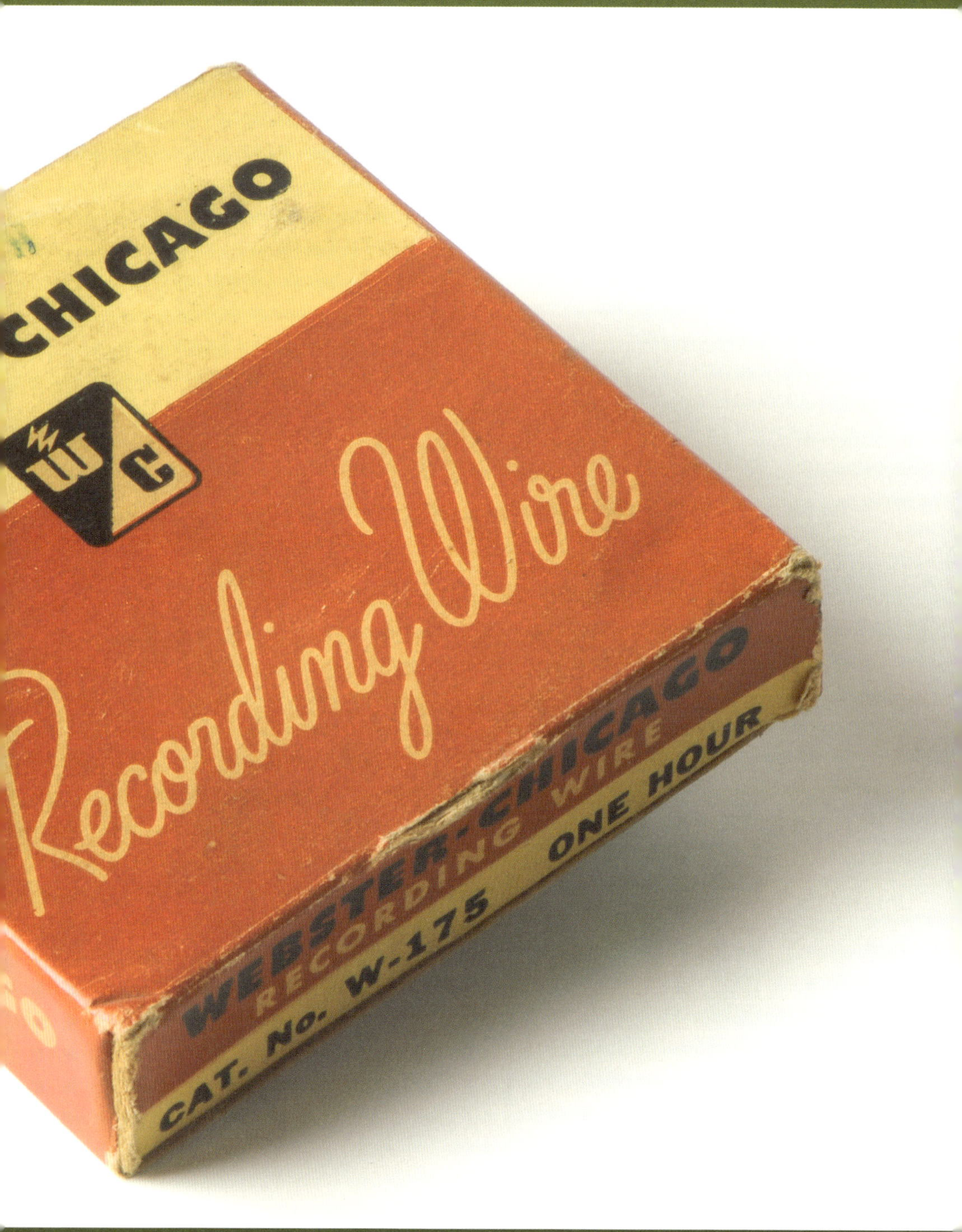
CHICAGO
WC
Recording Wire
WEBSTER-CHICAGO
RECORDING WIRE
CAT. No. W-175 ONE HOUR

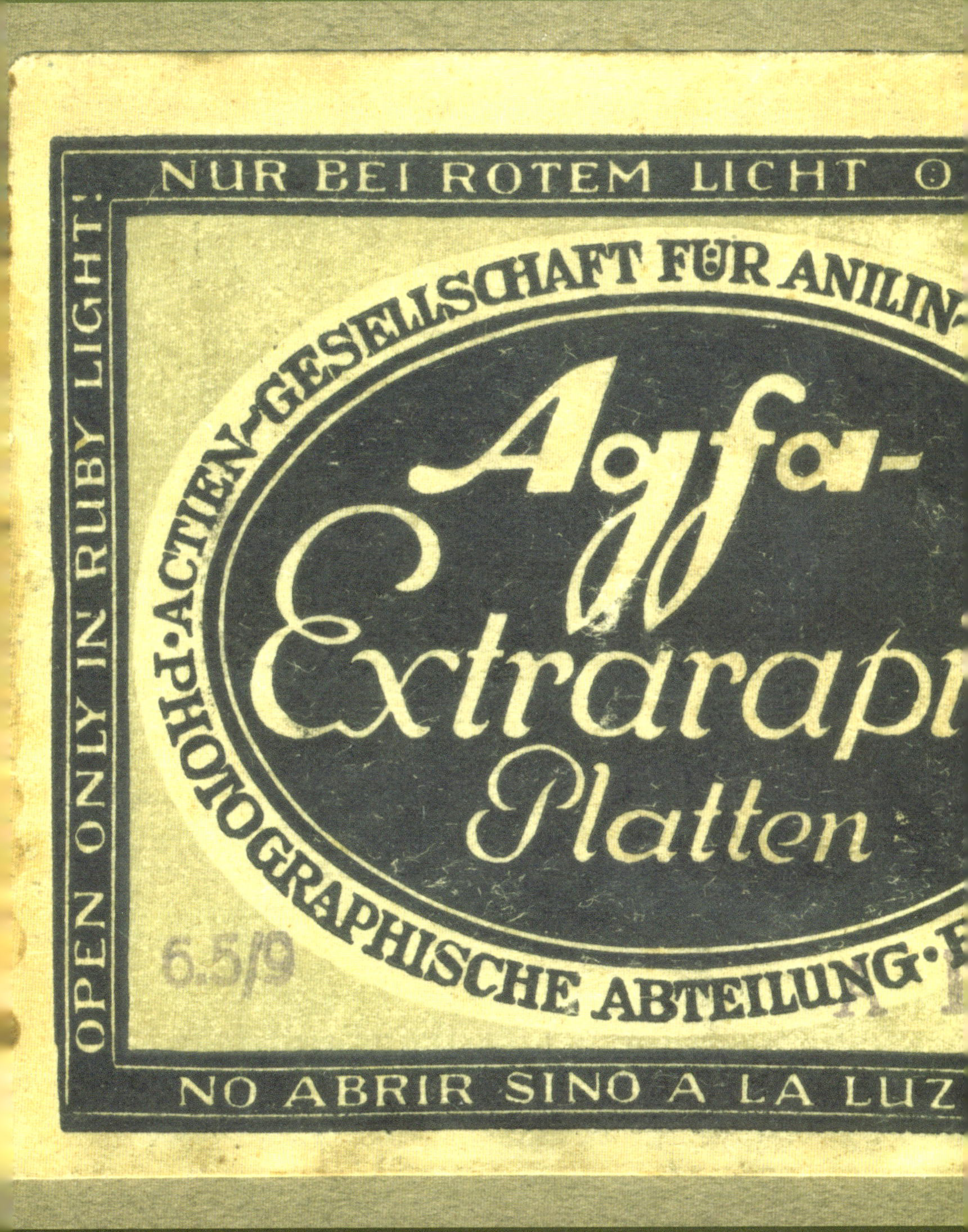
NUR BEI ROTEM LICHT
OPEN ONLY IN RUBY LIGHT!
ACTIEN-GESELLSCHAFT FÜR ANILIN-
Agfa-
Extrarapi
Platten
6.5/9
PHOTOGRAPHISCHE ABTEILUNG·
NO ABRIR SINO A LA LUZ

SONALC

RSONALCAREPE

PERSONAL CARE

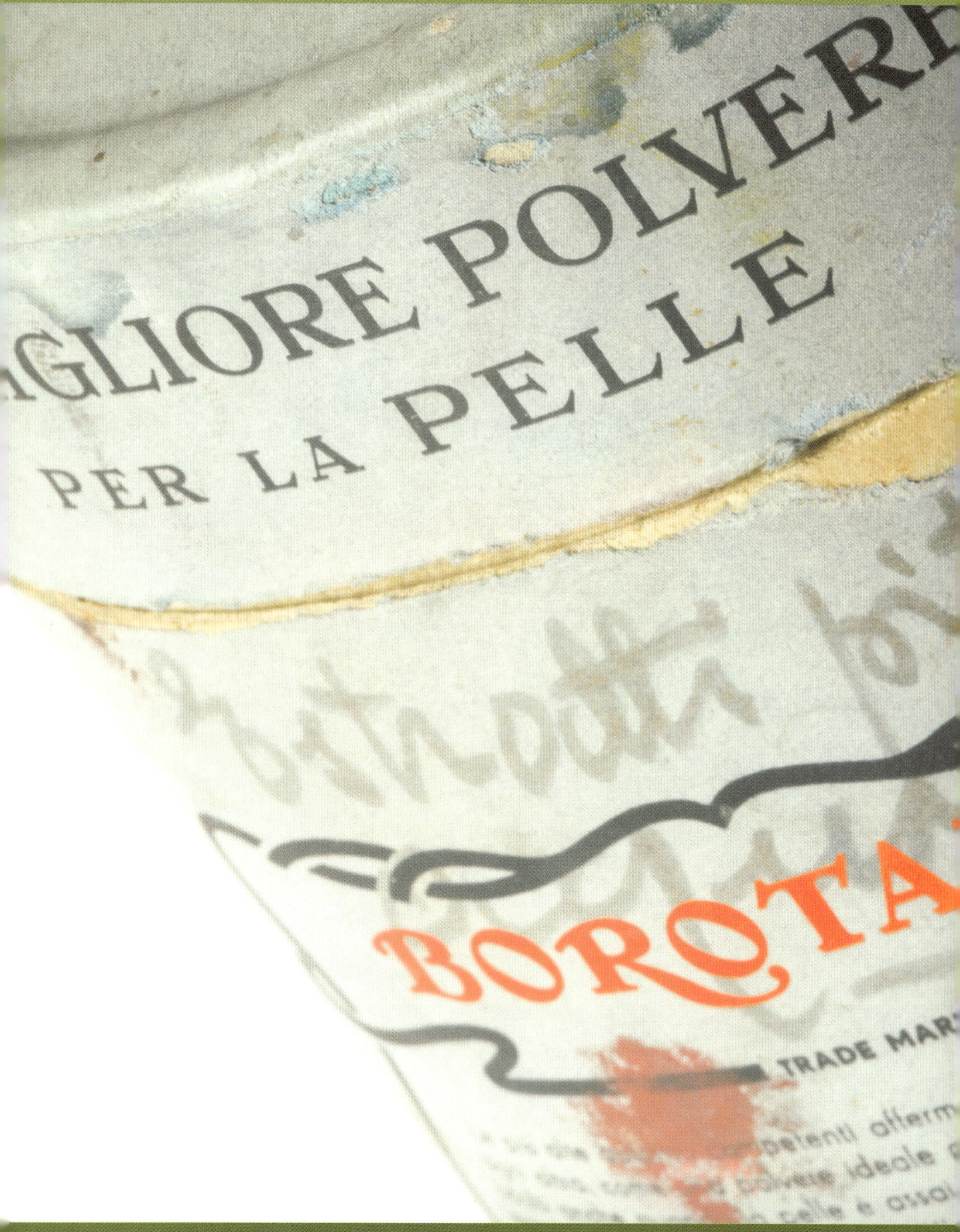
GLIORE POLVER
PER LA PELLE
BOROTA
TRADE MAR

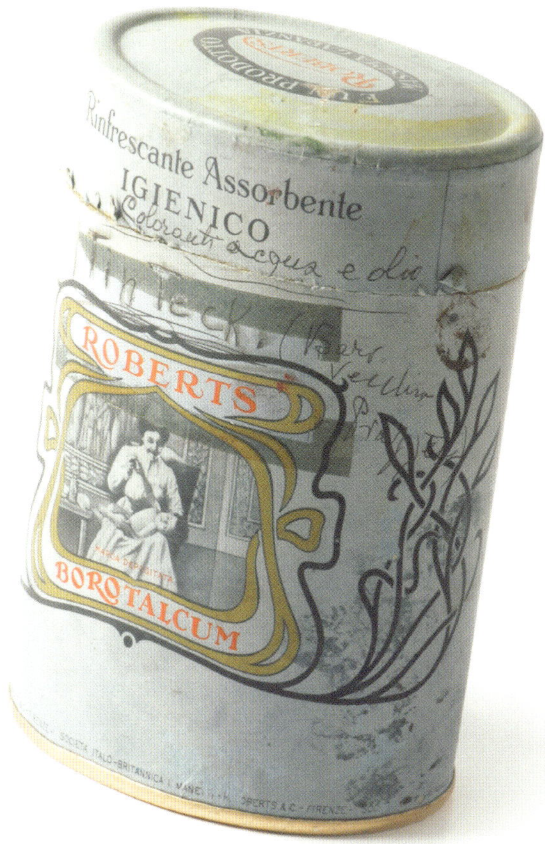
Rinfrescante Assorbente
IGIENICO
ROBERTS
BOROTALCUM

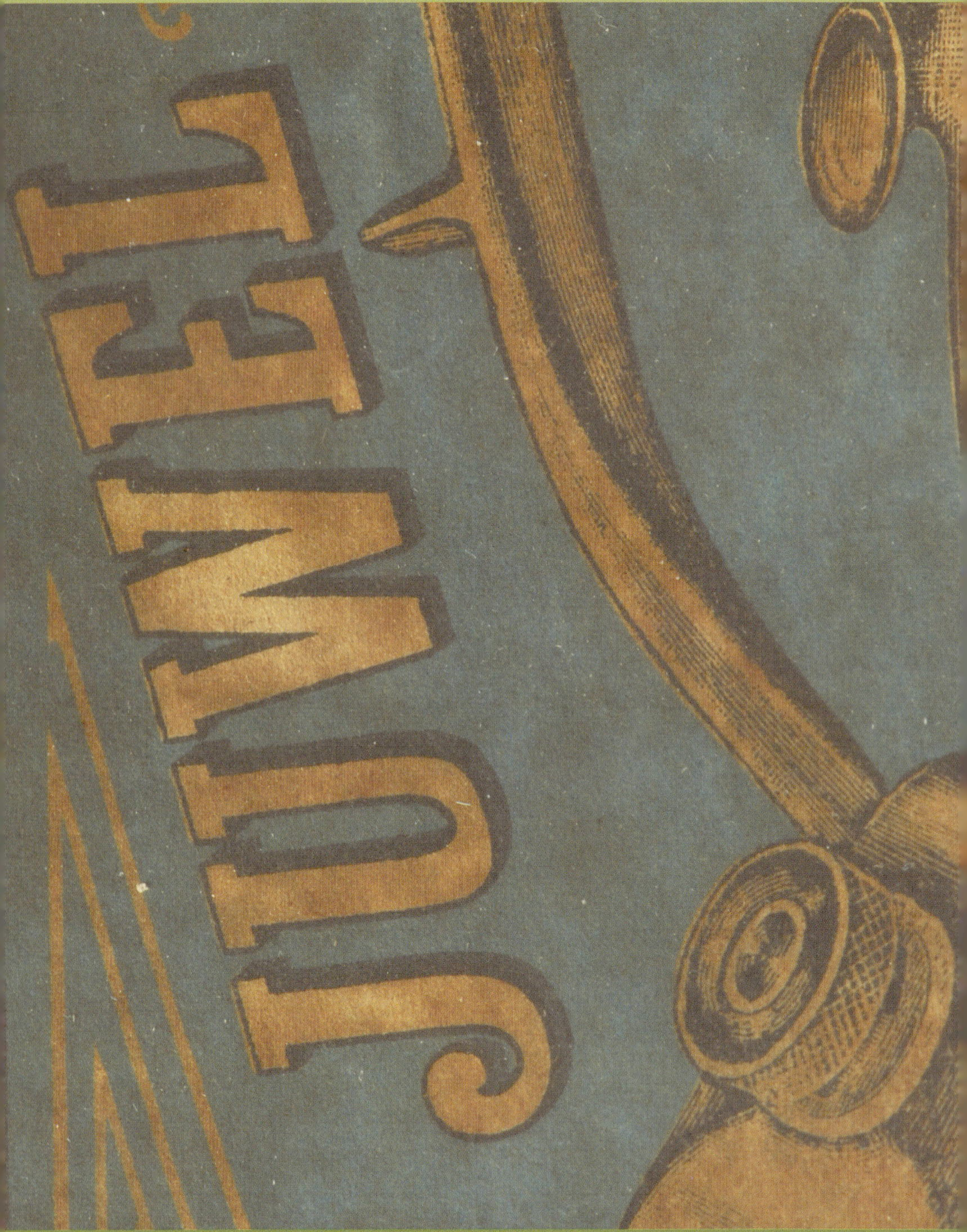
JUWEL

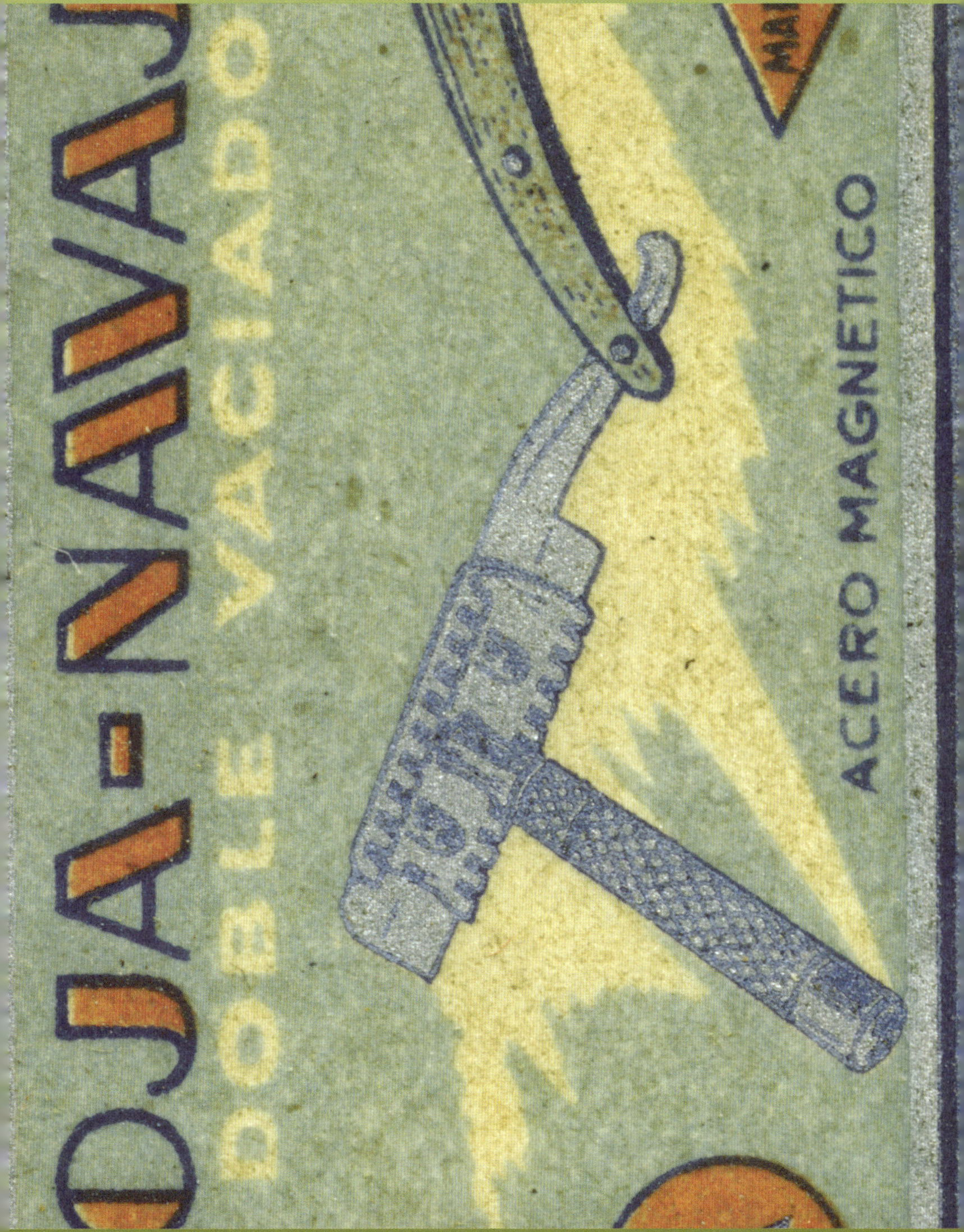
DOBLE VACIADO
ACERO MAGNETICO

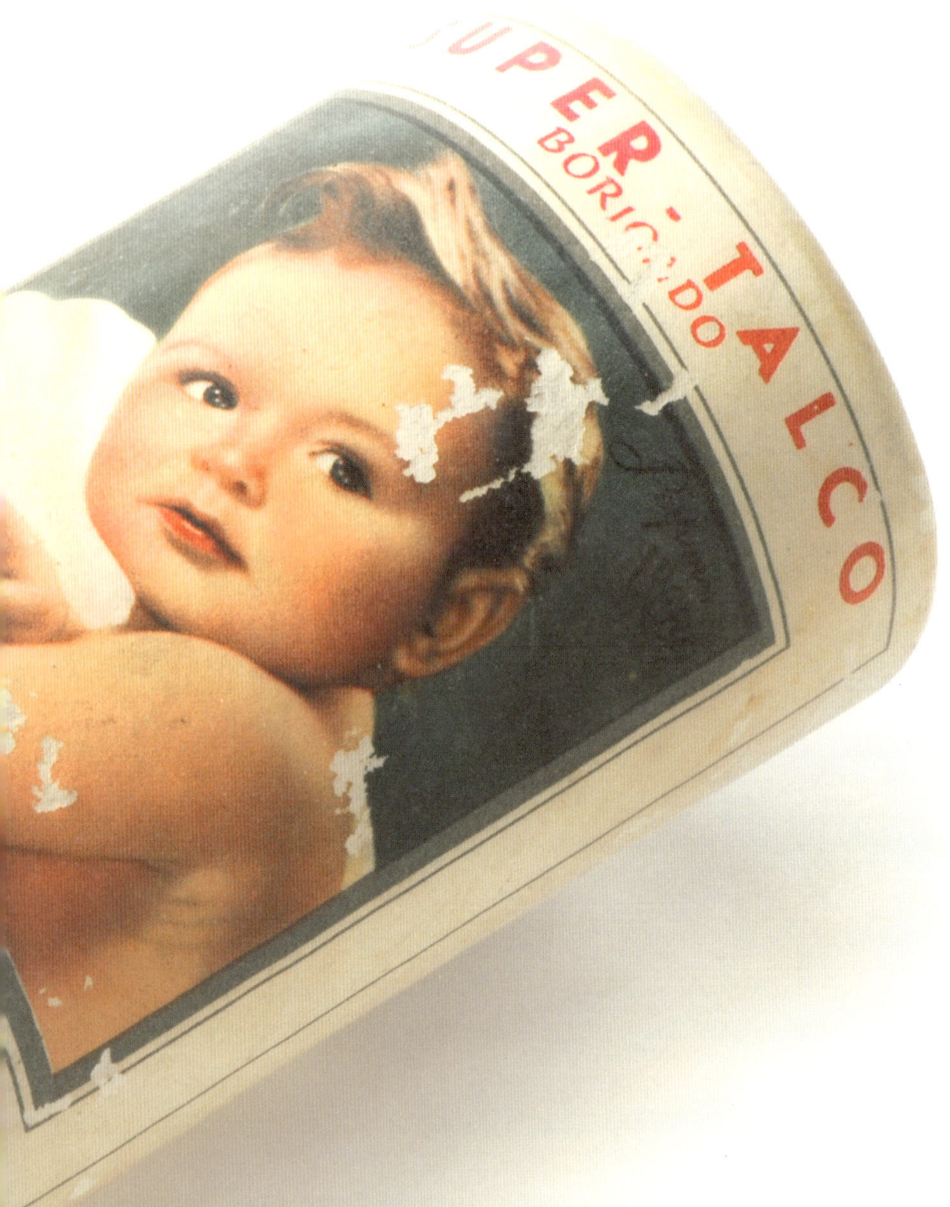
UPER·TALCO

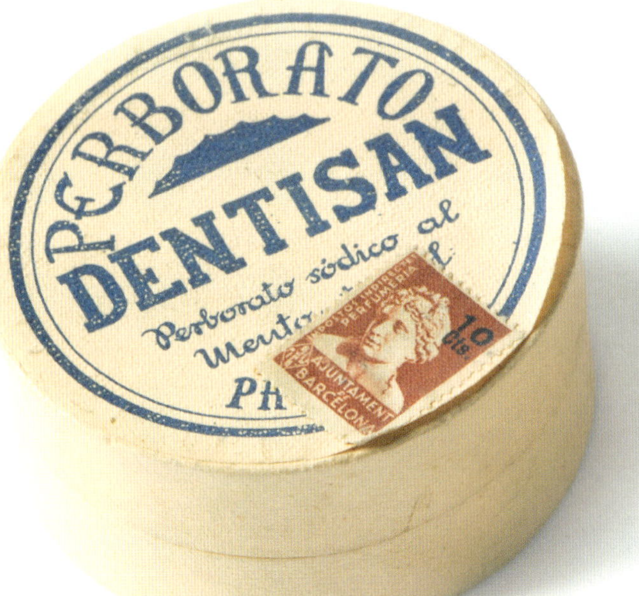
PERBORATO
DENTISAN
Perborato sódico al
PERFUMERIA
10
AJUNTAMENT
BARCELONA

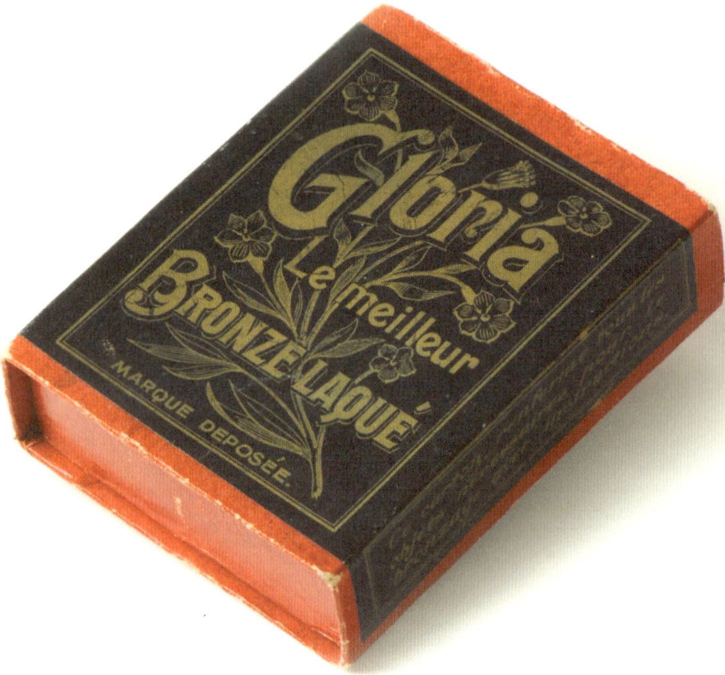
Gloria
Le meilleur
BRONZE LAQUÉ
MARQUE DEPOSÉE.

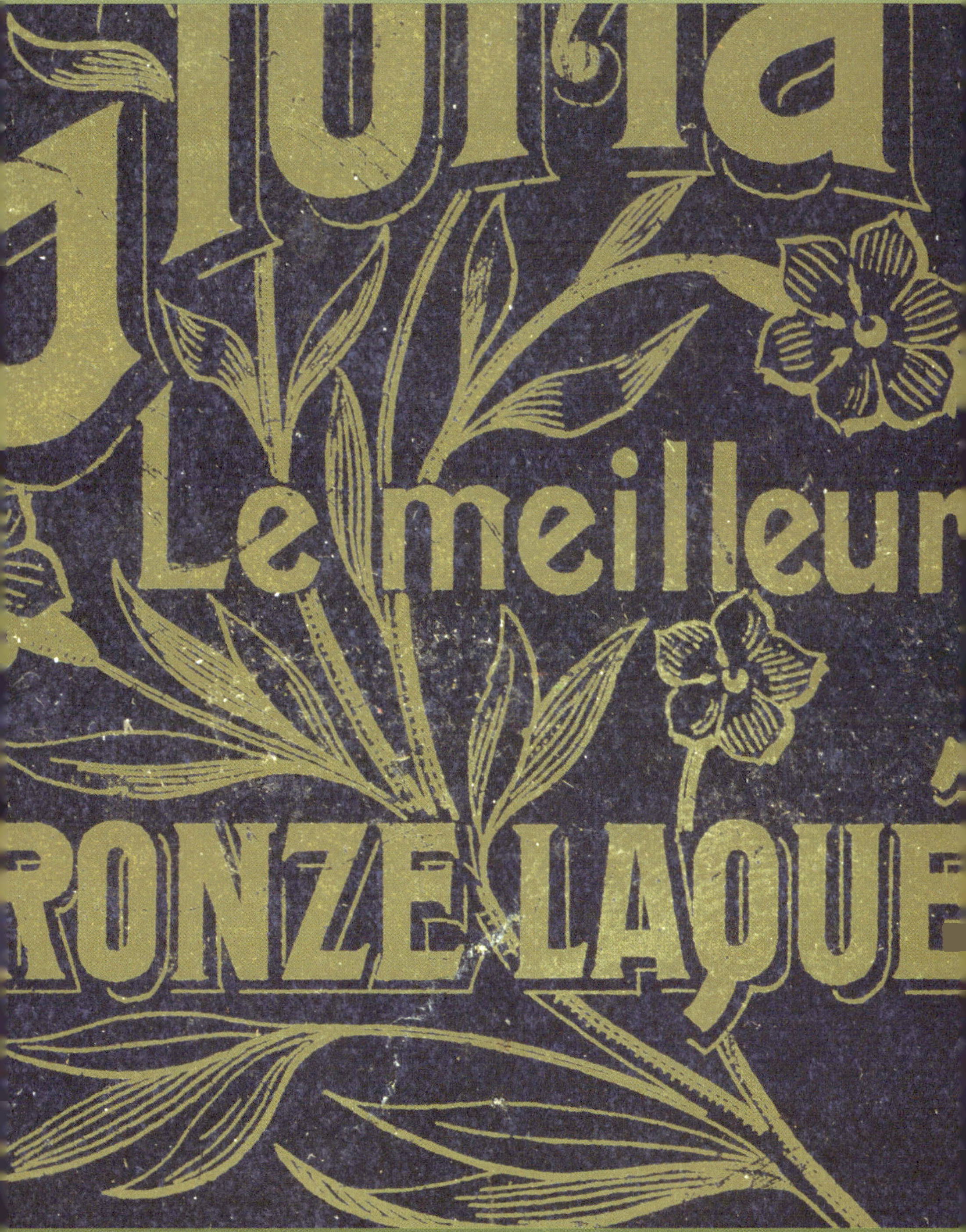
Le meilleur
RONZE LAQUE

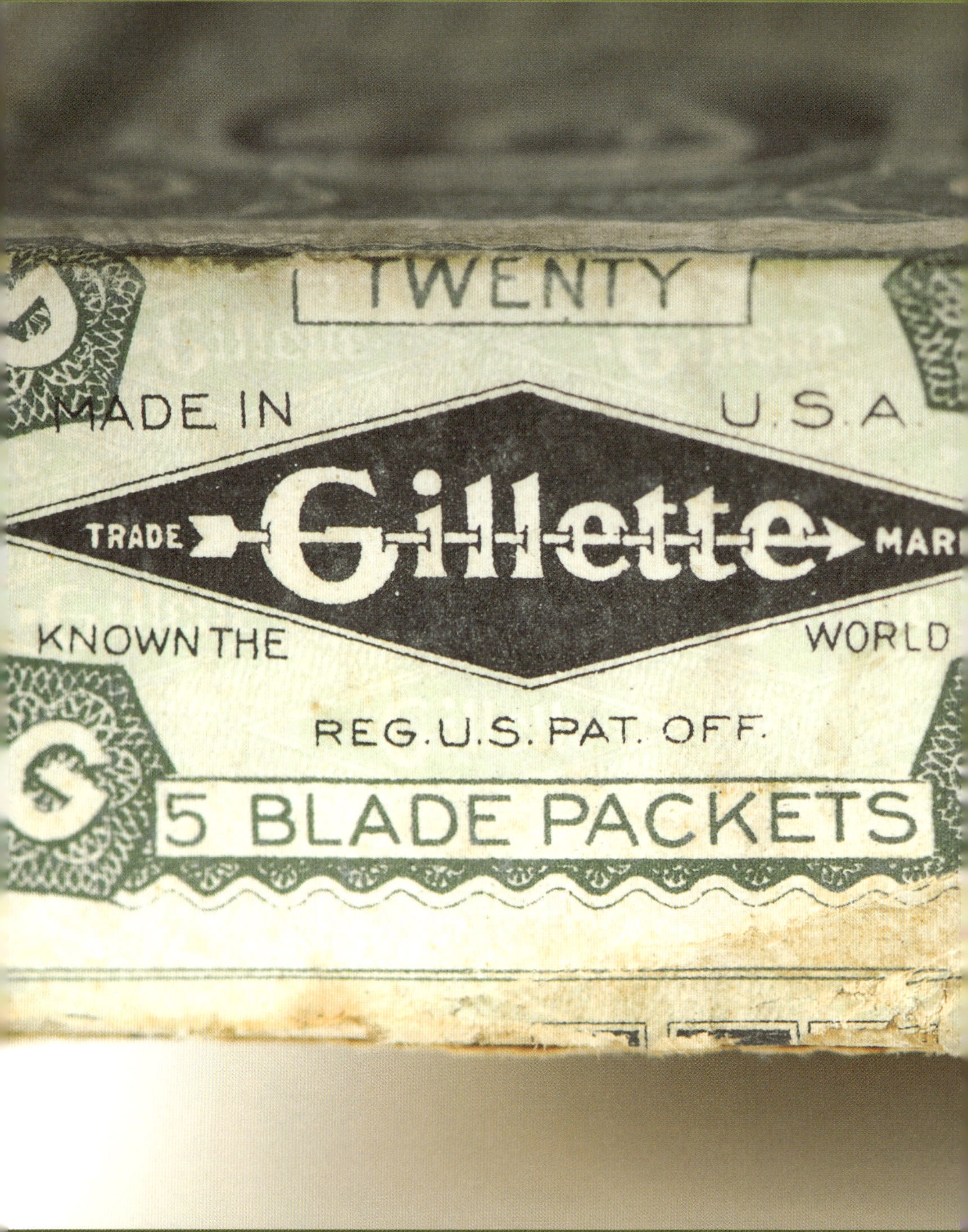
TWENTY
MADE IN
U.S.A.
TRADE
Gillette
KNOWN THE
WORLD
REG. U.S. PAT. OFF.
5 BLADE PACKETS

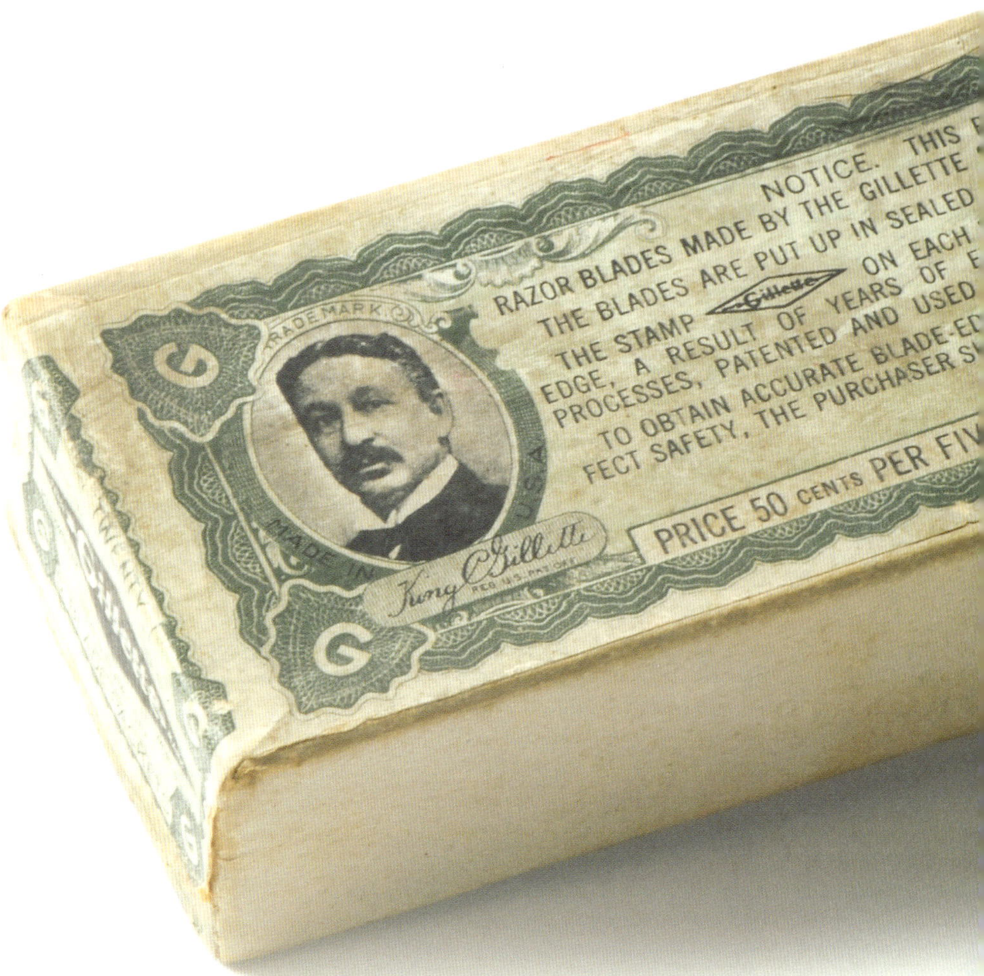
NOTICE. THIS
RAZOR BLADES MADE BY THE GILLETTE
THE BLADES ARE PUT UP IN SEALED
THE STAMP Gillette ON EACH
EDGE, A RESULT OF YEARS OF
PROCESSES, PATENTED AND USED
TO OBTAIN ACCURATE
FECT SAFETY, THE PURCHASER
PRICE 50 CENTS PER
MADE IN U.S.A.
King C. Gillette
REG. U.S. PAT. OFF.
G
G

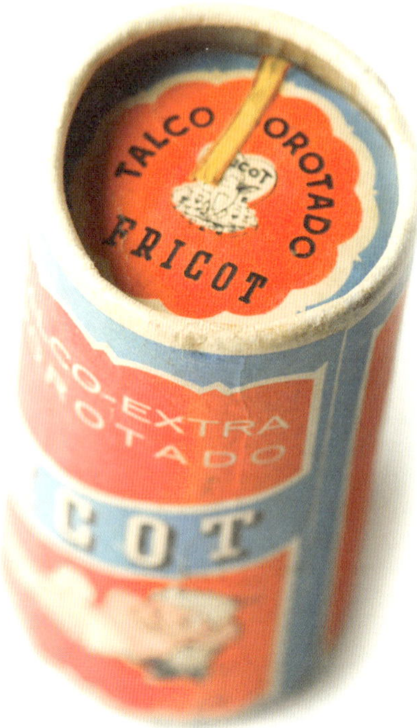
TALCO BOROTADO
FRICOT
TALCO-EXTRA
BOROTADO
COT

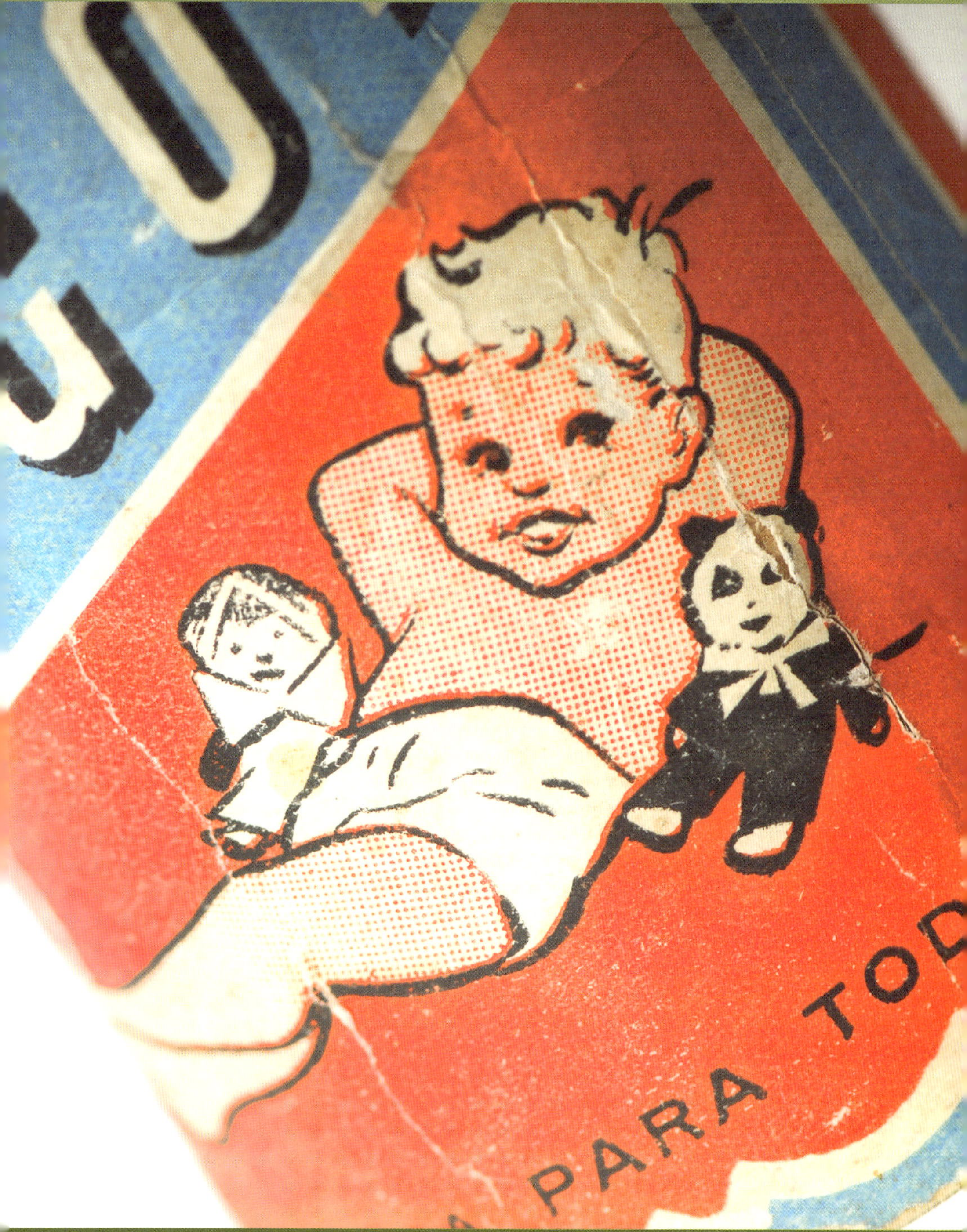
PARA TOD

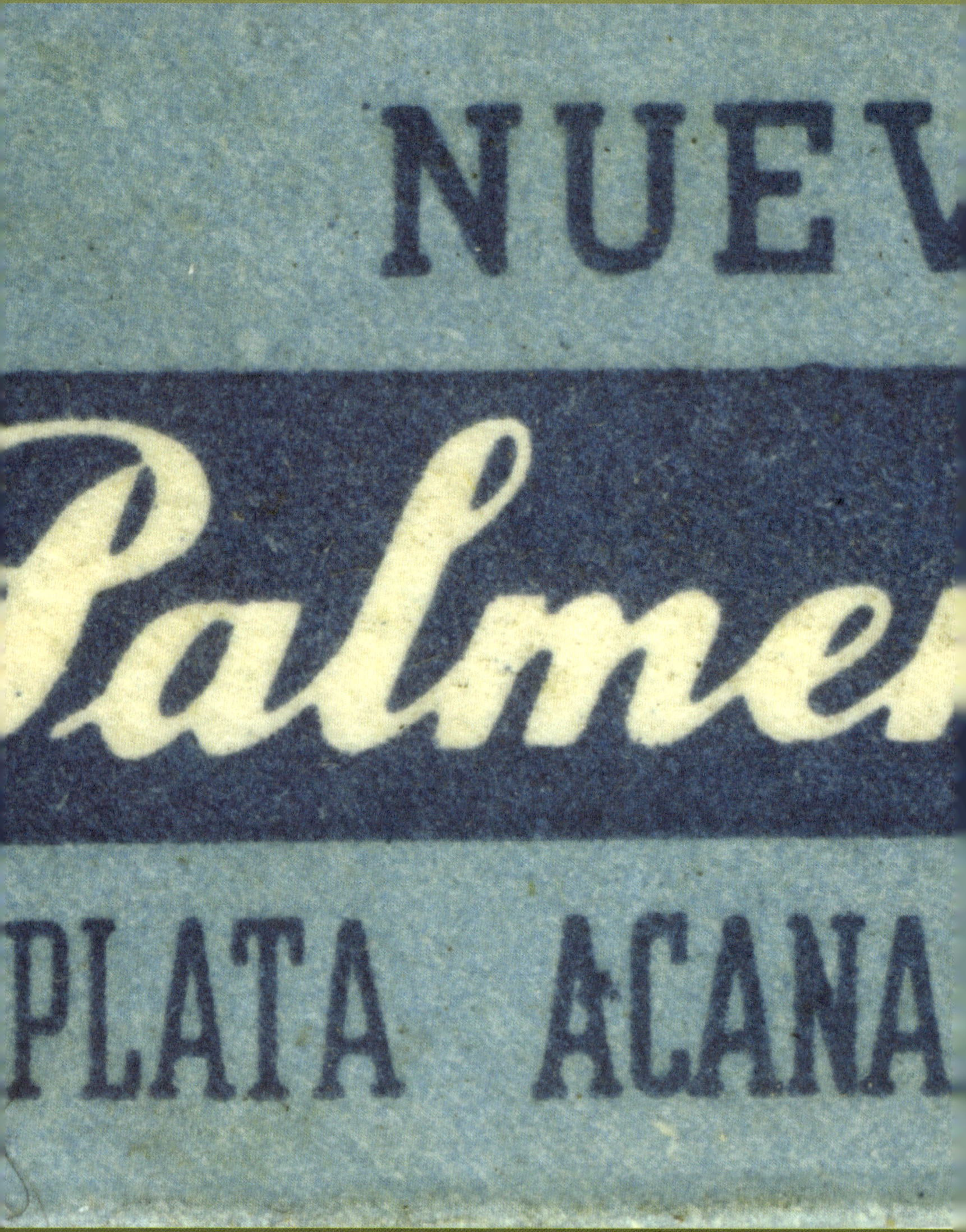
NUEV
Palme
PLATA ACANA

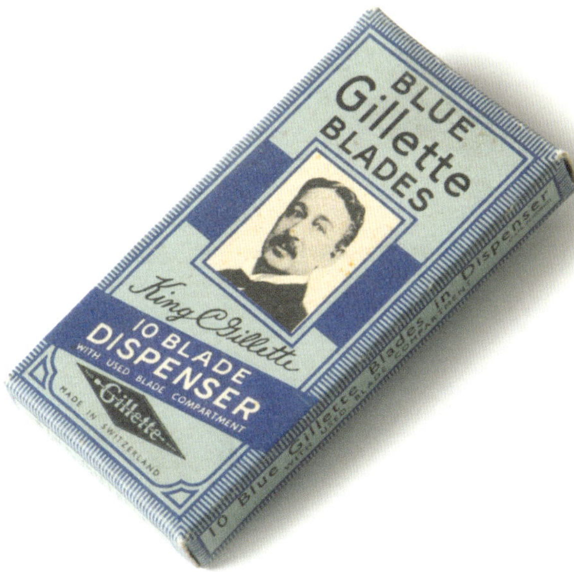
BLUE
Gillette
BLADES
King C. Gillette
10 BLADE
DISPENSER
WITH USED BLADE COMPARTMENT
Gillette
MADE IN SWITZERLAND
10 Blue Gillette Blades in Dispenser

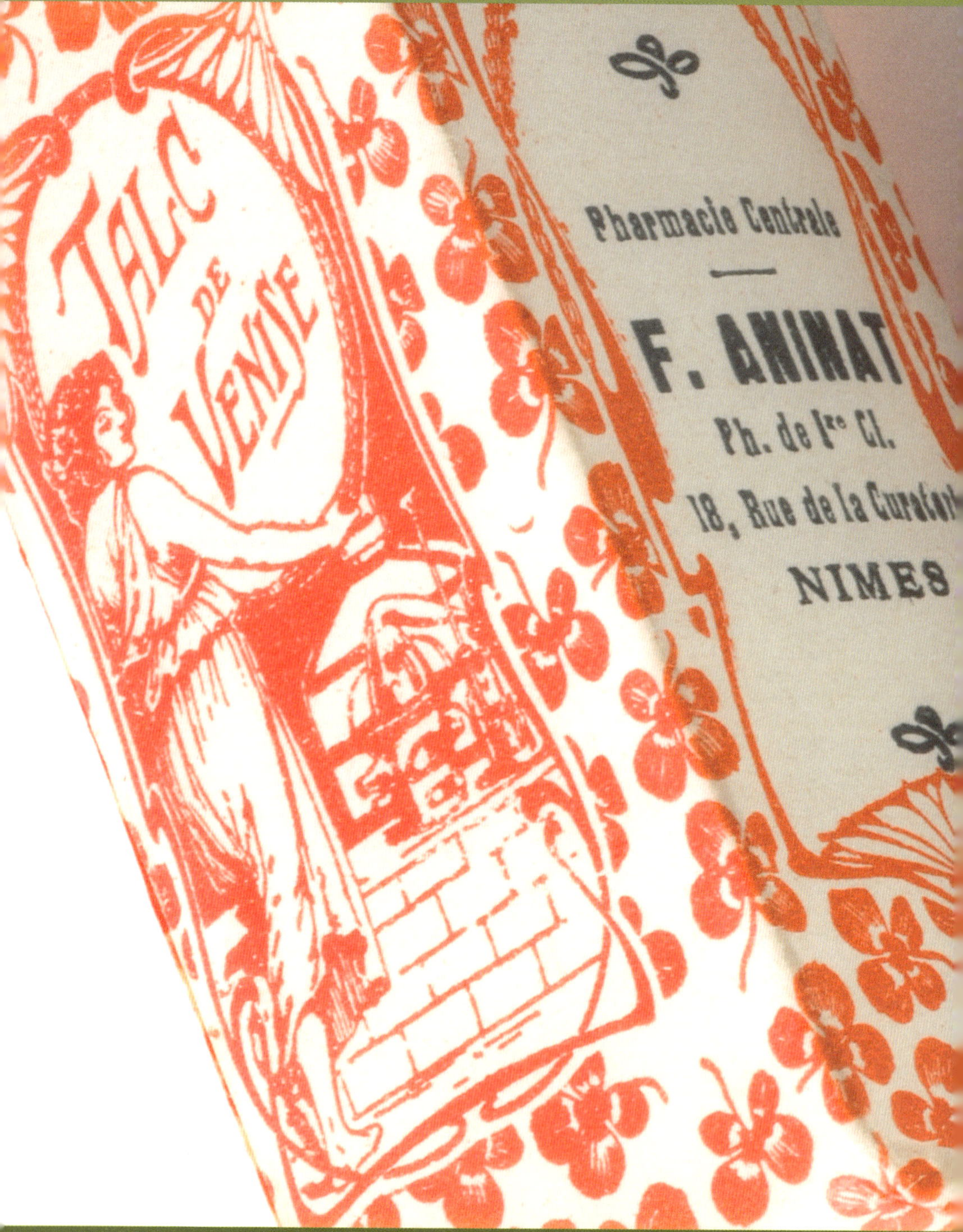
TALC DE VENISE
Pharmacie Centrale
F. ANIMAT
Ph. de 1re Cl.
18, Rue de la Curaterie
NIMES

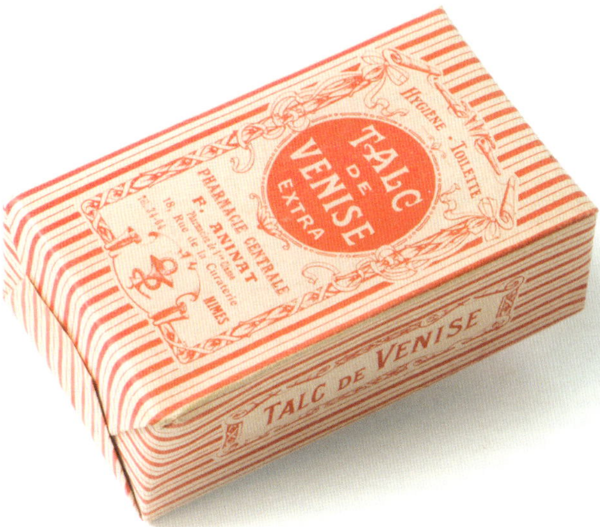
HYGIÈNE · TOILETTE
TALC DE VENISE EXTRA
PHARMACIE CENTRALE
F. ANINAT
NIMES
TALC DE VENISE

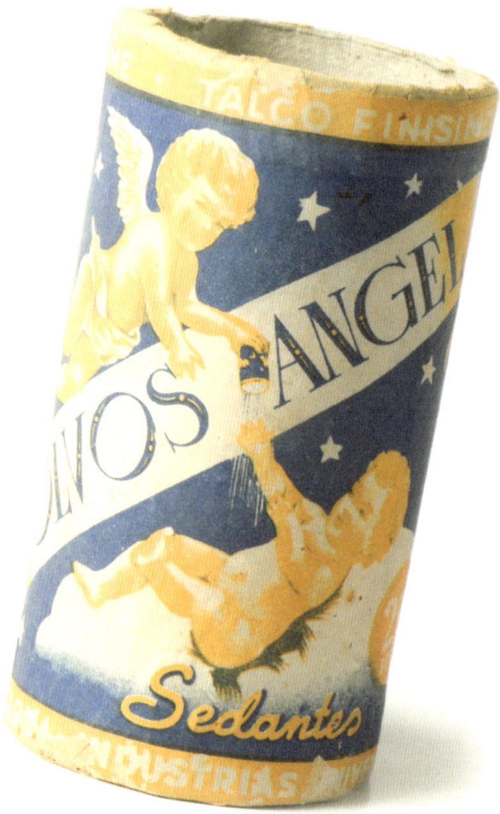
TALCO
Sedantes

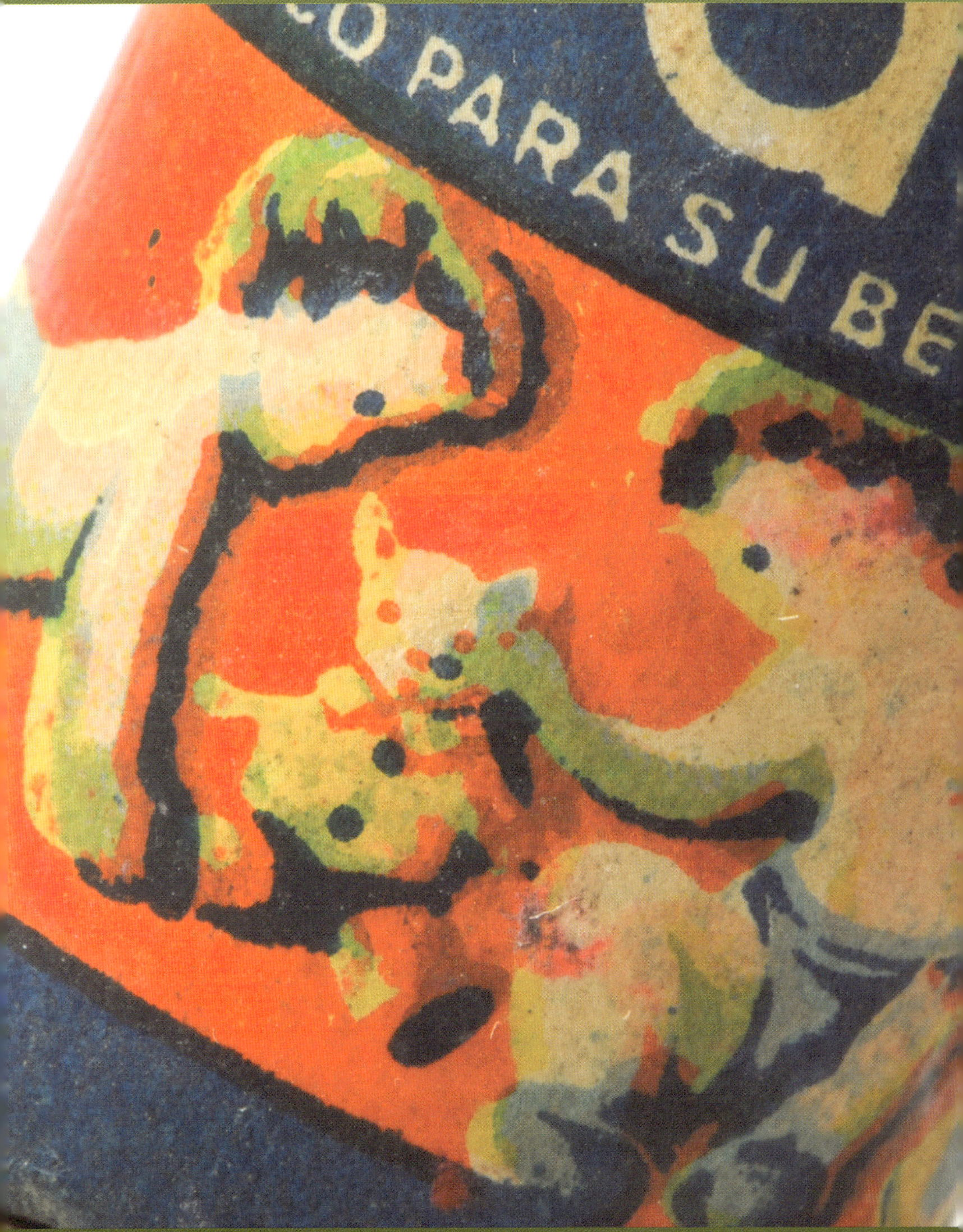
CO PARA SUBE

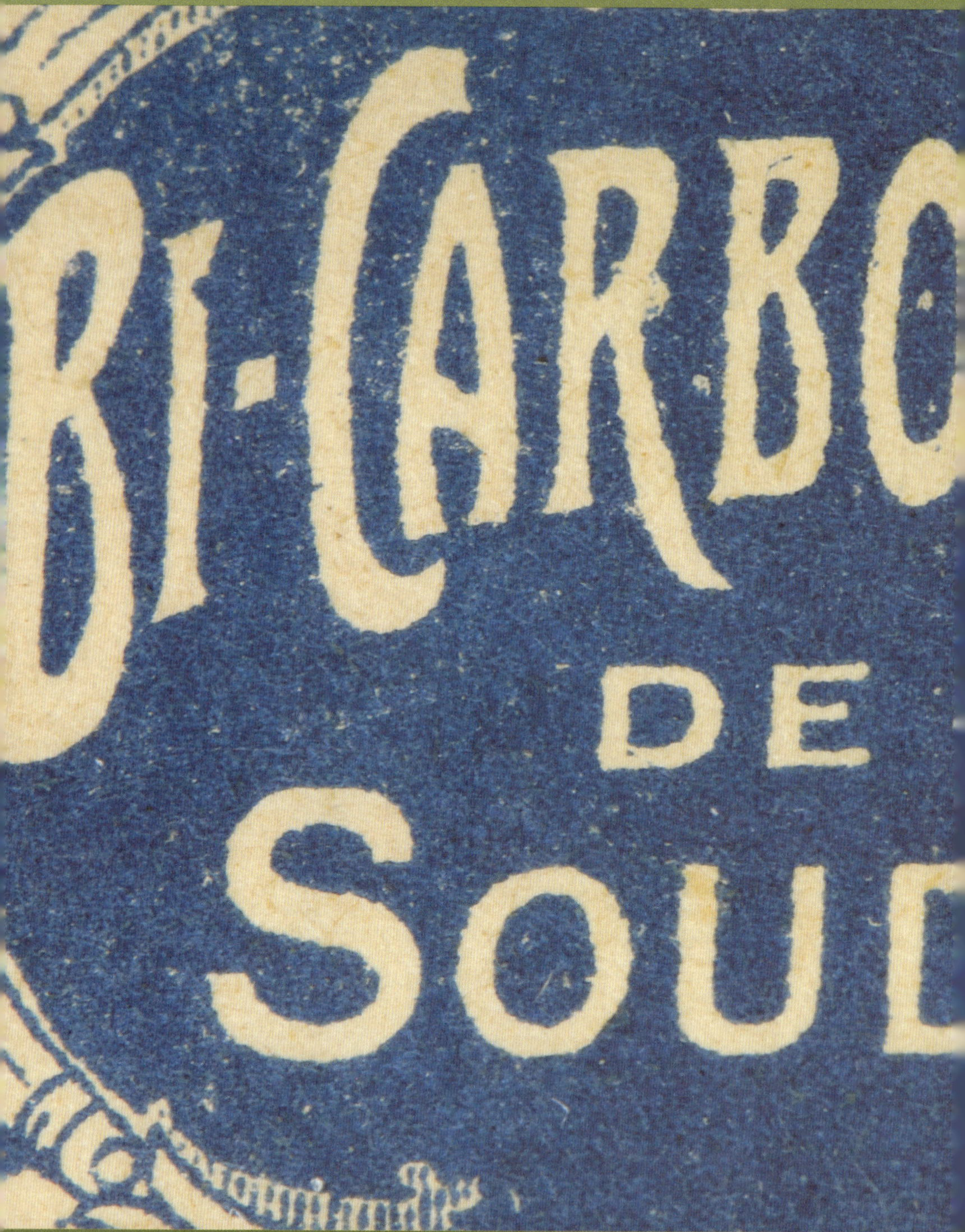
BI-CARBO
DE
SOUD

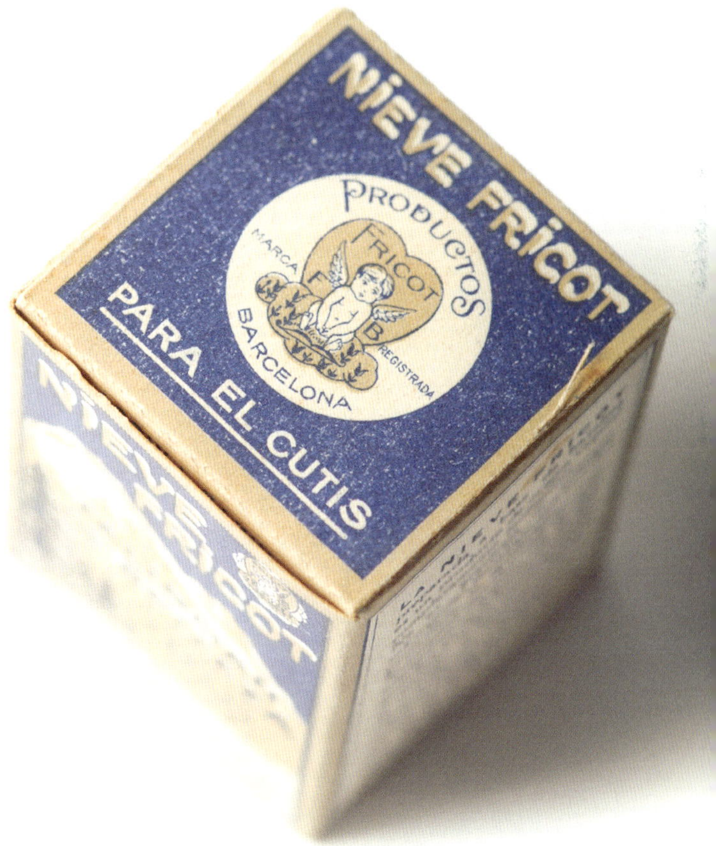
NIEVE FRICOT
PRODUCTOS
FRICOT
MARCA
F
B
BARCELONA
REGISTRADA
PARA EL CUTIS

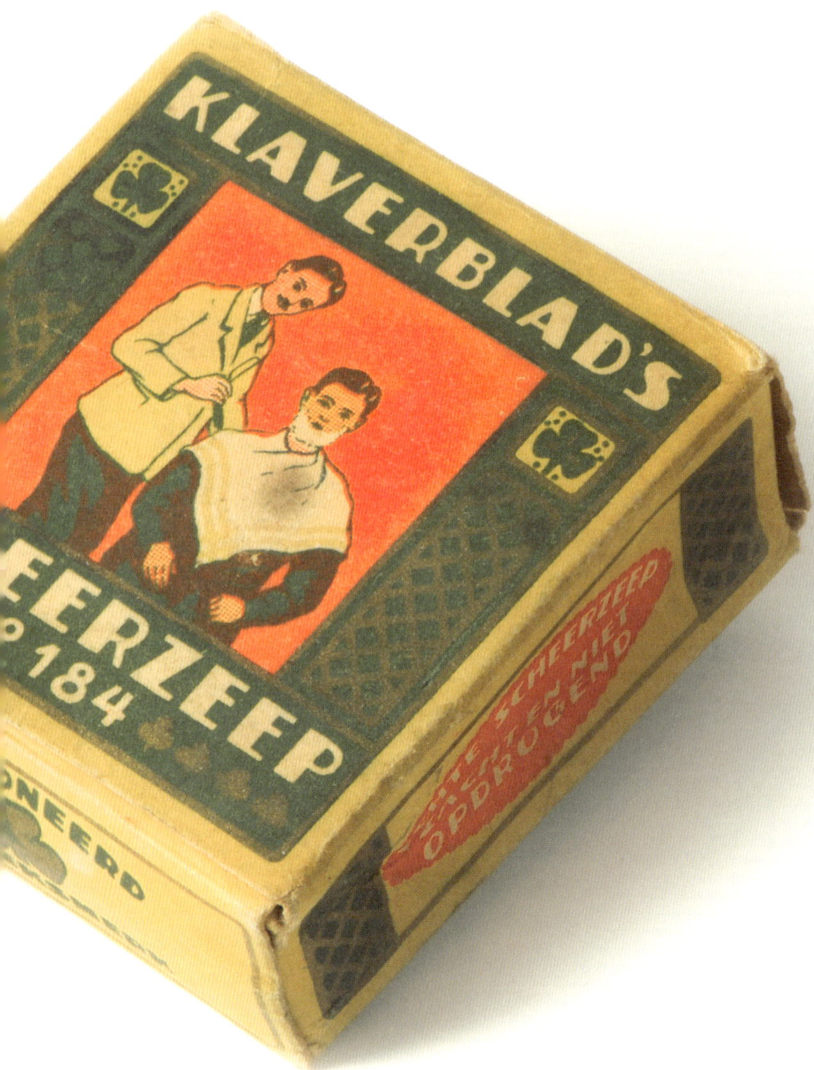
KLAVERBLAD'S
SCHEERZEEP
OPDROGEND

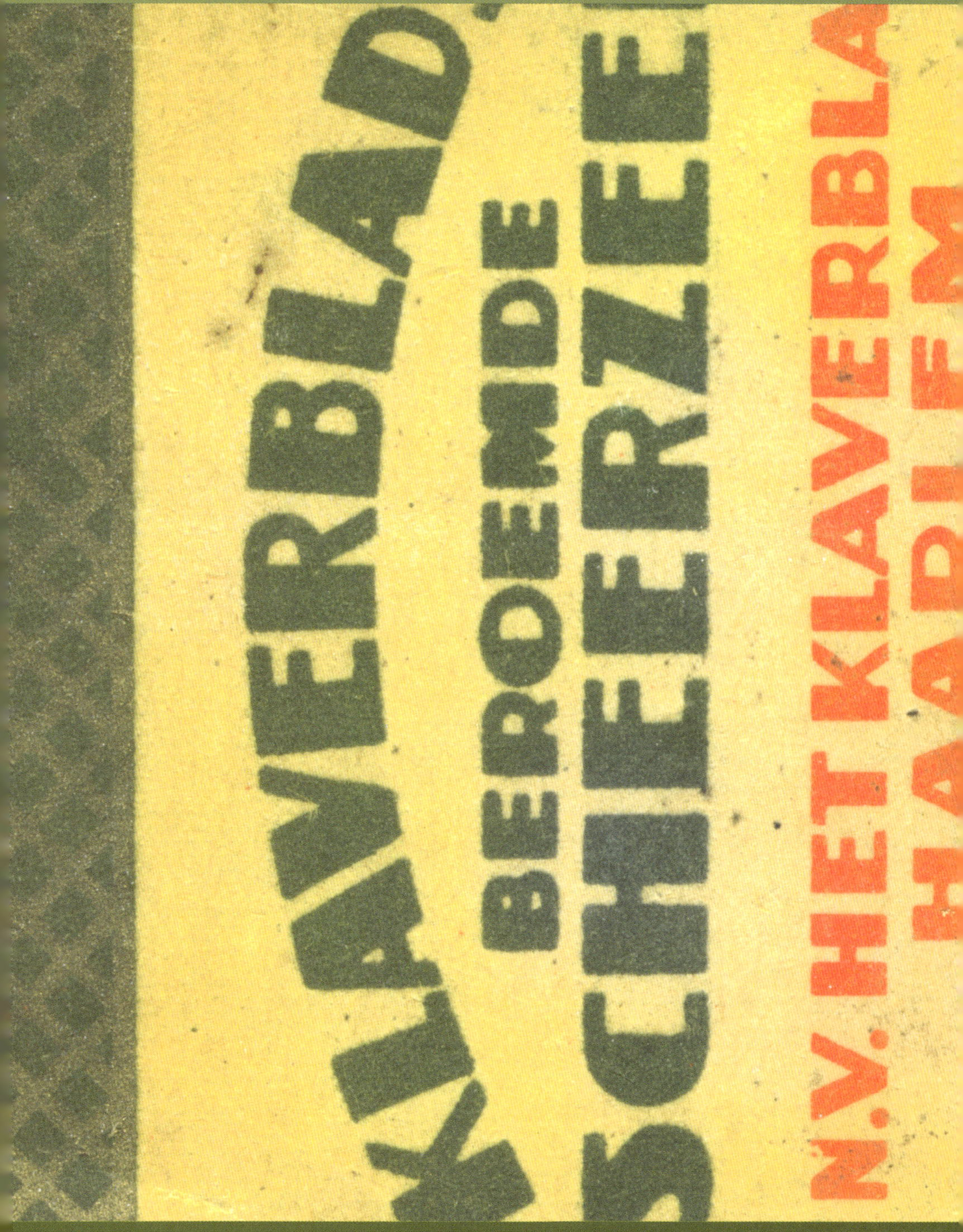
KLAVERBLAD
BEROEMDE
N.V. HET KLAVERBLA

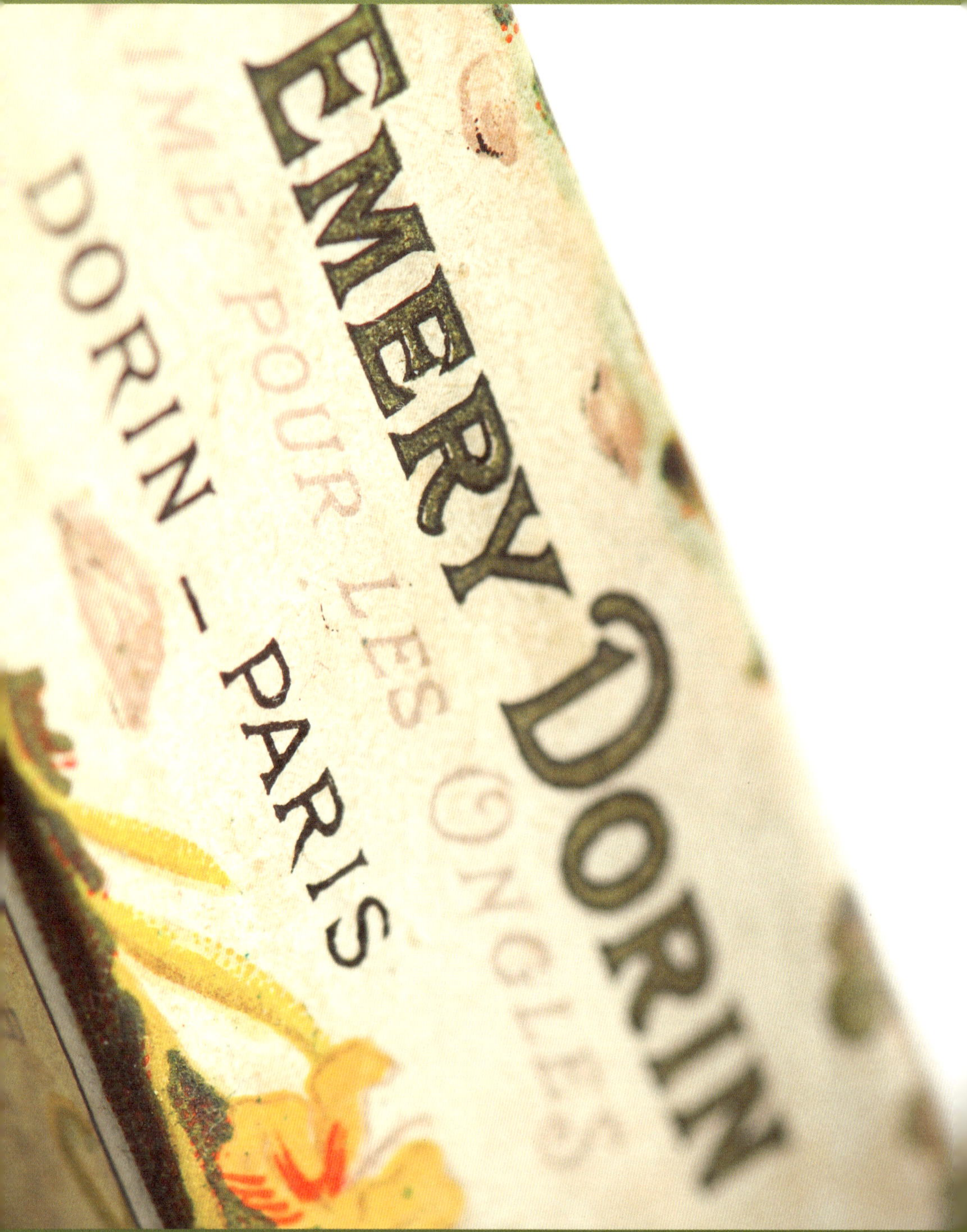
EMERY DORIN
POUR LES ONGLES
DORIN — PARIS

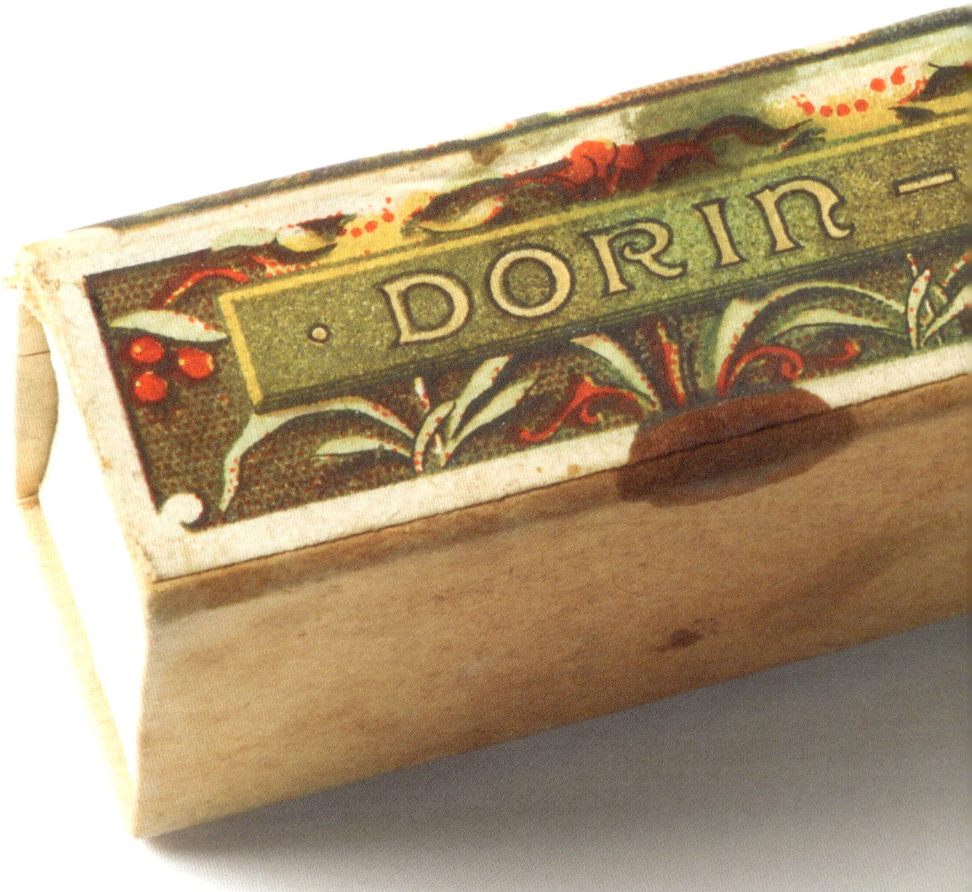
DORIN

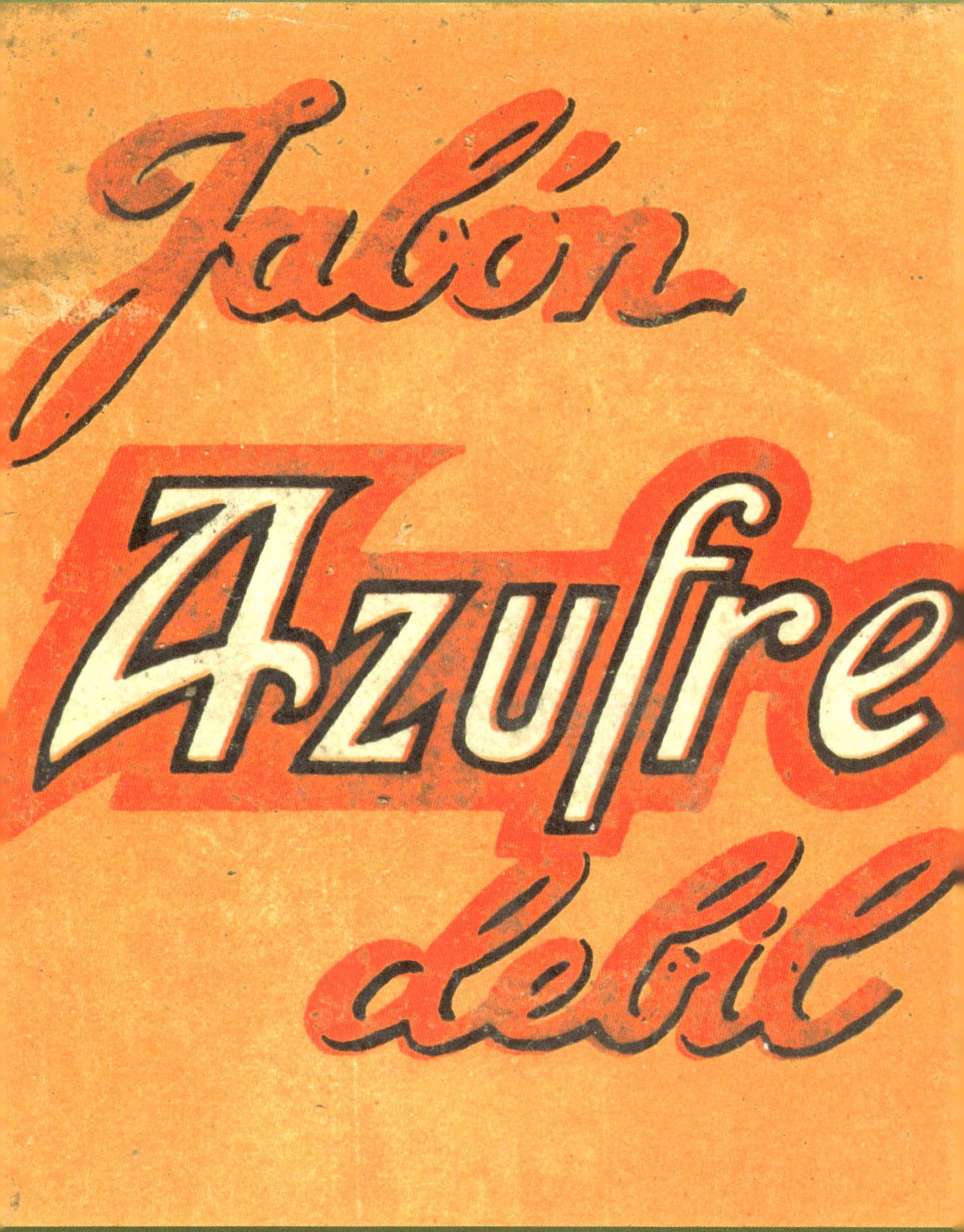
Jabón
Azufre
débil

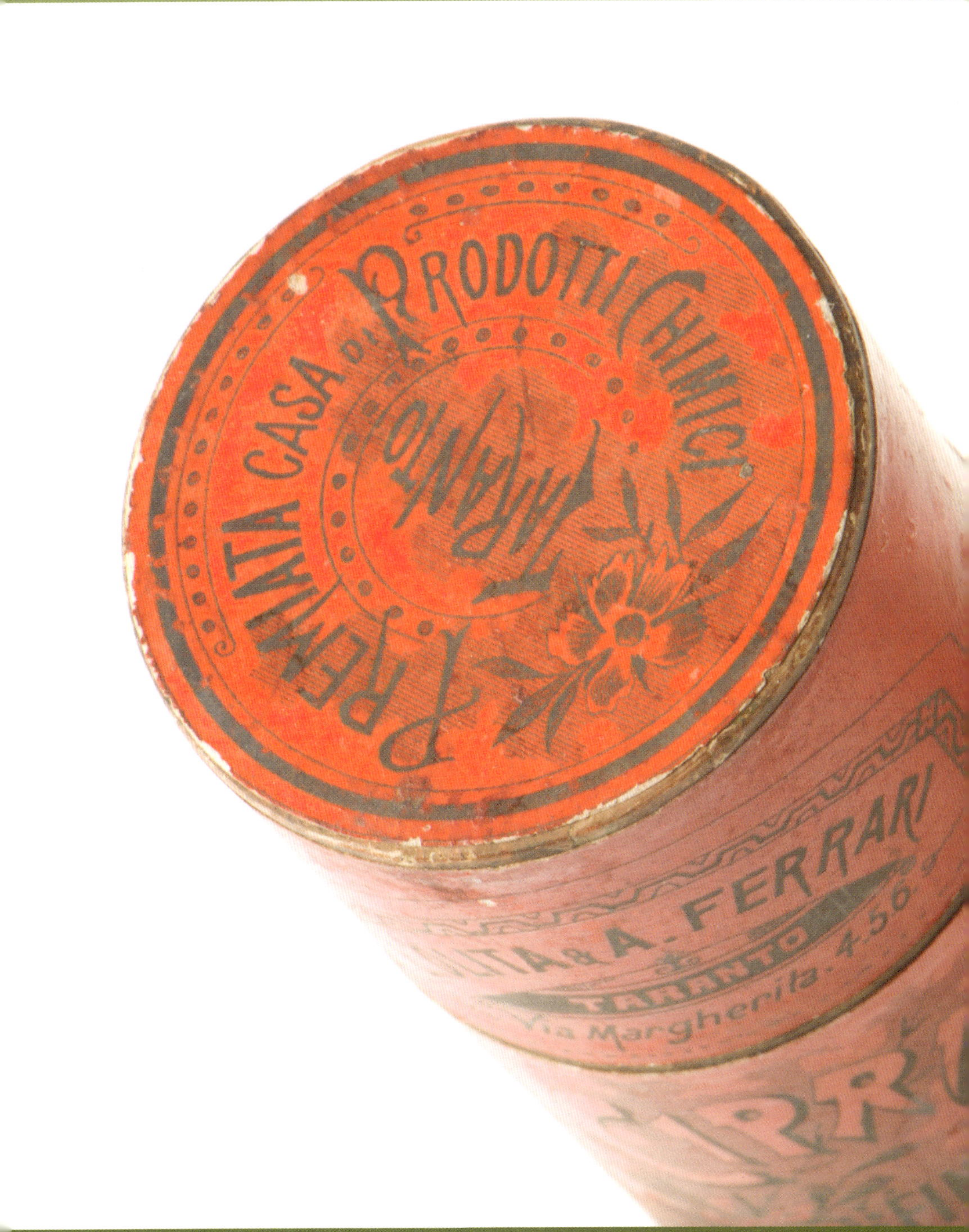
PREMIATA CASA DI PRODOTTI CHIMICI
A. FERRARI
TARANTO
Via Margherita 4.56

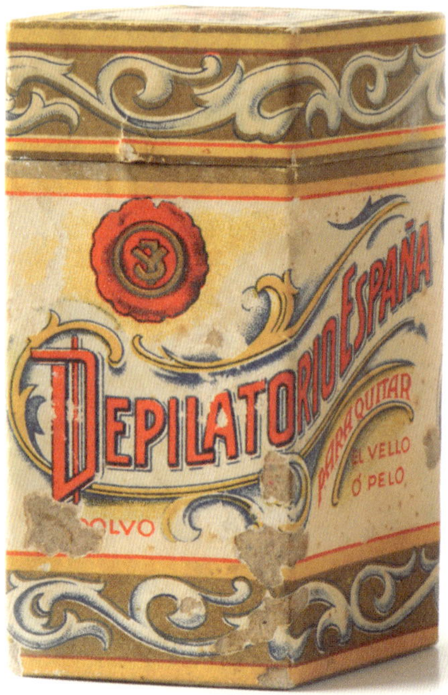
DEPILATORIO ESPAÑA
PARA QUITAR
EL VELLO
Ó PELO
POLVO

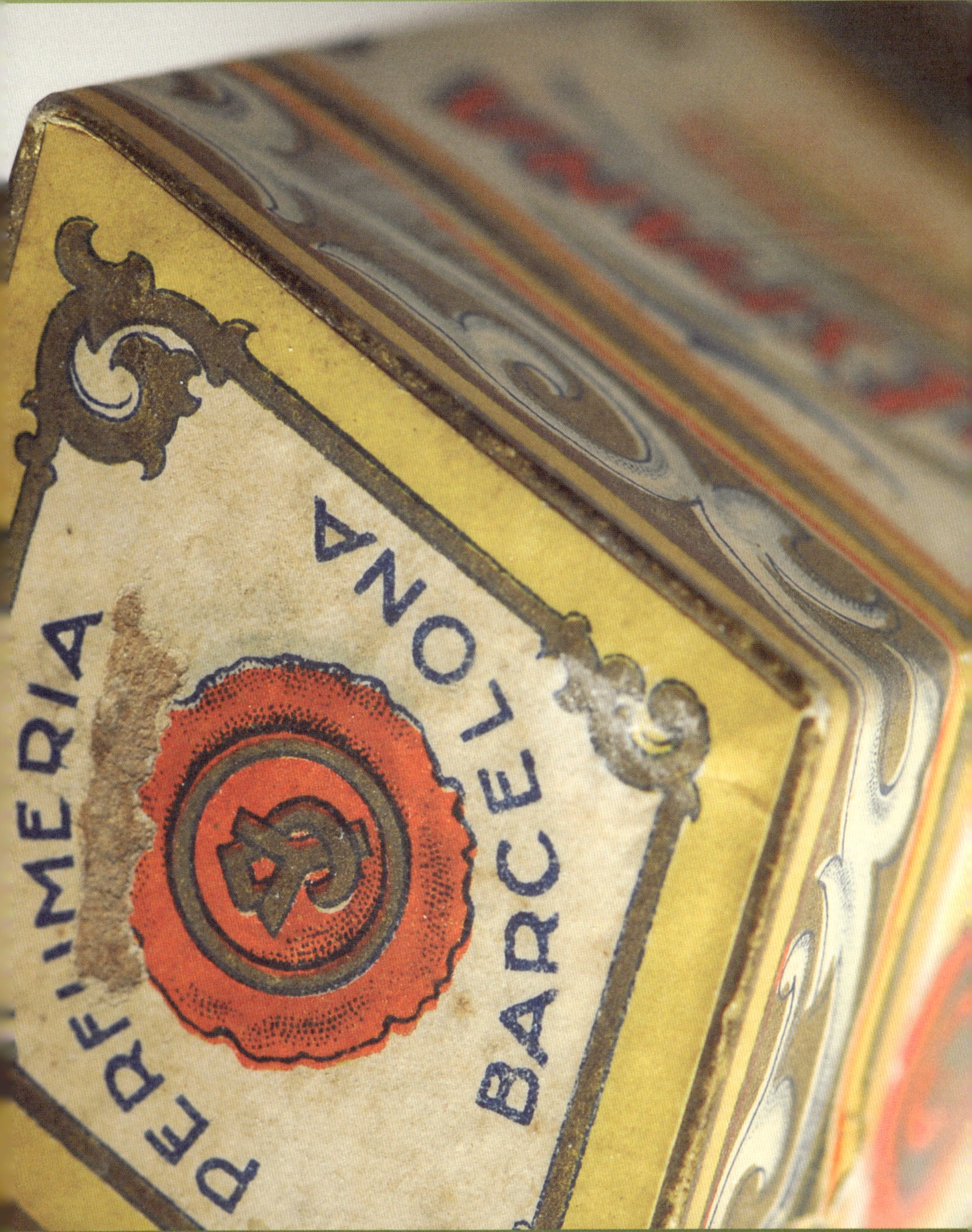
PERFUMERIA
BARCELONA

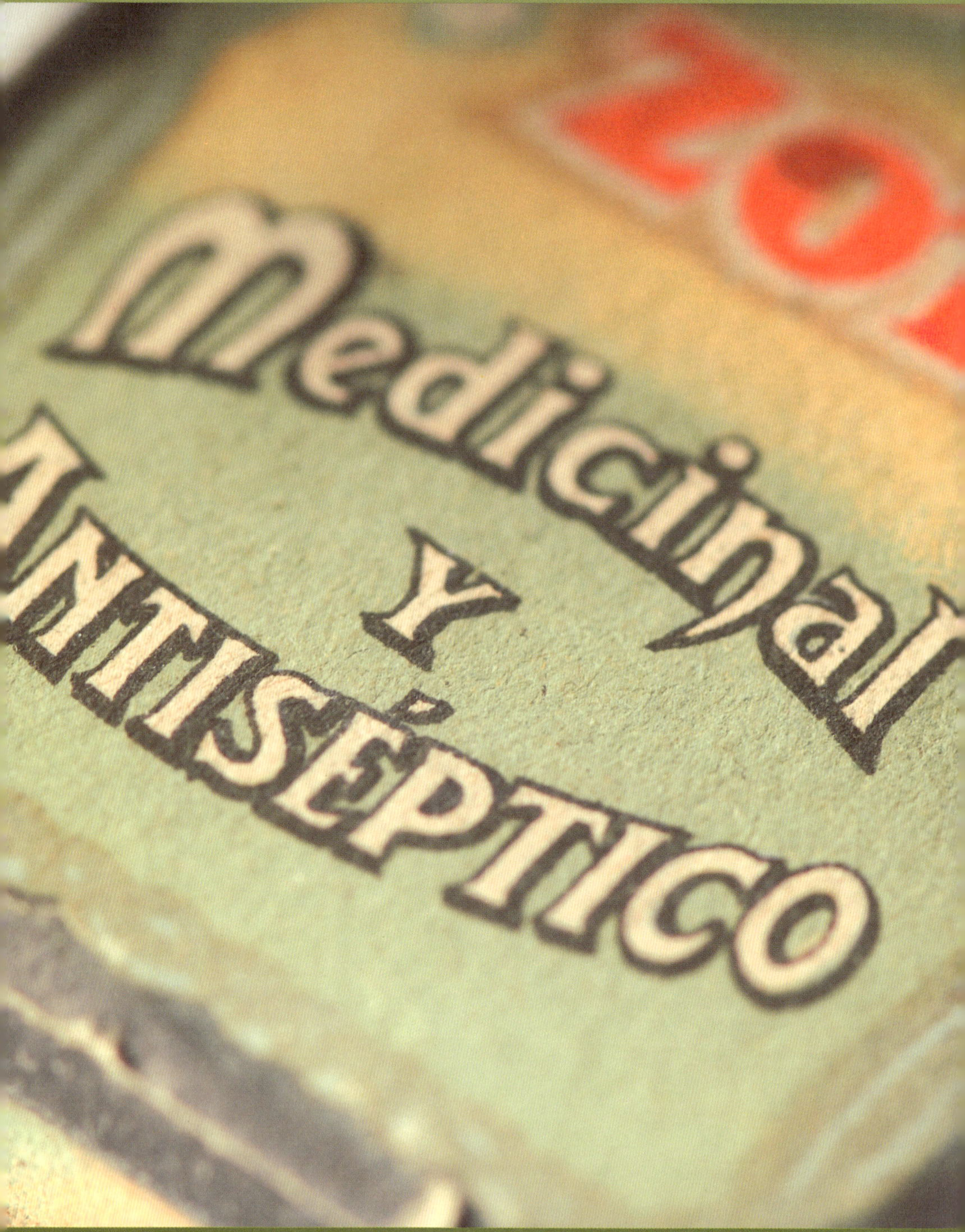
Medicinal
Y
ANTISÉPTICO

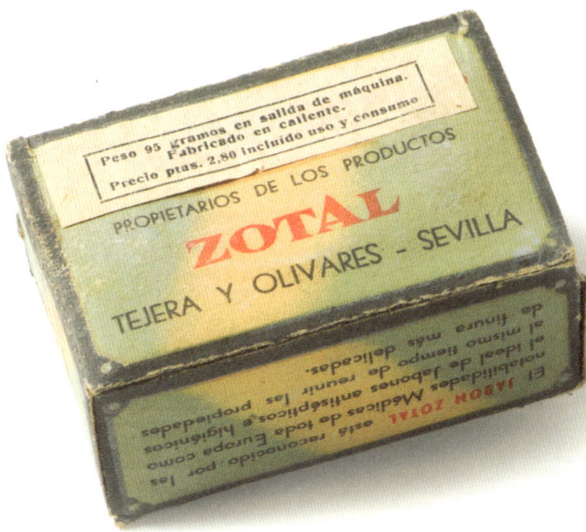
Peso 95 gramos en salida de máquina.
Fabricado en caliente.
Precio ptas. 2,80 incluido uso y consumo
PROPIETARIOS DE LOS PRODUCTOS
ZOTAL
TEJERA Y OLIVARES - SEVILLA
El JABON ZOTAL está reconocido por las
notabilidades Médicas de toda Europa como
el ideal de Jabones antisépticos, e higiénicos
al mismo tiempo de reunir las propiedades
de finura más delicadas.

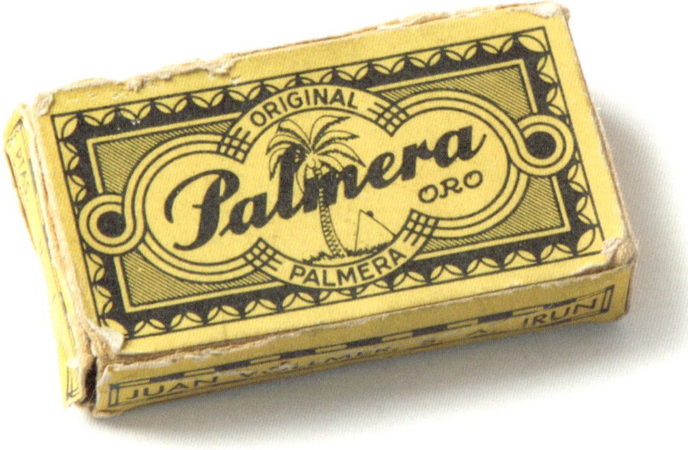
ORIGINAL
Palmera
ORO
PALMERA
IRUN

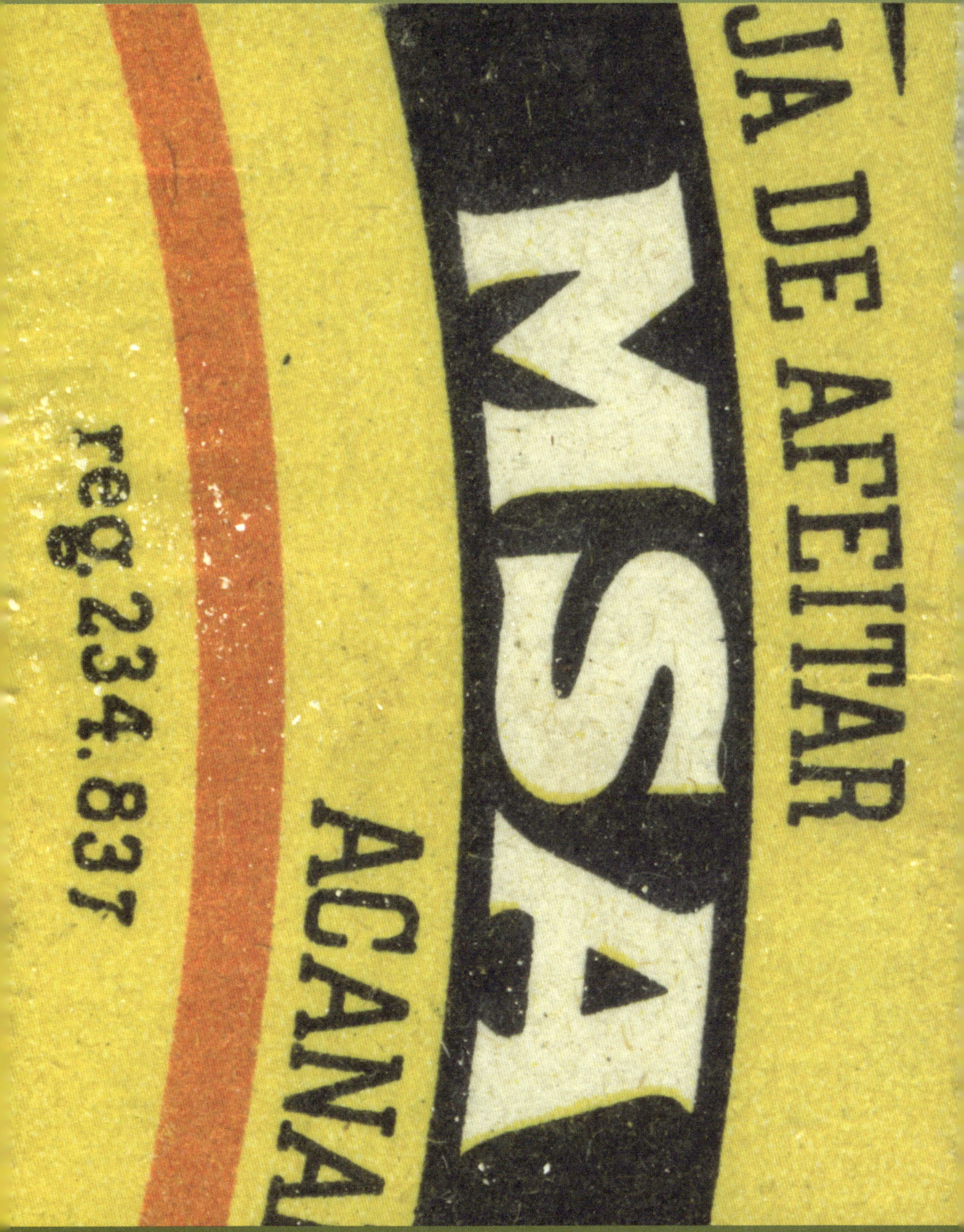
JA DE AFEITAR
MSA
ACANA
reg. 234.837

ESSMAK

ESSMAKINGDRES

INGDRE

MAKINGDRESSM

DRESSMAKING

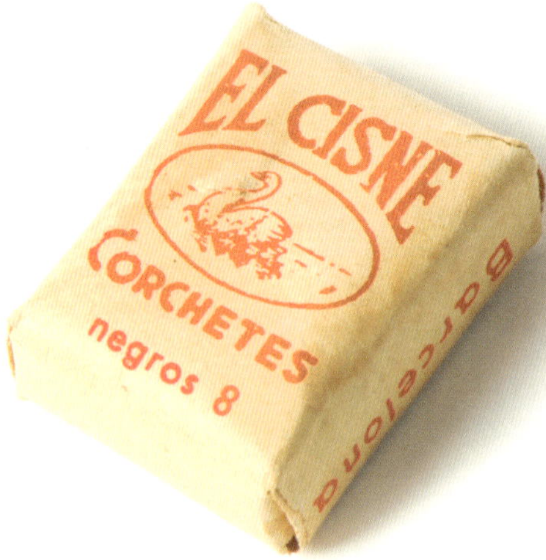
EL CISNE
CORCHETES
negros 8
Barcelona

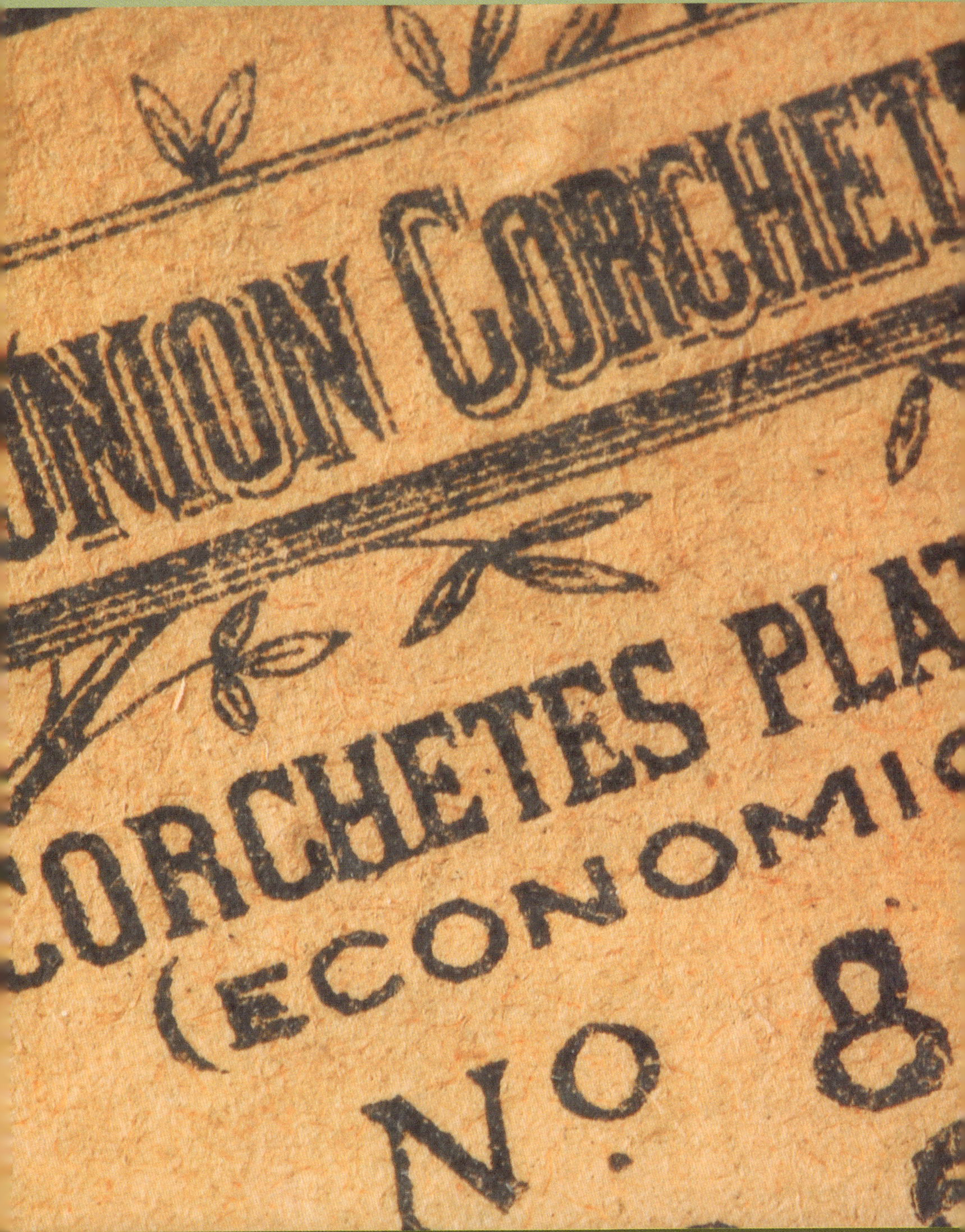
CORCHETES
No 8

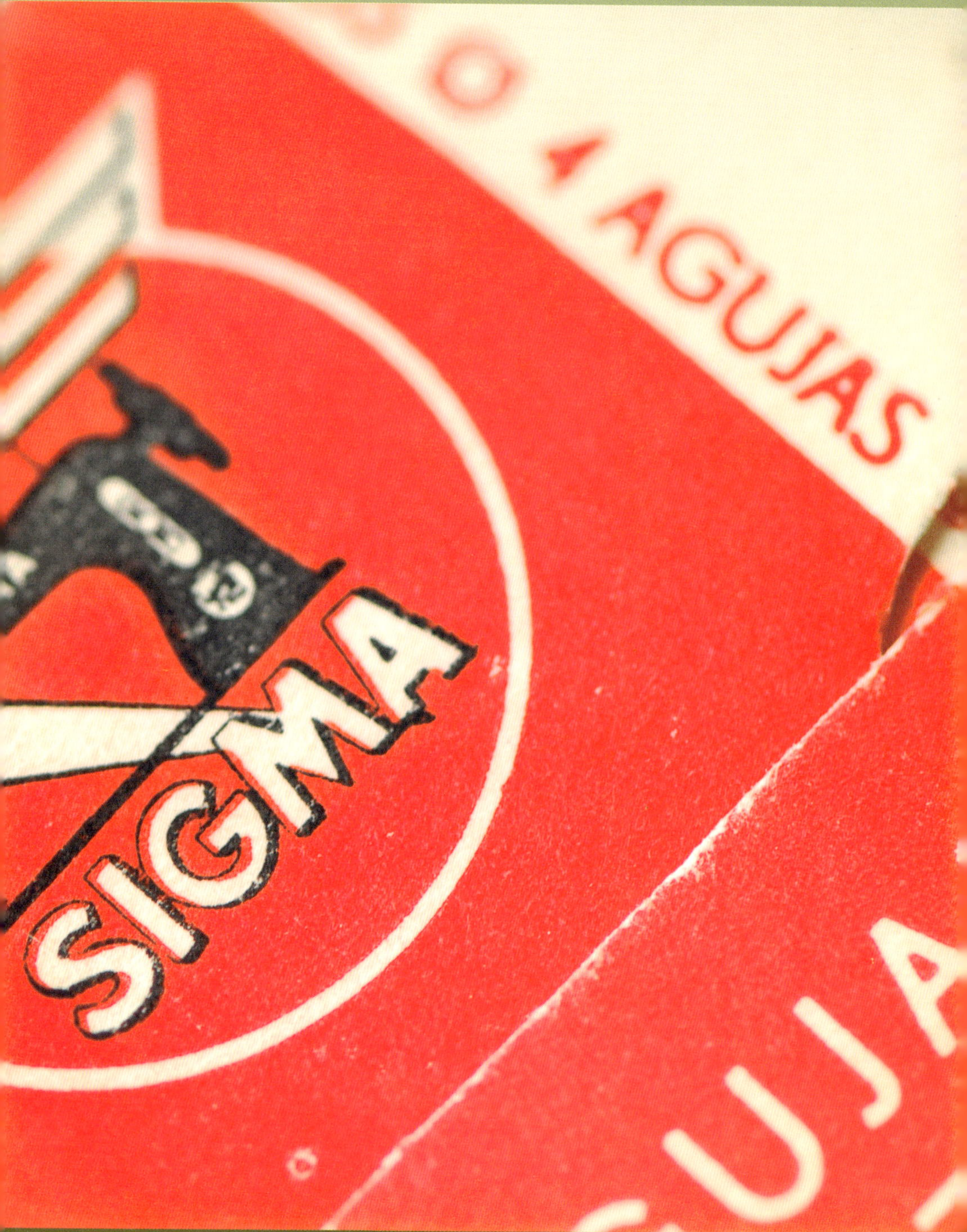
4 AGUJAS
SIGMA

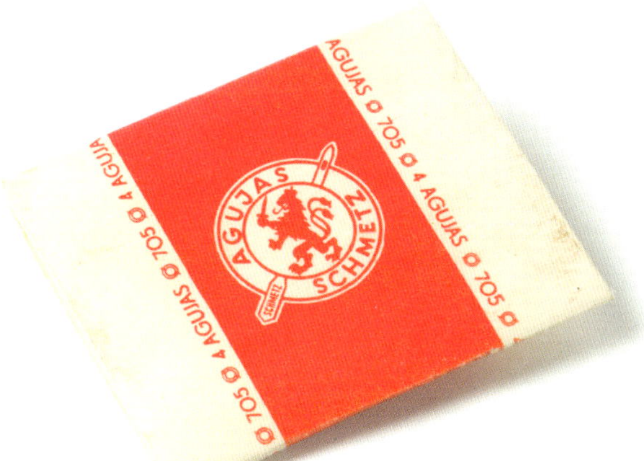
AGUJAS SCHMETZ
4 AGUJAS 705
4 AGUJAS 705

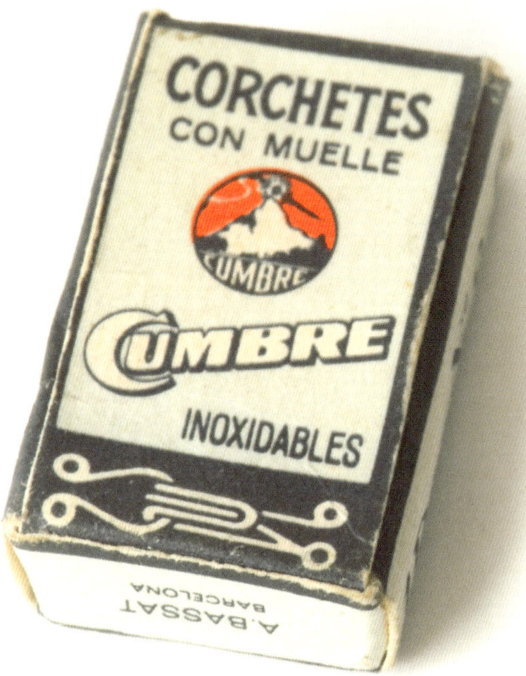
CORCHETES
CON MUELLE
CUMBRE
CUMBRE
INOXIDABLES
A.BASSAT
BARCELONA

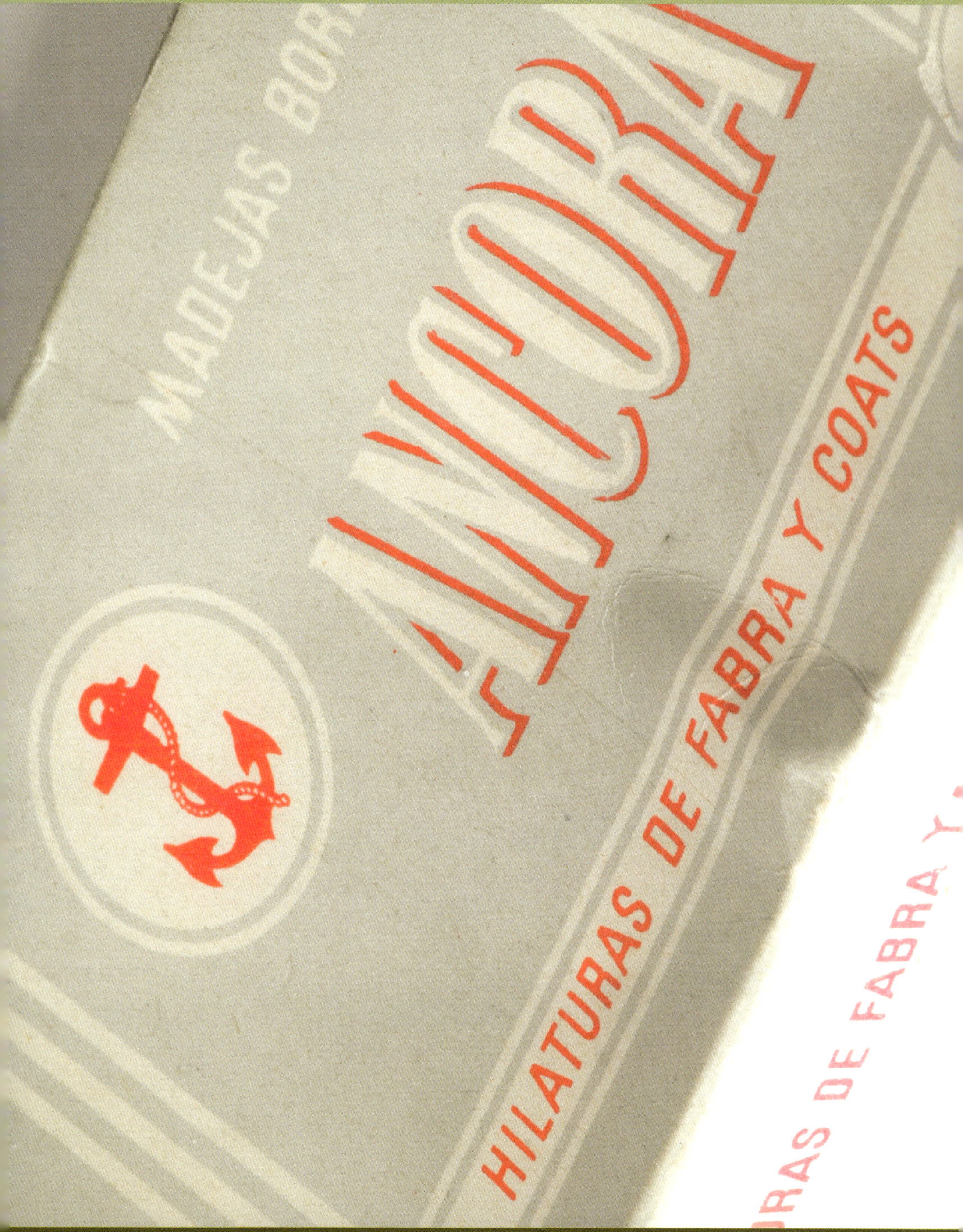
HILATURAS DE FABRA Y COATS

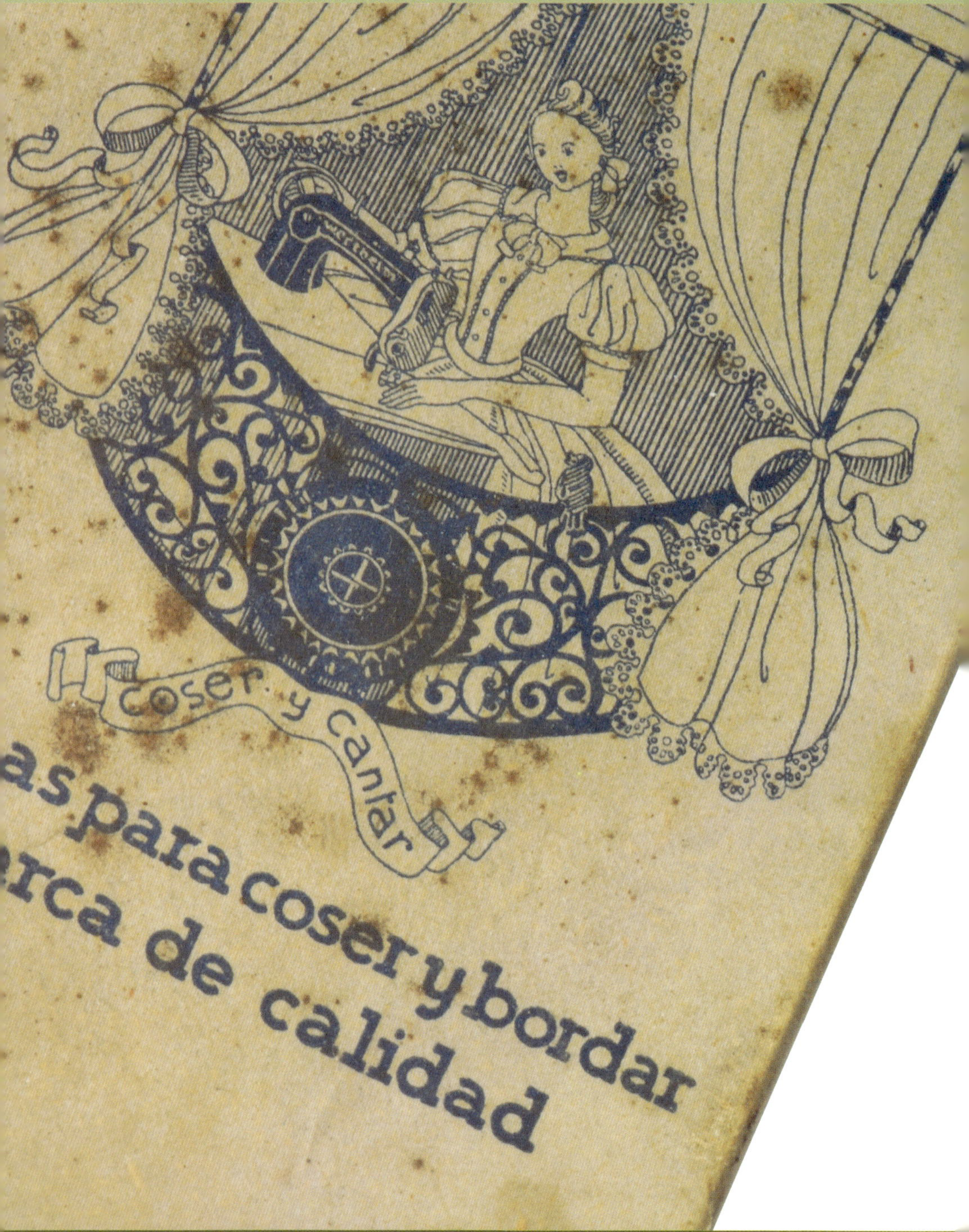
coser y cantar
as para coser y bordar
erca de calidad

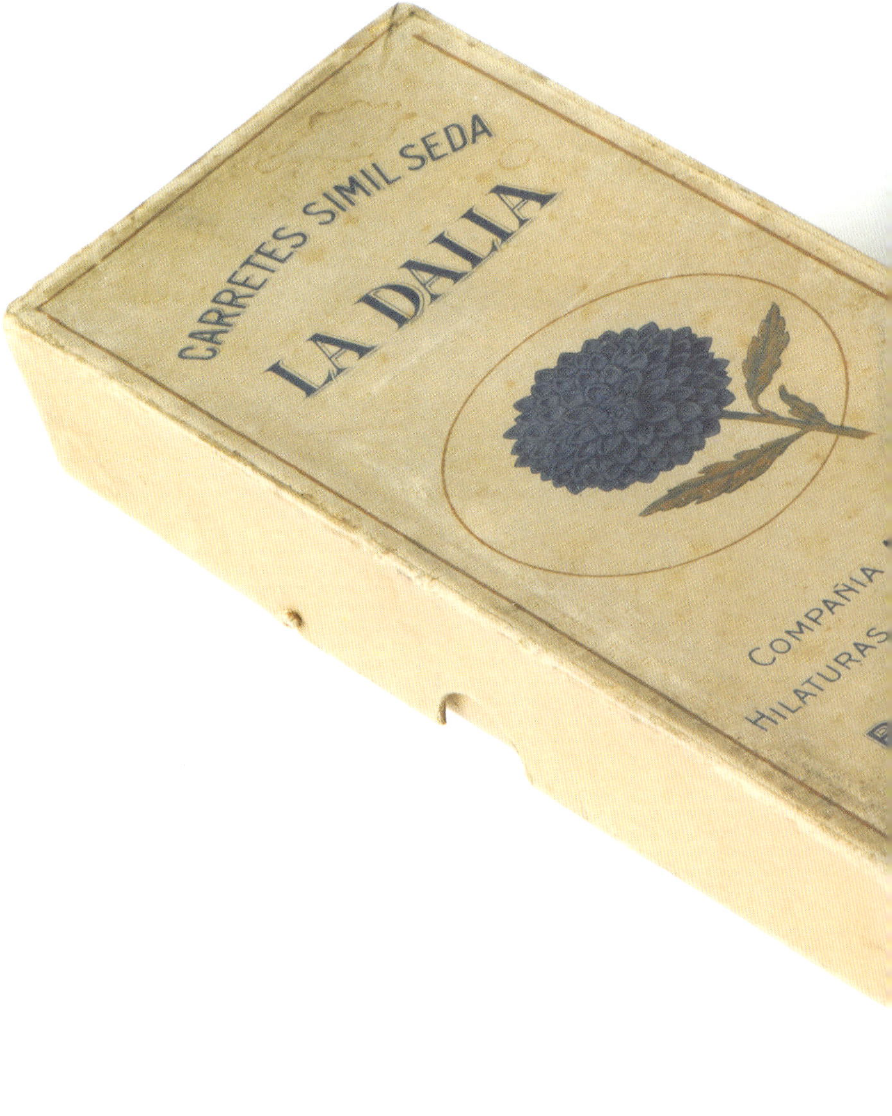
CARRETES SIMIL SEDA
LA DALIA

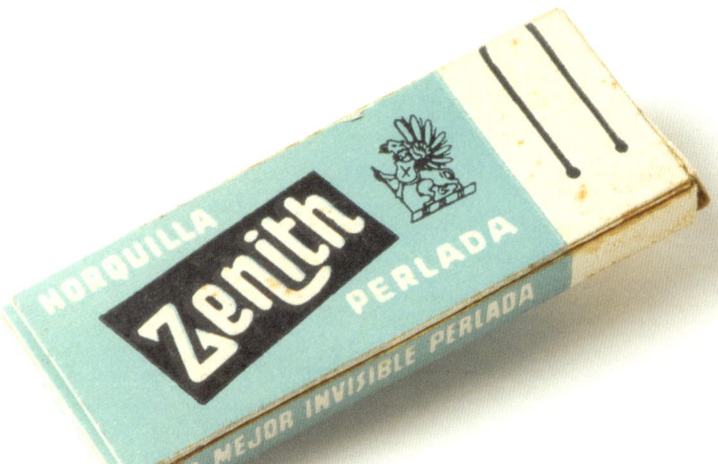
HORQUILLA
Zenith
PERLADA
LA MEJOR INVISIBLE PERLADA

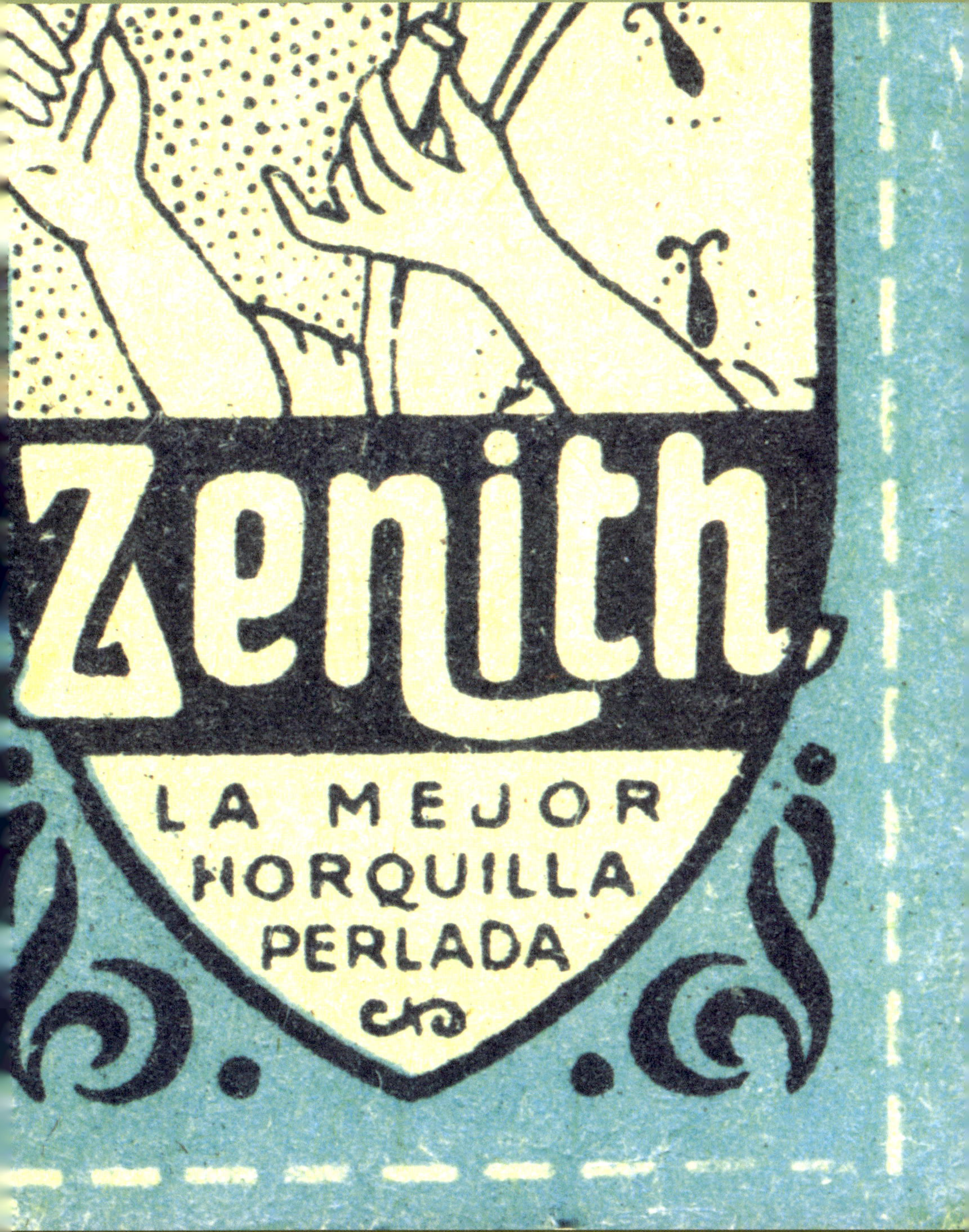
Zenith
LA MEJOR
HORQUILLA
PERLADA

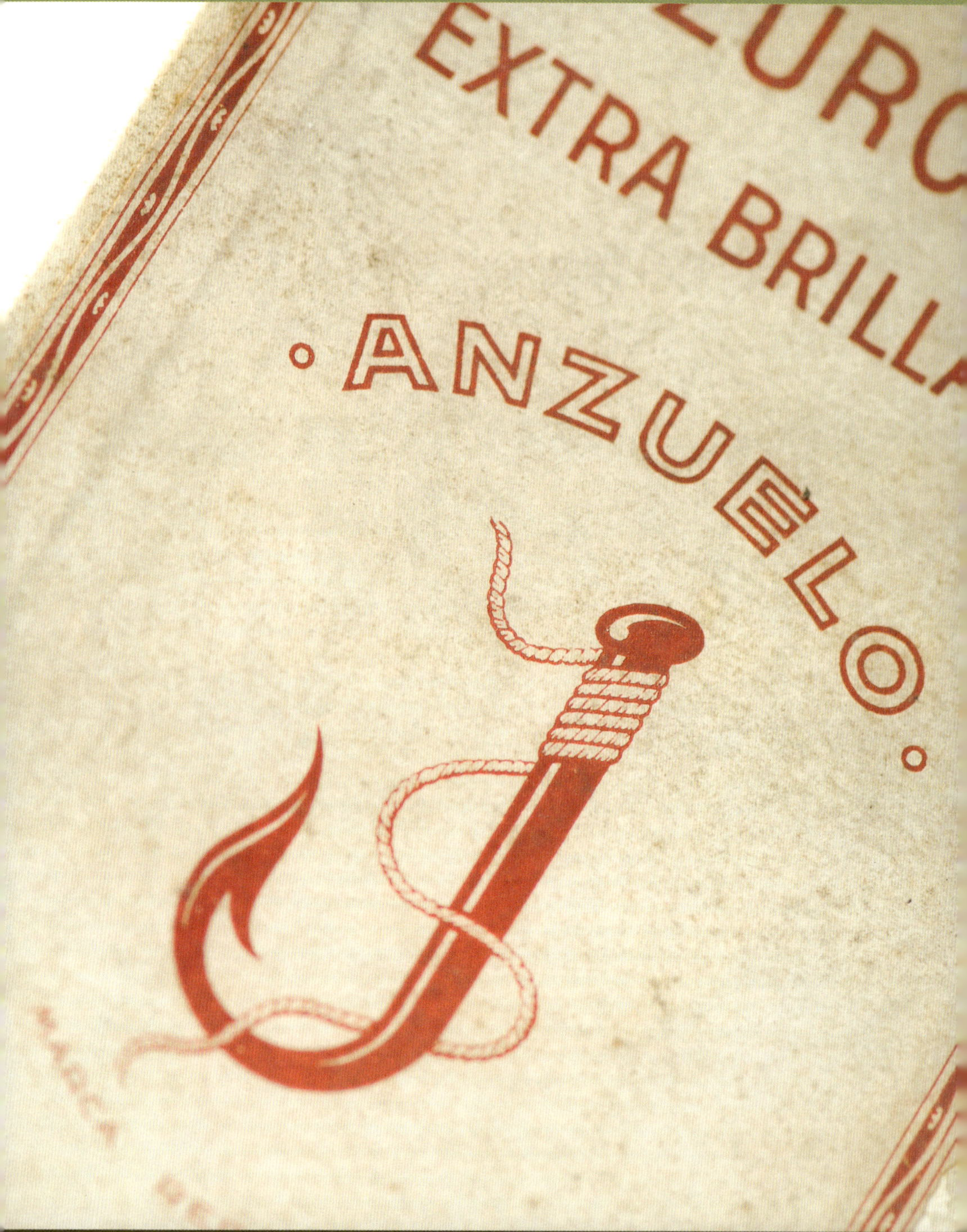
EXTRA BRILL
·ANZUELO·

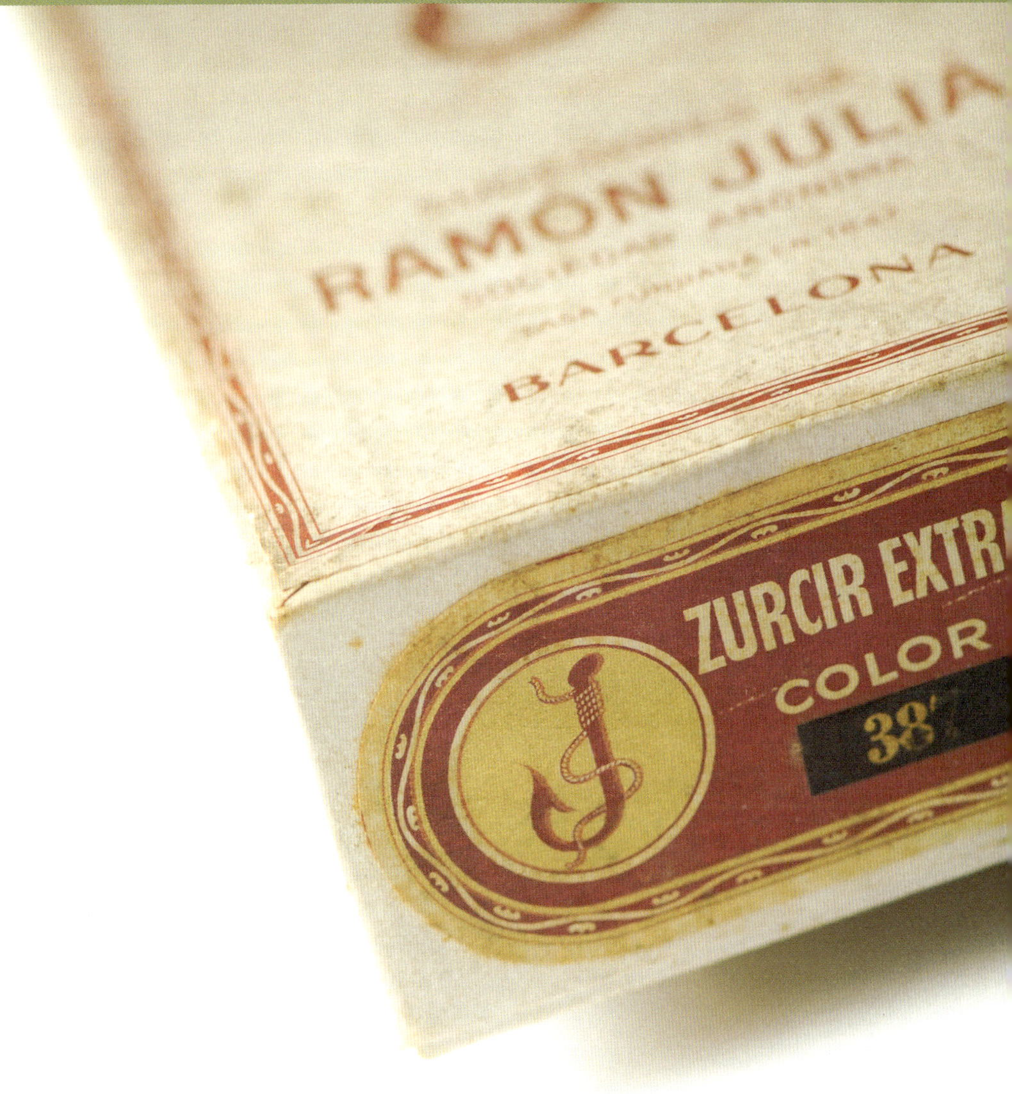
RAMON JULIA
BARCELONA
ZURCIR
COLOR

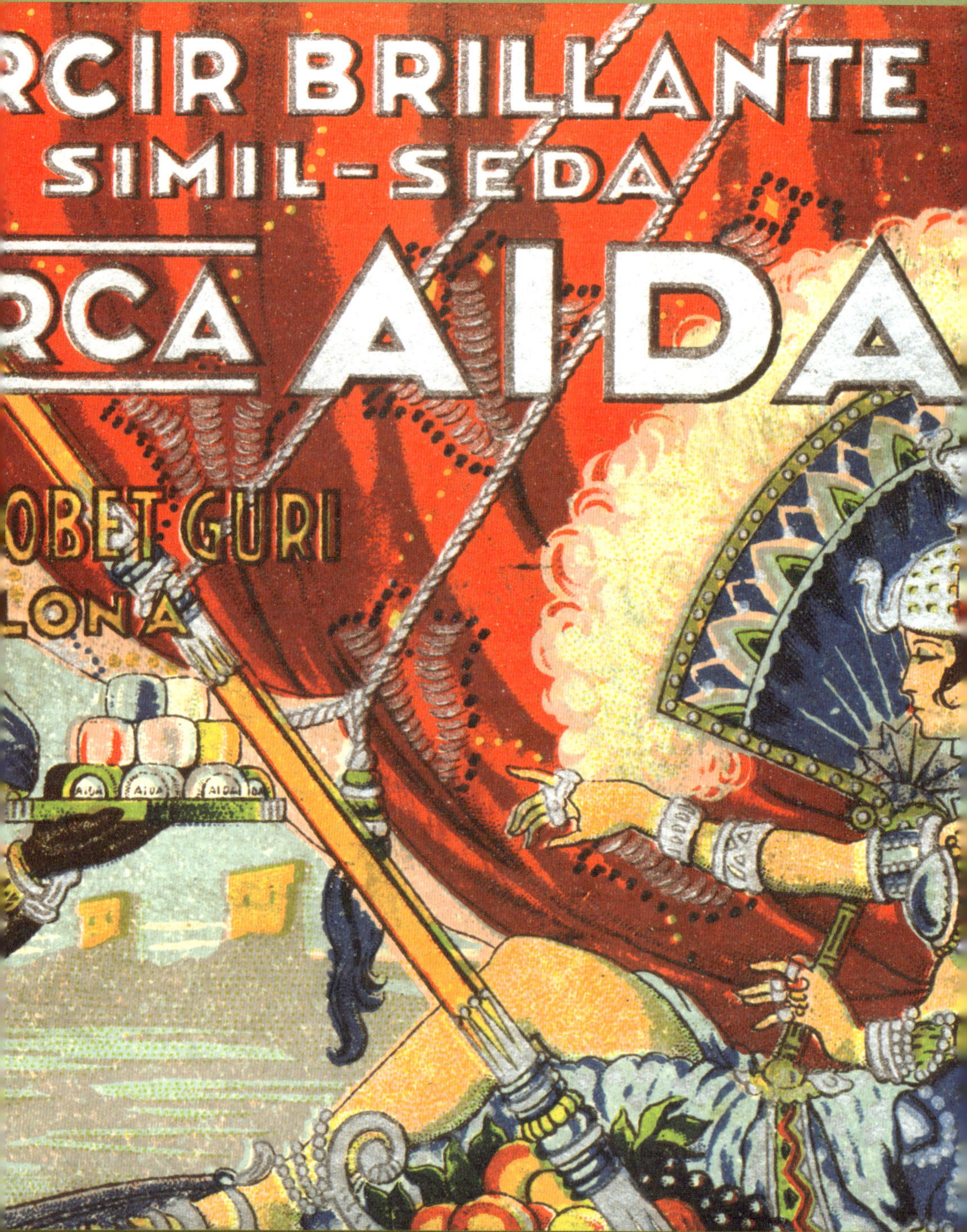
RCIR BRILLANTE
SIMIL-SEDA
RCA AIDA
OBET GURI
LONA
AIDA

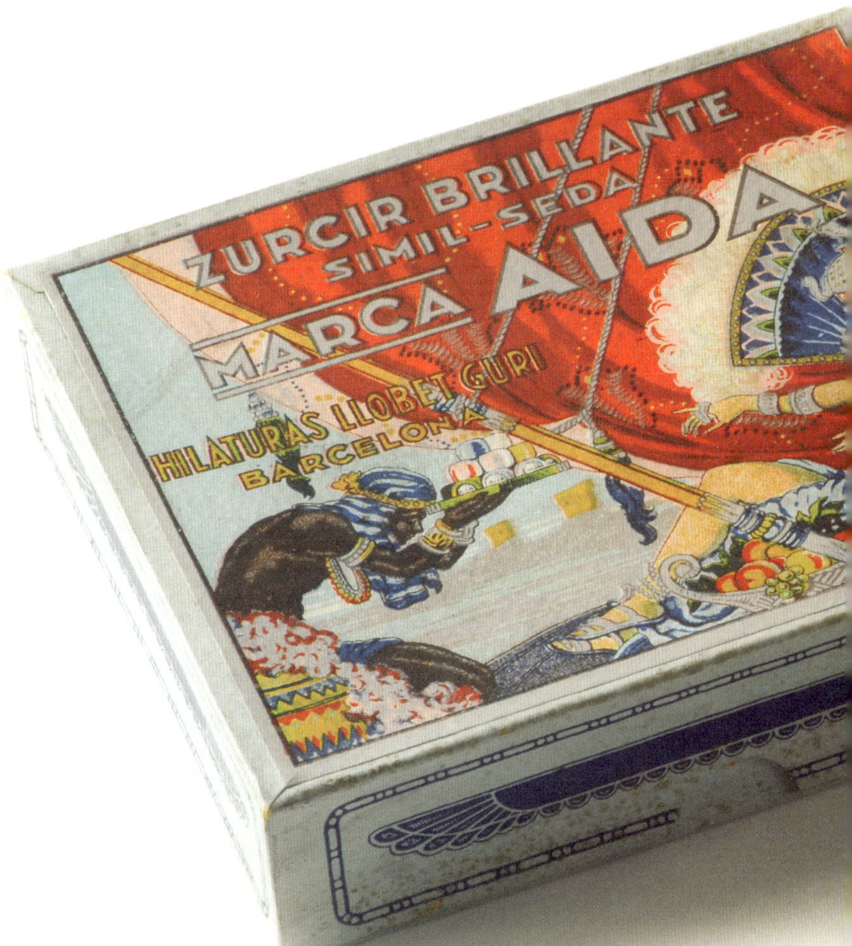
ZURCIR BRILLANTE
SIMIL-SEDA
MARCA AIDA
HILATURAS LLOBET GURI
BARCELONA

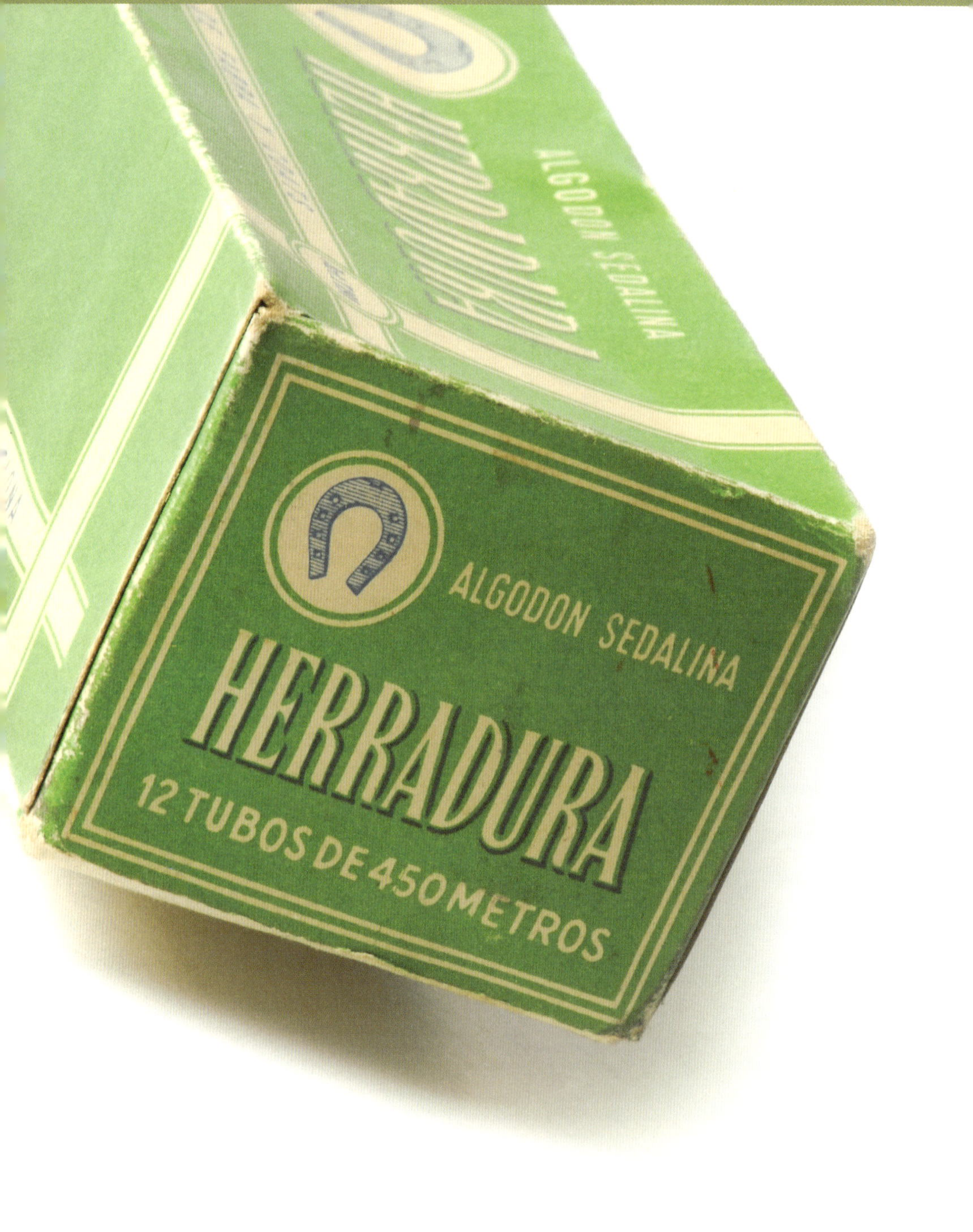
ALGODON SEDALINA
HERRADURA
12 TUBOS DE 450 METROS

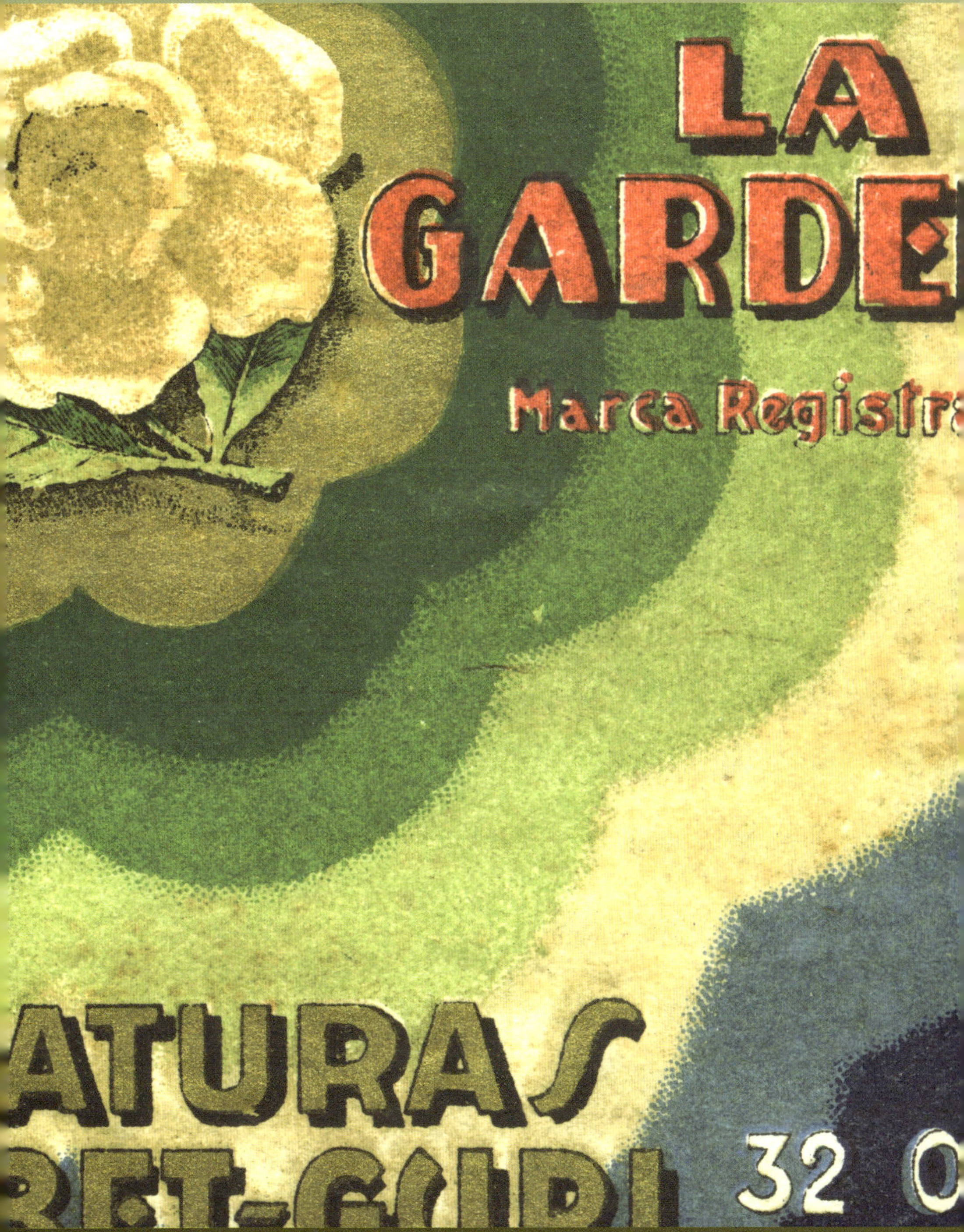
LA
GARDE
Marca Regist
ATURAS
32

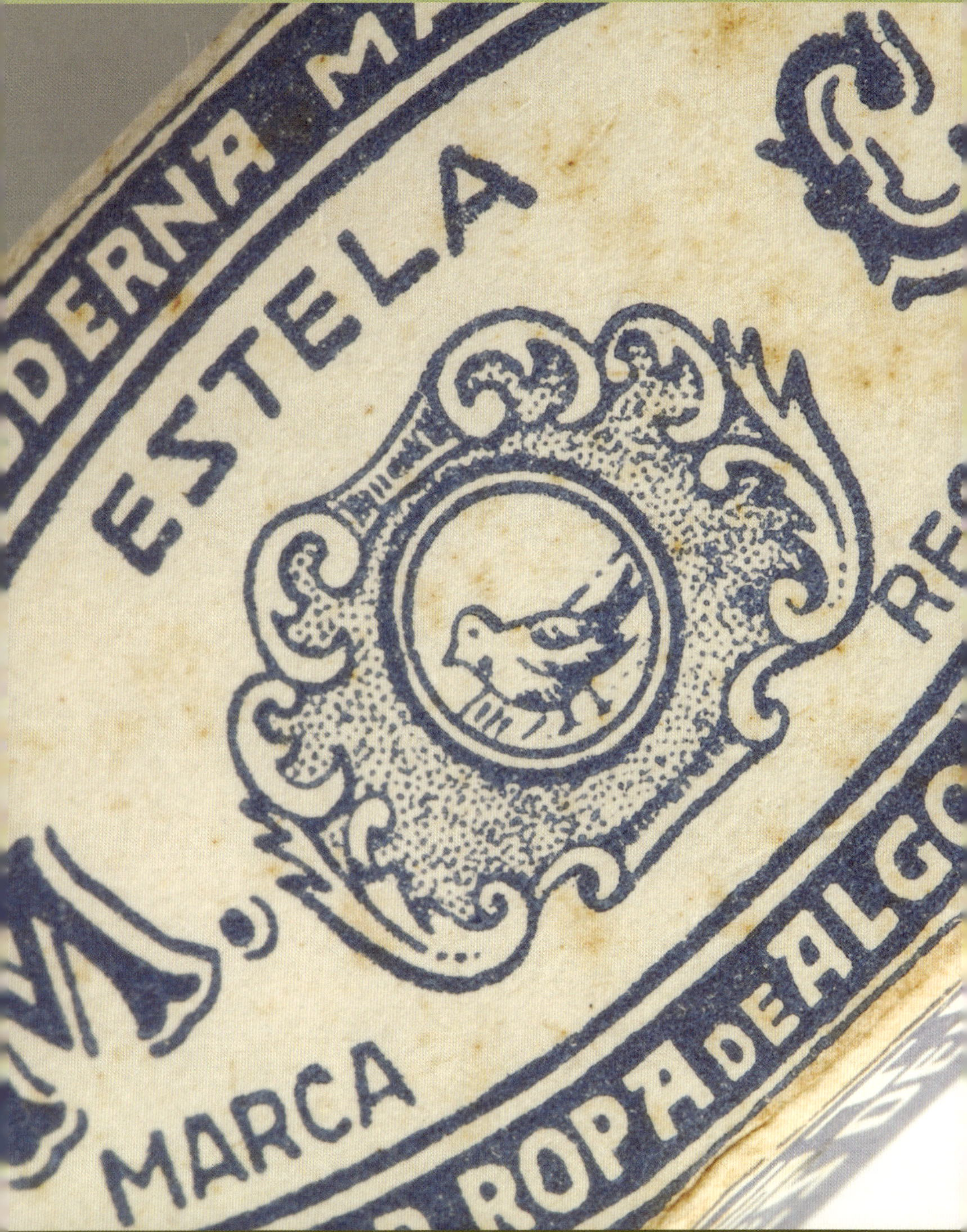
ESTELA
MARCA
ROPA

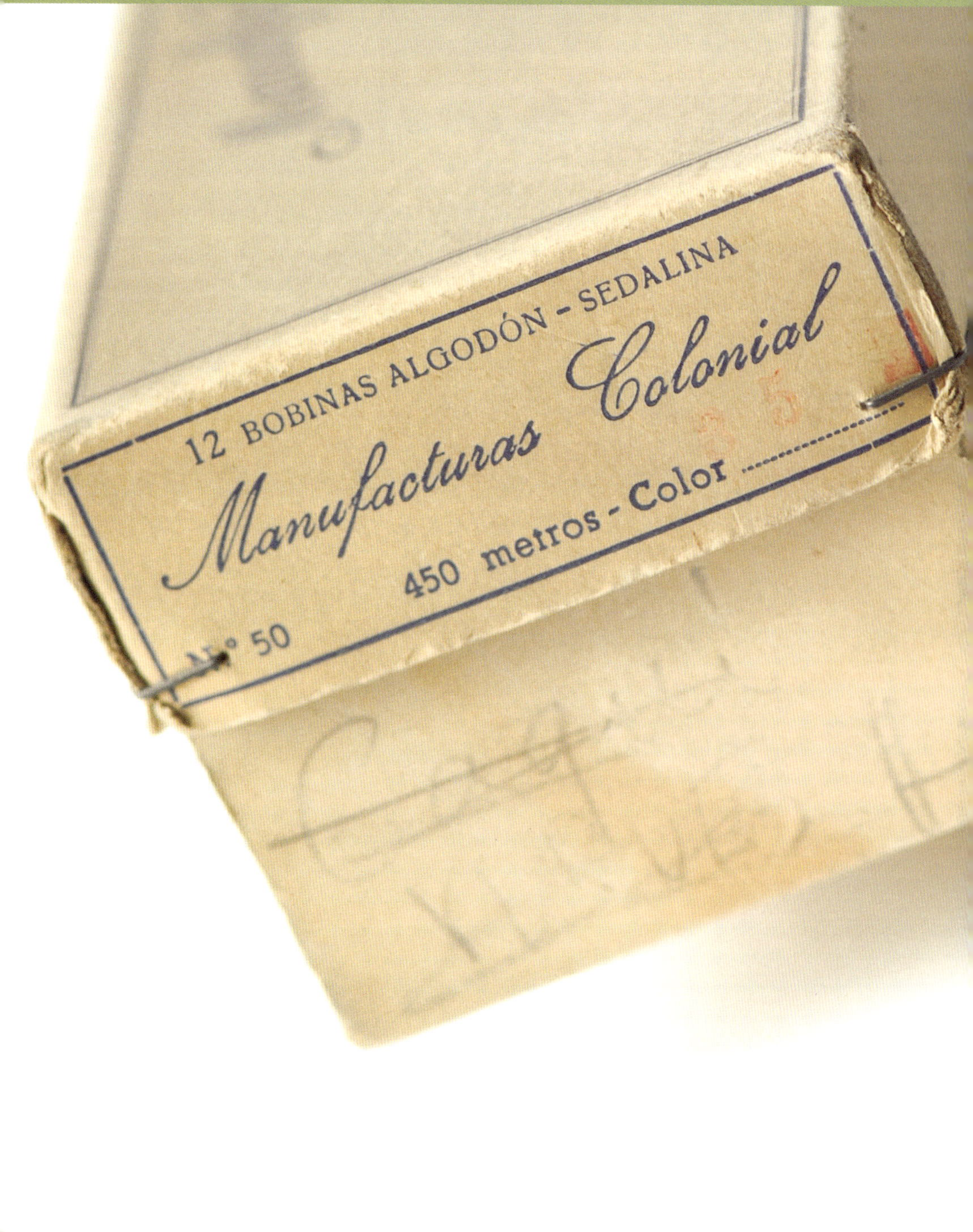
12 BOBINAS ALGODÓN - SEDALINA
Manufacturas Colonial
450 metros - Color
Nº 50

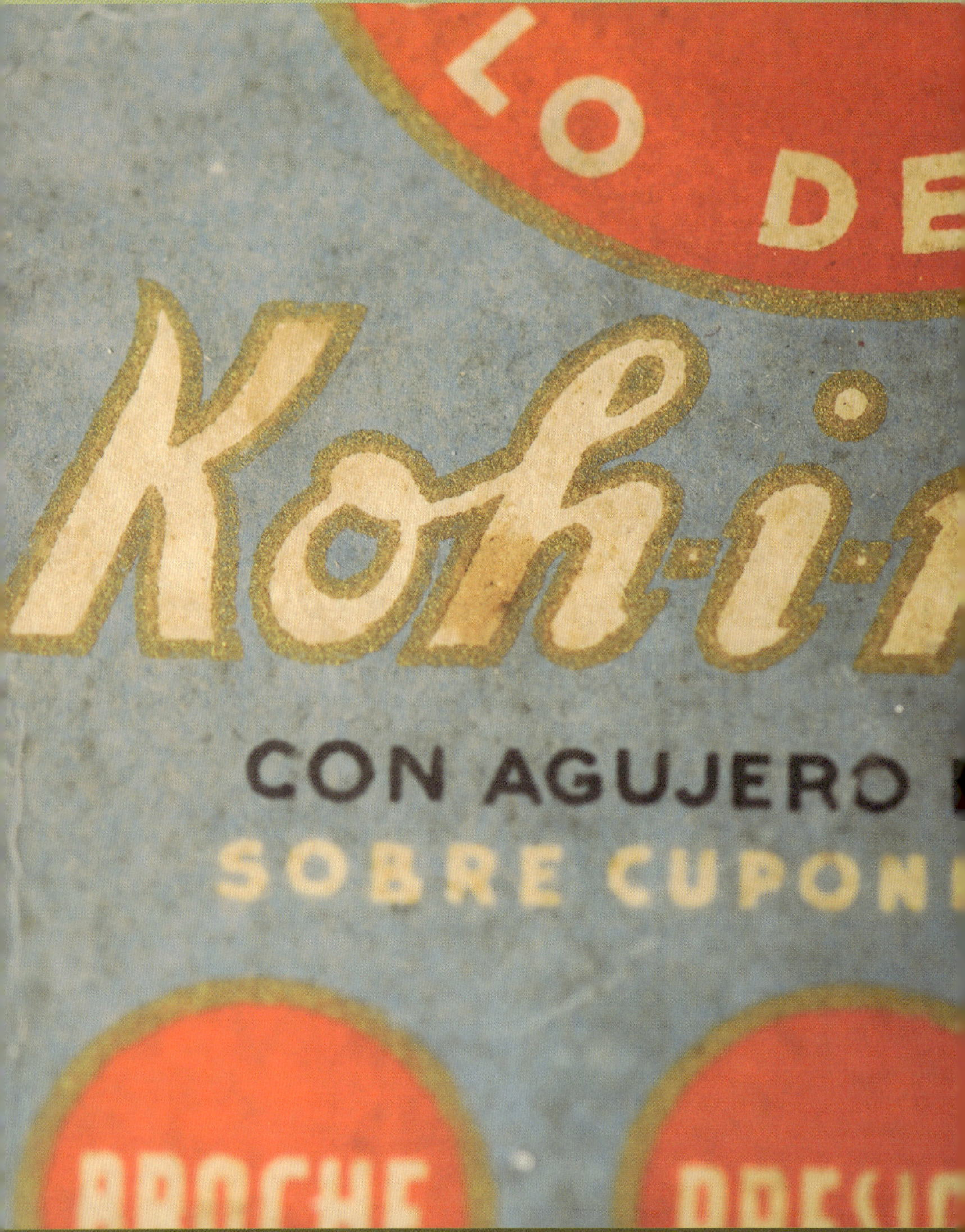
LO DE
CON AGUJERO
SOBRE CUPON

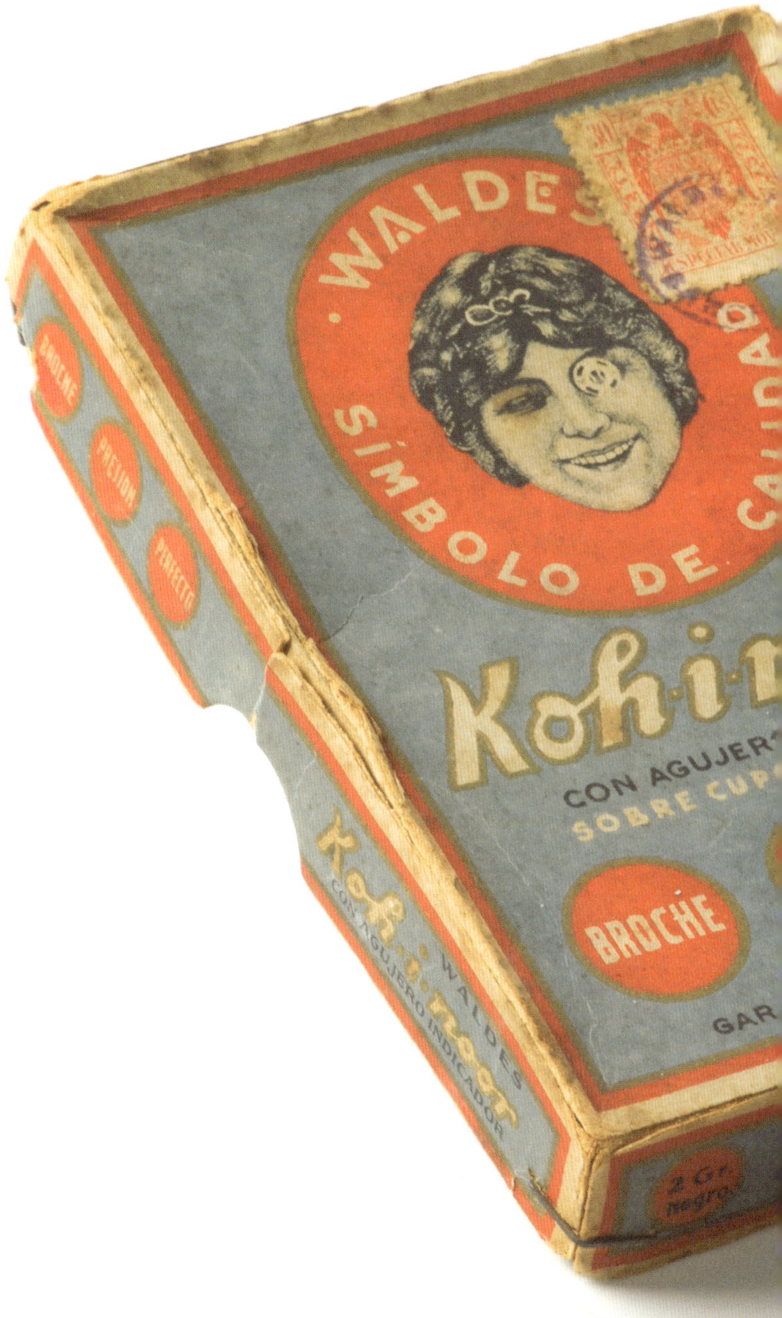
SÍMBOLO DE CALIDAD
BROCHE
PRESION
PERFECTO
BROCHE
WALDES
CON AGUJERO INDICADOR

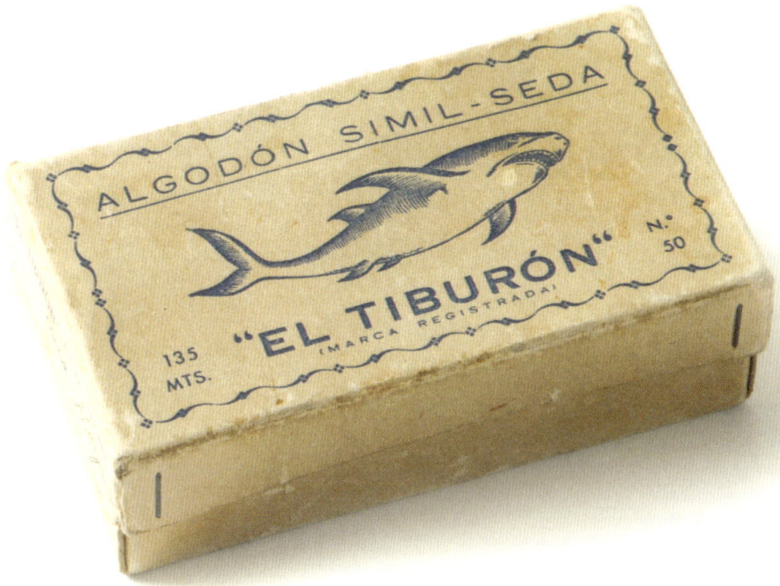
ALGODÓN SIMIL-SEDA
N.° 50
"EL TIBURÓN"
(MARCA REGISTRADA)
135 MTS.

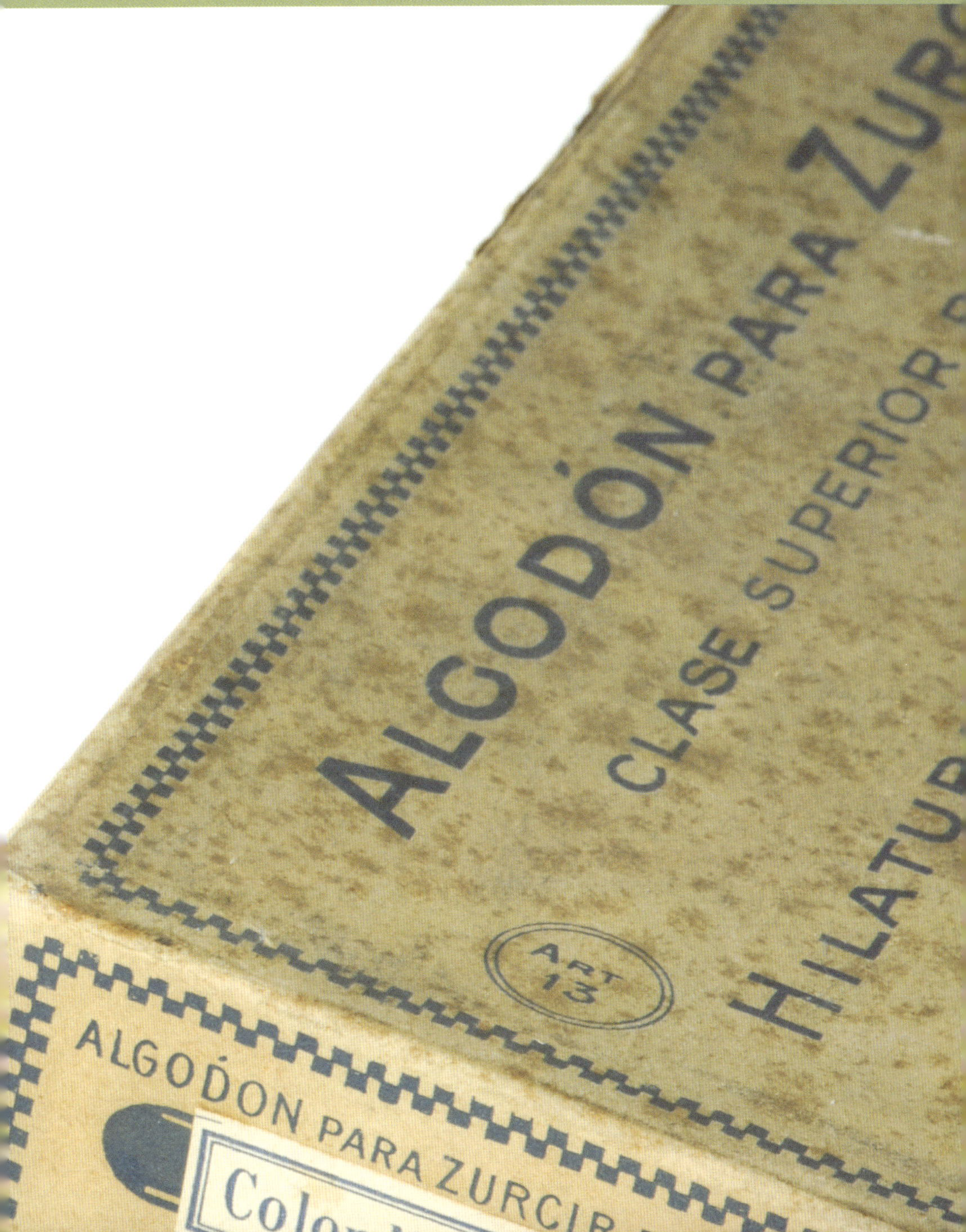
ALGODÓN PARA
CLASE SUPERIOR
ART 13
ALGODON PARA ZURCIR

FICEOFI

FICEOFFICEOFFIC

OFFICE

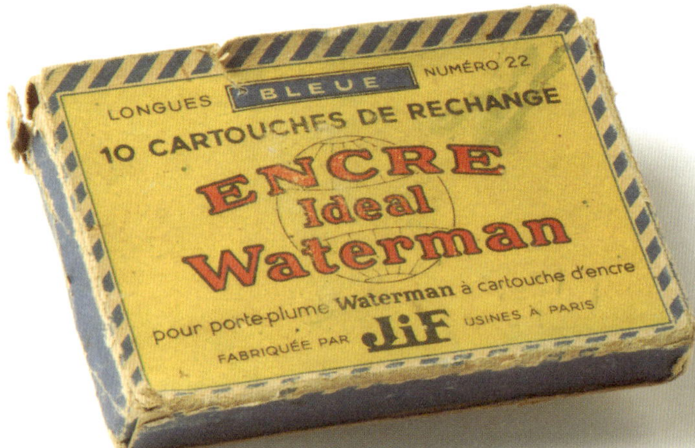
BLEUE
NUMÉRO 22
LONGUES
10 CARTOUCHES DE RECHANGE
ENCRE
Ideal
Waterman
pour porte-plume Waterman à cartouche d'encre
FABRIQUÉE PAR JiF USINES À PARIS

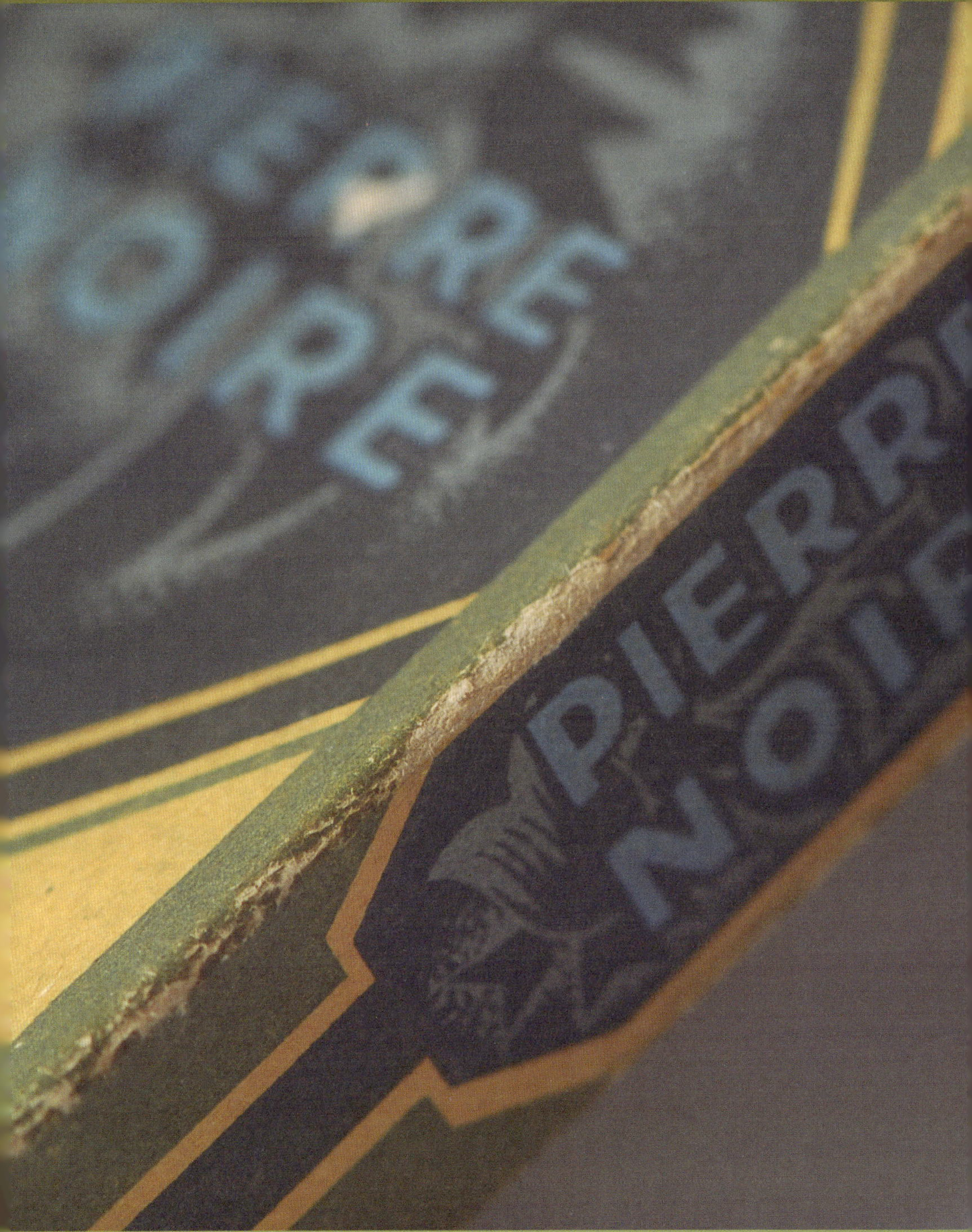
PIERR
NOIR

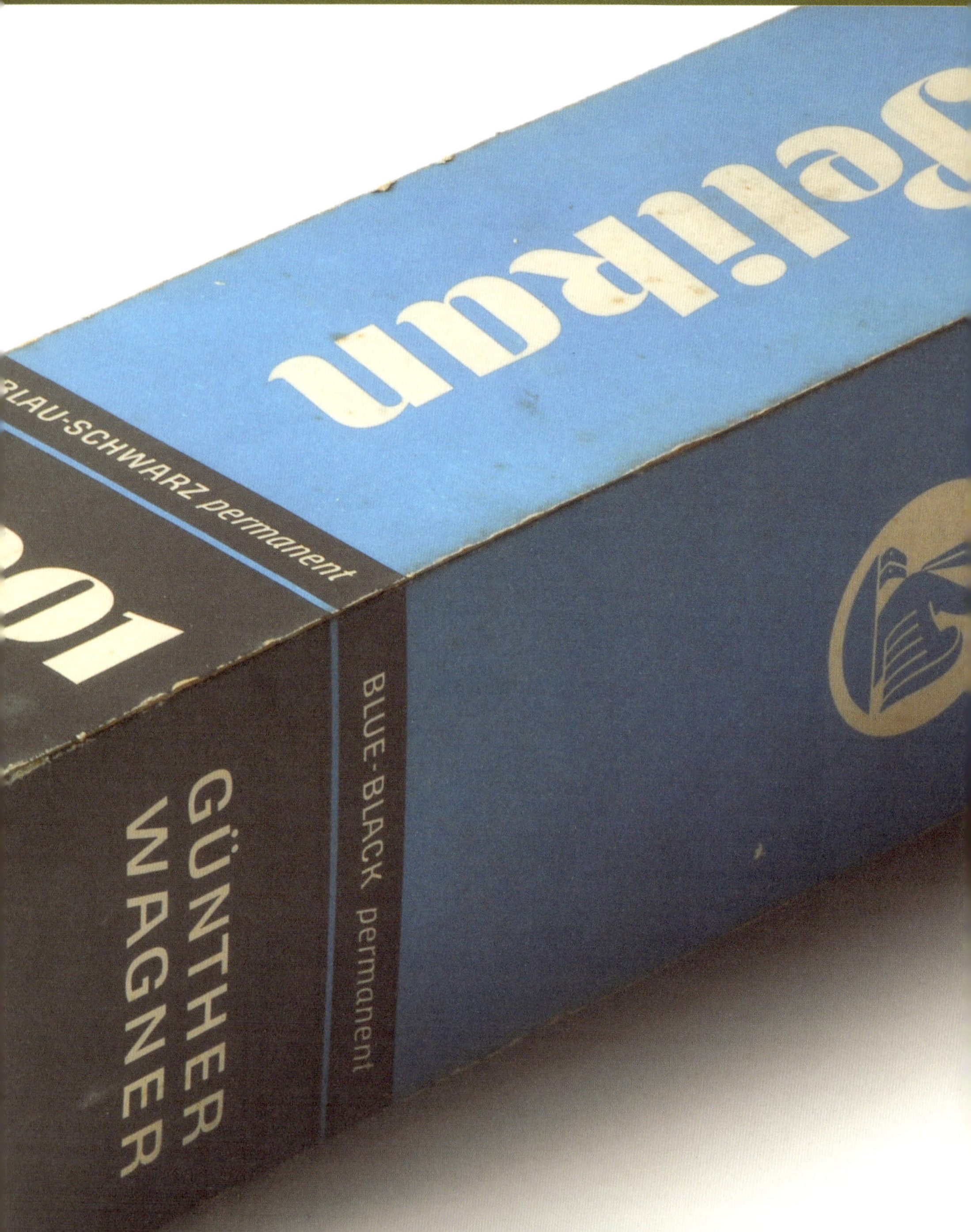
Pelikan
BLAU-SCHWARZ permanent
BLUE-BLACK permanent
GÜNTHER WAGNER

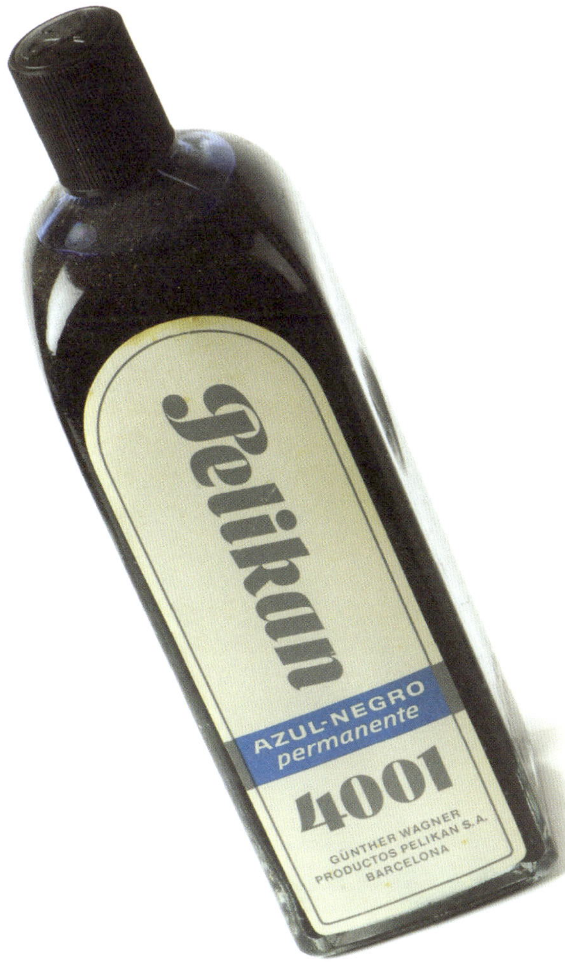
Pelikan
AZUL-NEGRO
permanente
4001
GÜNTHER WAGNER
PRODUCTOS PELIKAN S.A.
BARCELONA

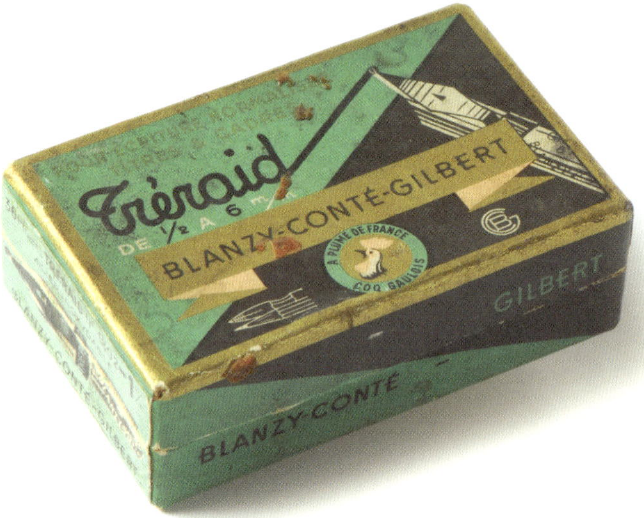
Tréroïd
BLANZY-CONTÉ-GILBERT
PLUME DE FRANCE
COQ GAULOIS
GILBERT
BLANZY-CONTÉ

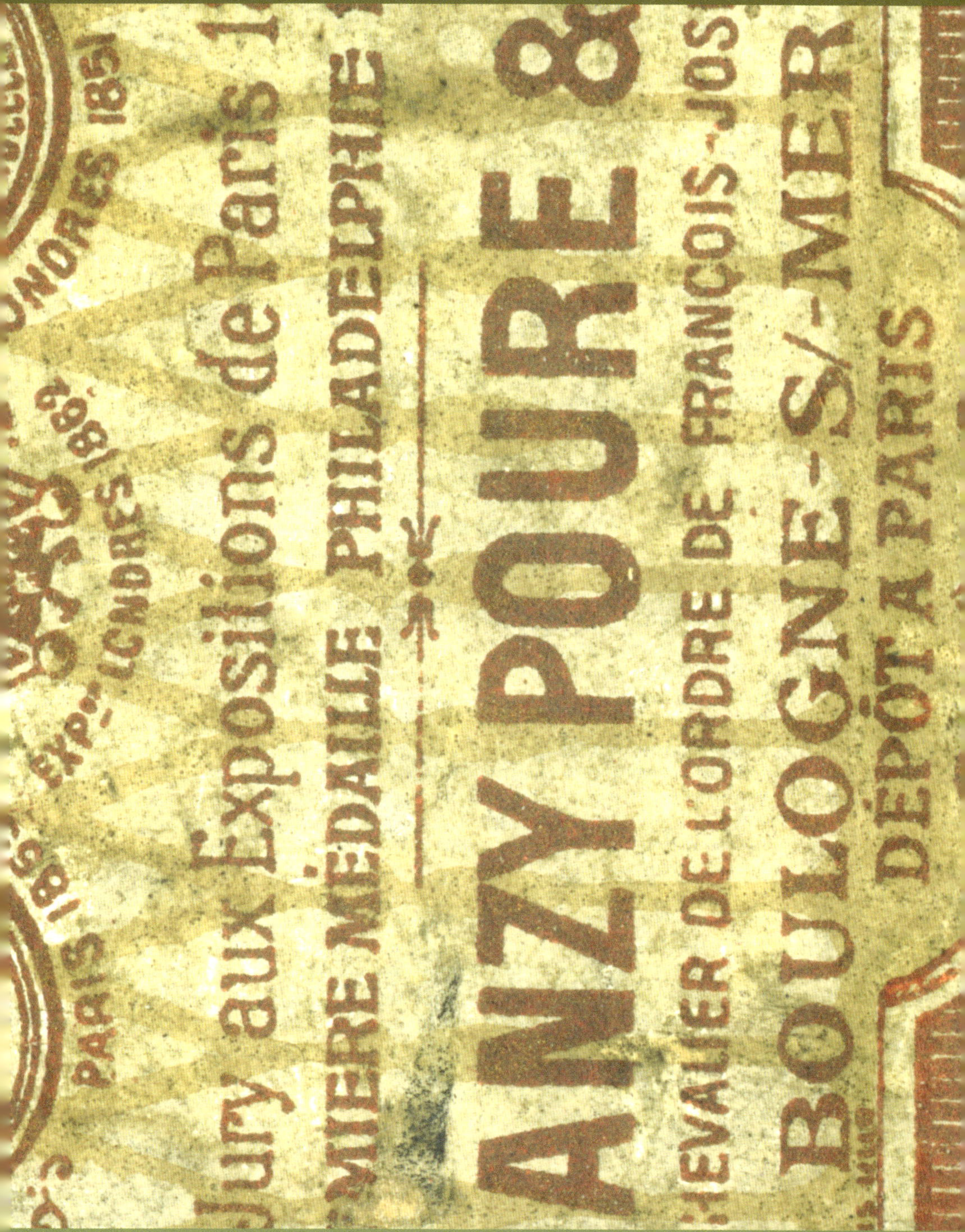
EXPo. LONDRES 1862
LONDRES 1851
Jury aux Expositions de Paris
MIERE MÉDAILLE PHILADELPHIE
ANZY POURE
EVALIER DE L'ORDRE DE FRANÇOIS-JOS
BOULOGNE-S/-MER
DÉPÔT A PARIS

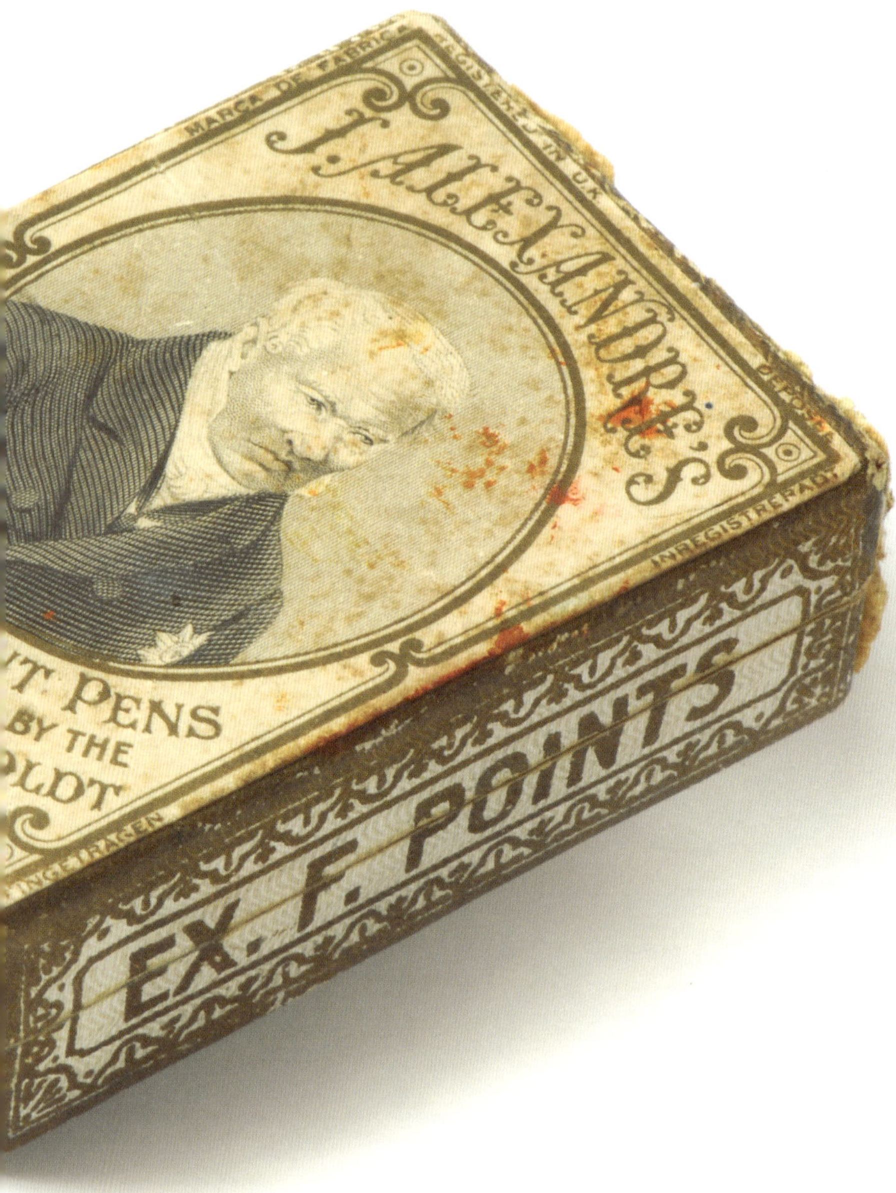
J. ALEXANDRE'S
PENS
BY THE
EX. F. POINTS

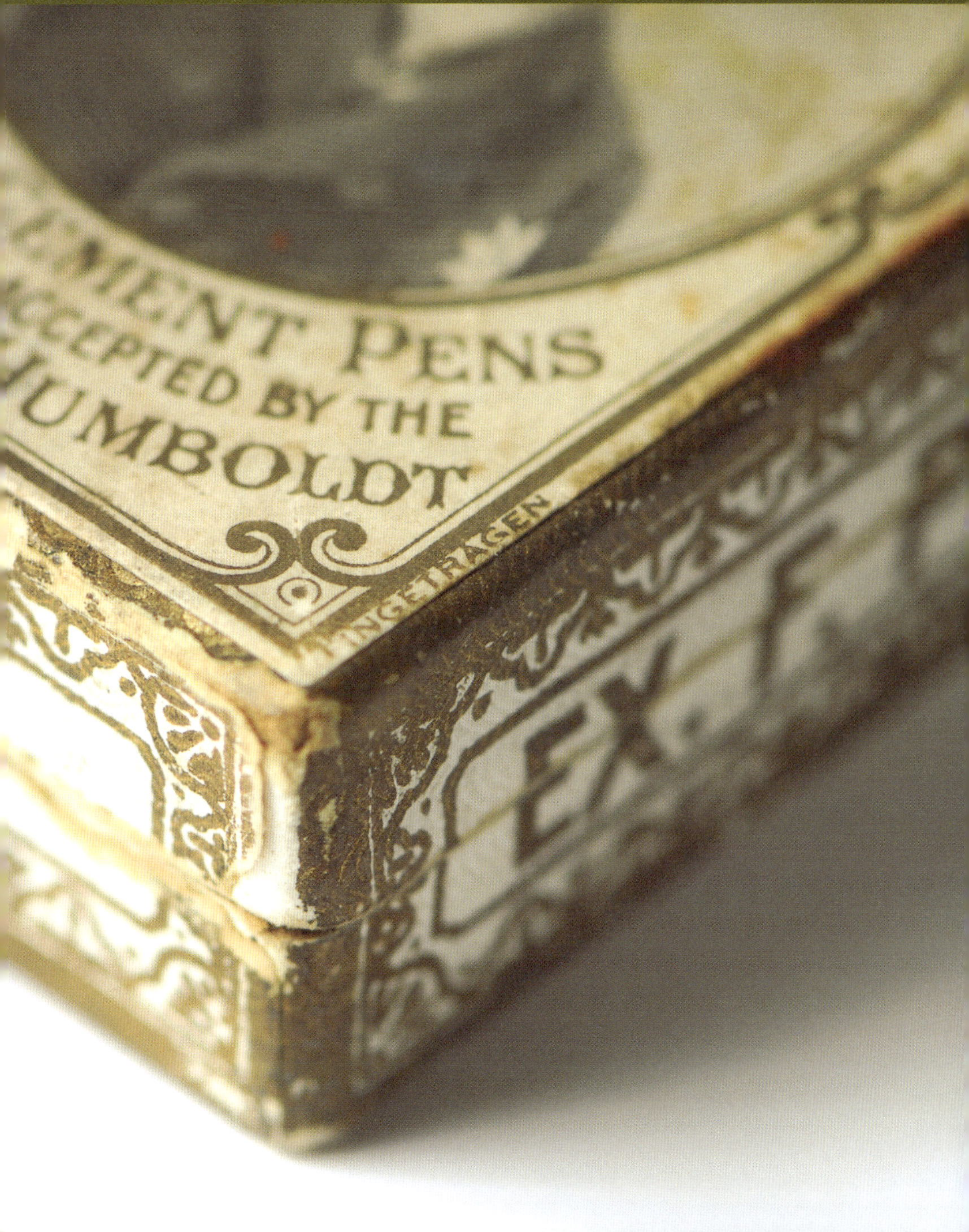
MENT PENS
CCEPTED BY THE
UMBOLDT

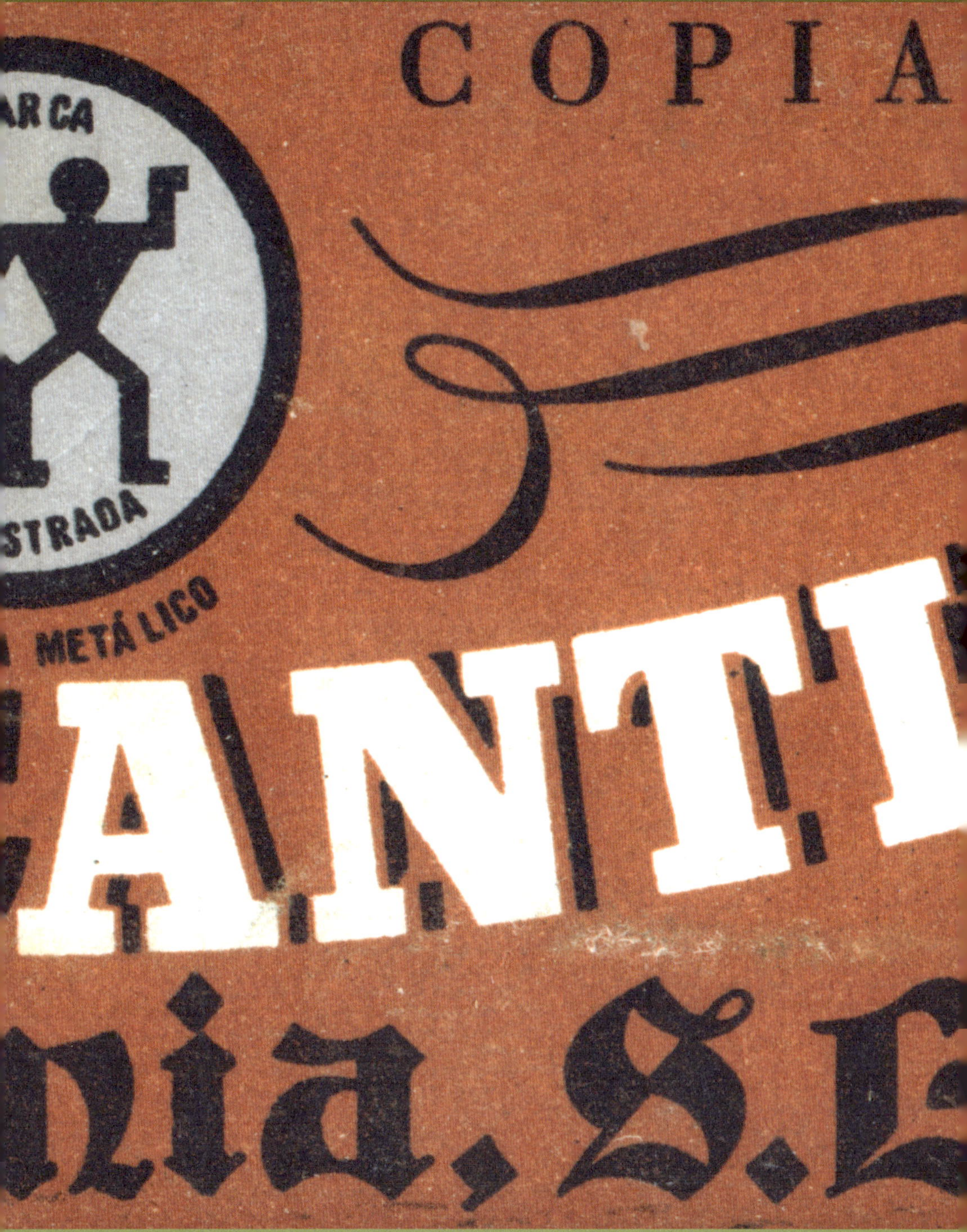
COPIA
METÁ LICO

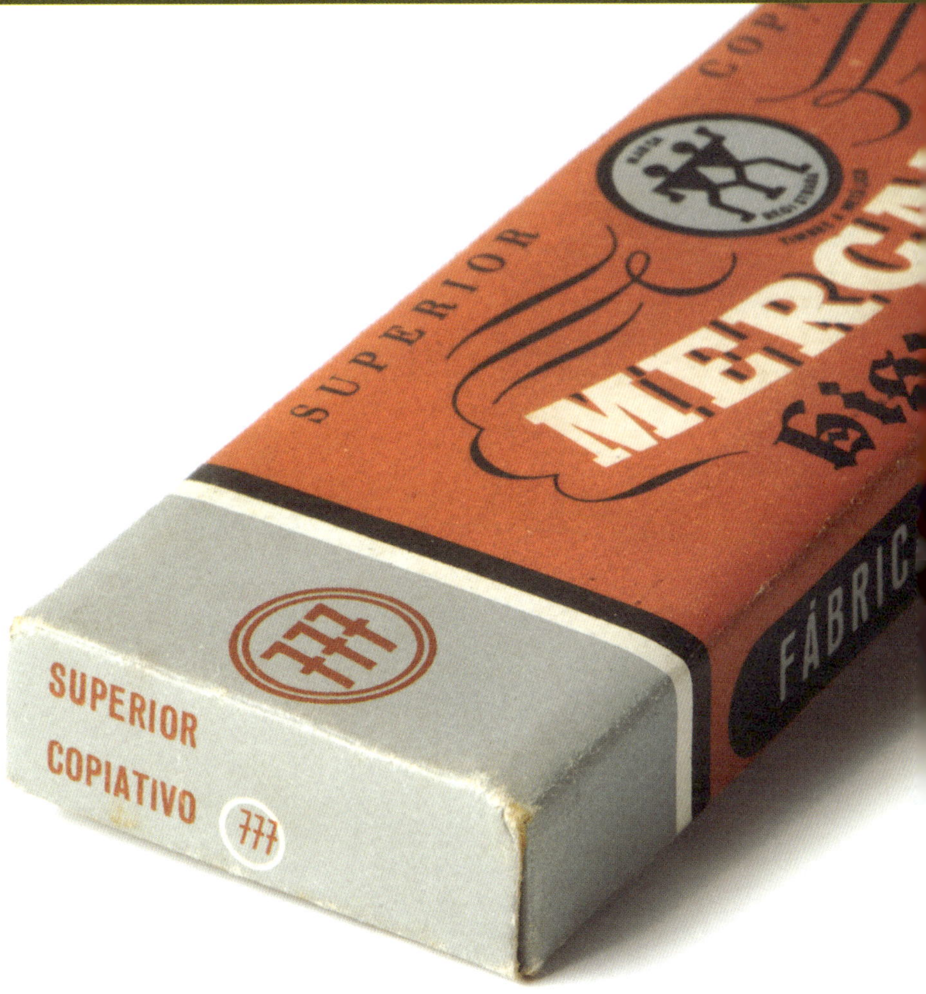
SUPERIOR
SUPERIOR
COPIATIVO
777

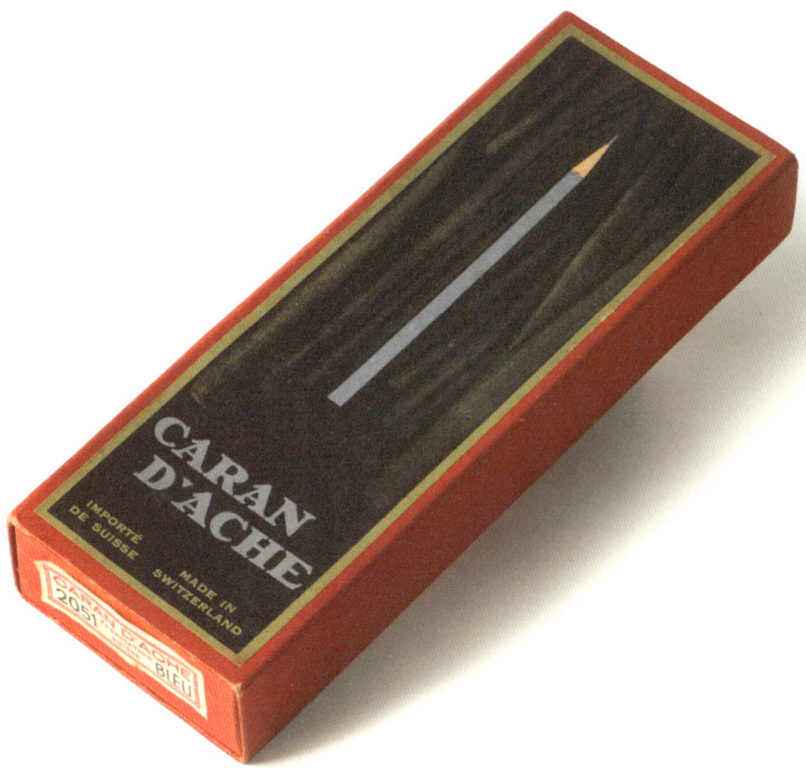
CARAN
D'ACHE
IMPORTÉ
DE SUISSE
MADE IN
SWITZERLAND
CARAN D'ACHE
2051
BLEU

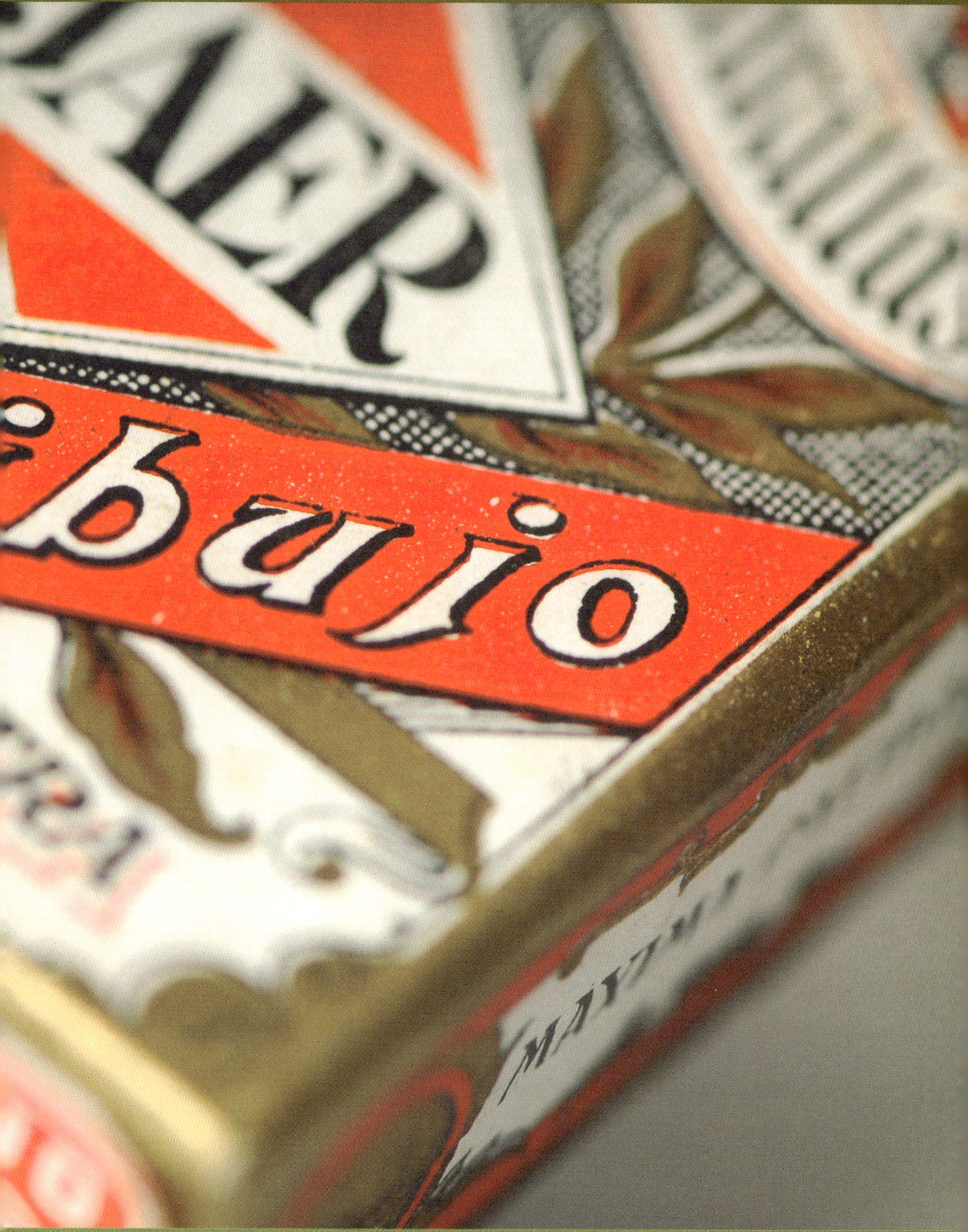

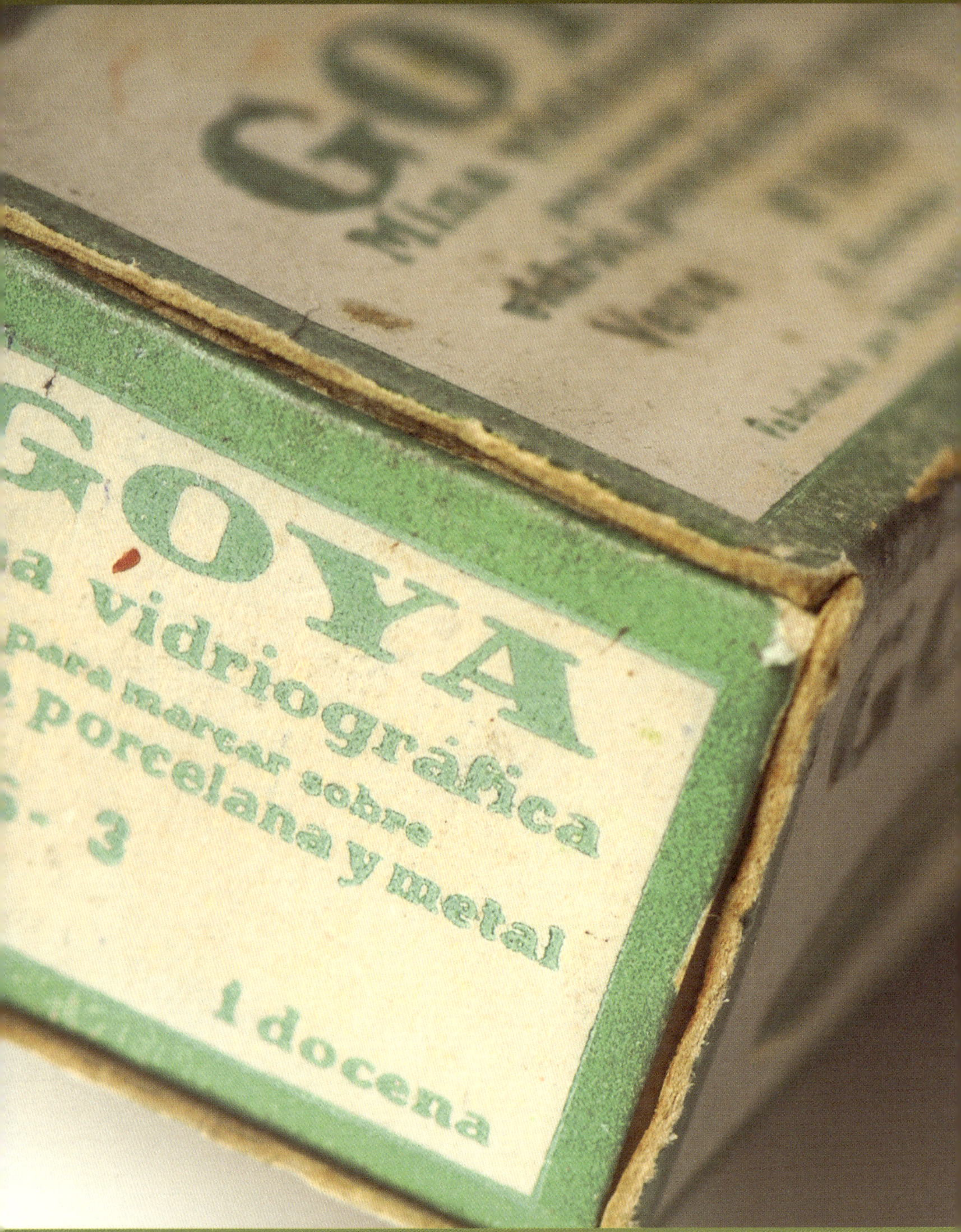
GOYA
vidriográfica
para marcar sobre
porcelana y metal
1 docena

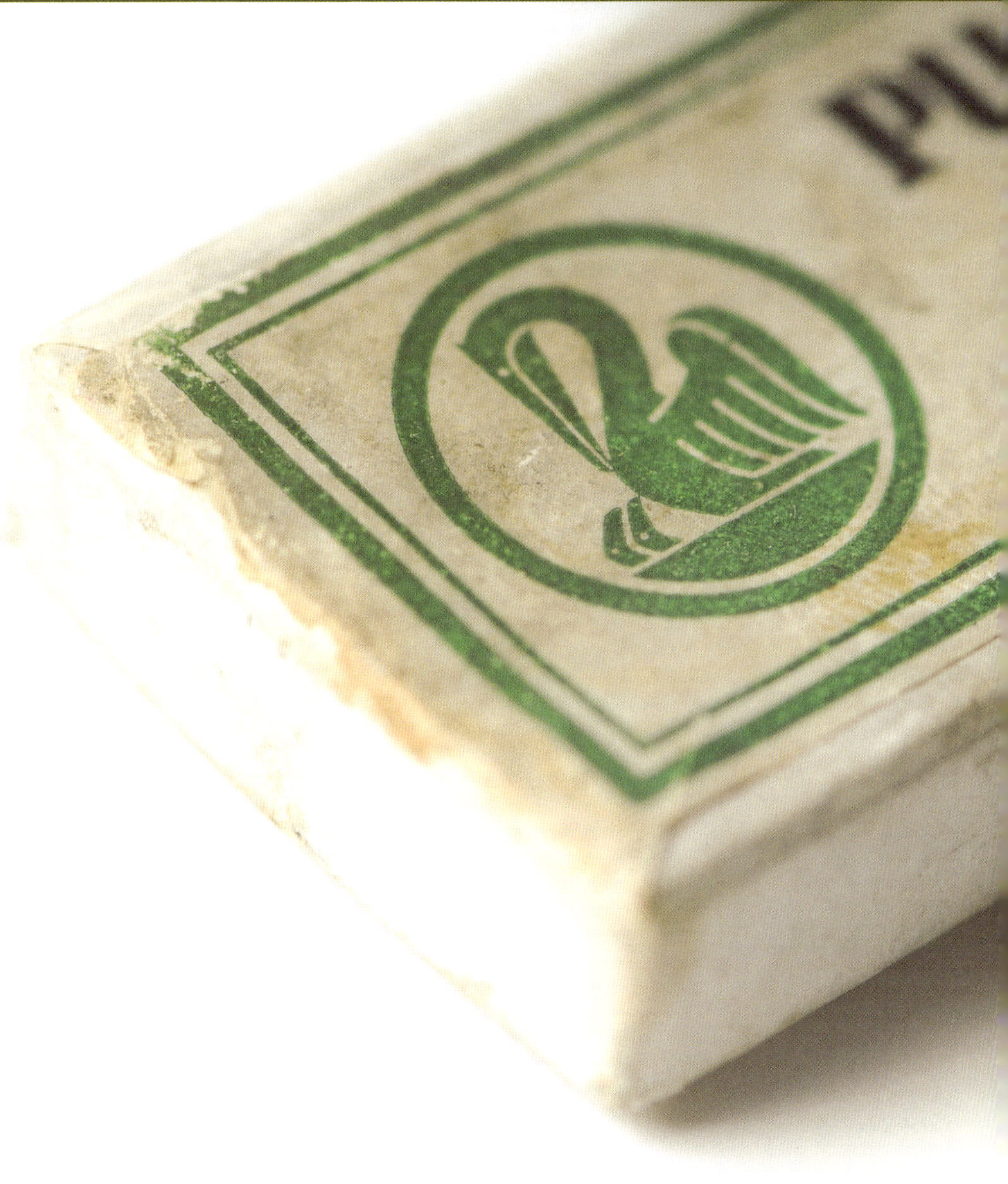

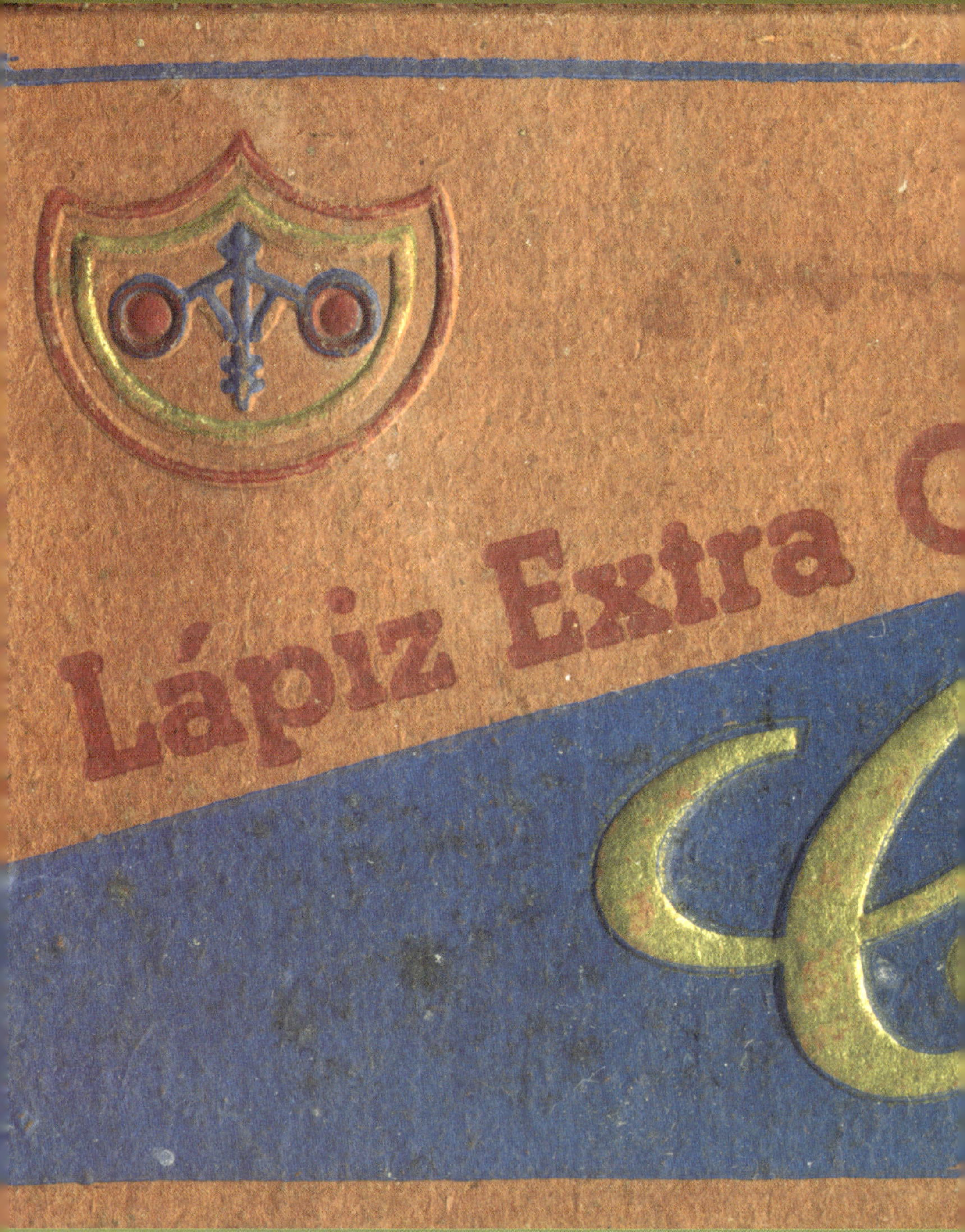
Lápiz Extra

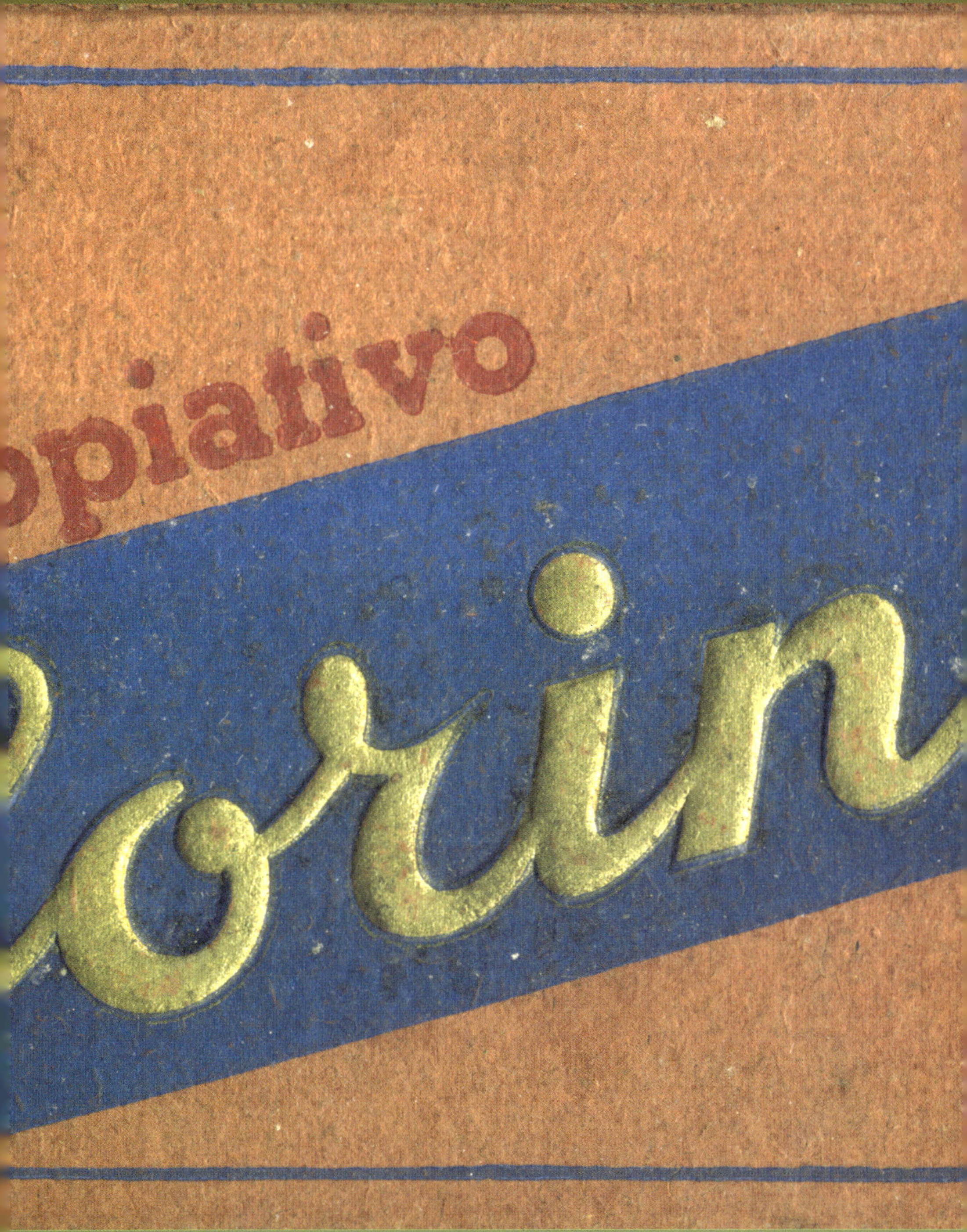
opiativo
orin

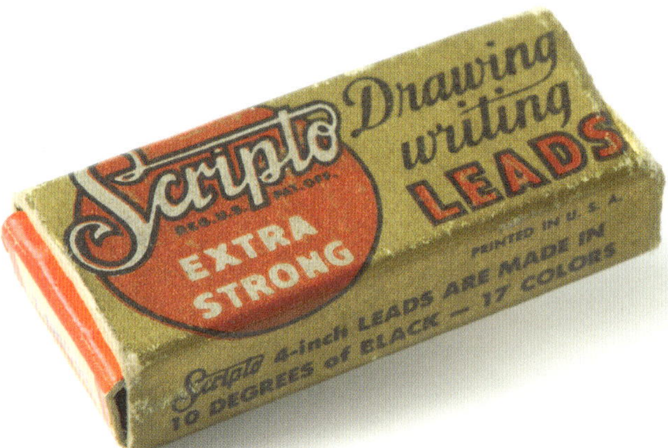
Scripto
REG. U.S. PAT. OFF.
EXTRA STRONG
Drawing writing LEADS
PRINTED IN U. S. A.
Scripto 4-inch LEADS ARE MADE IN
10 DEGREES of BLACK — 17 COLORS

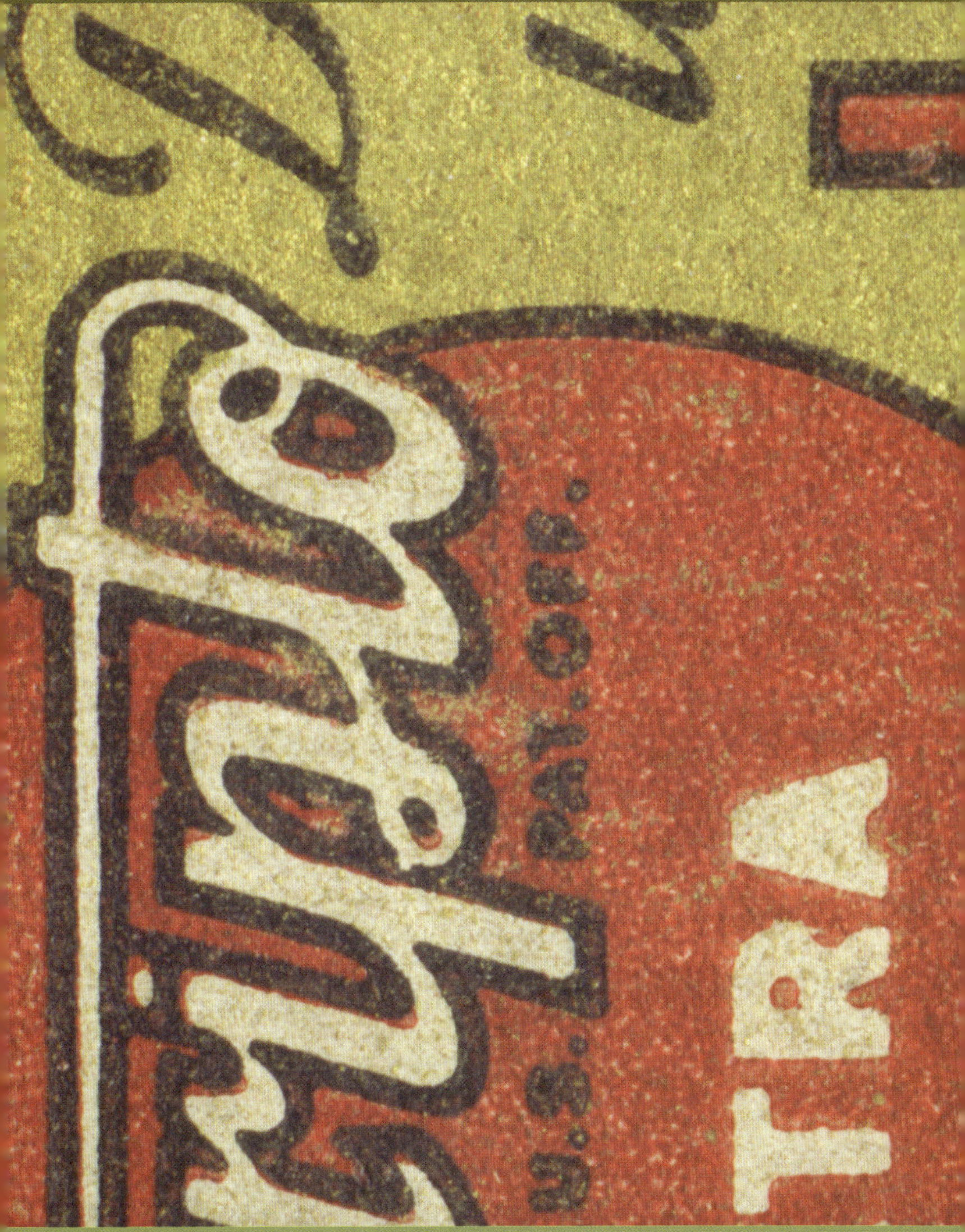
U.S. PAT. OFF.
TRA

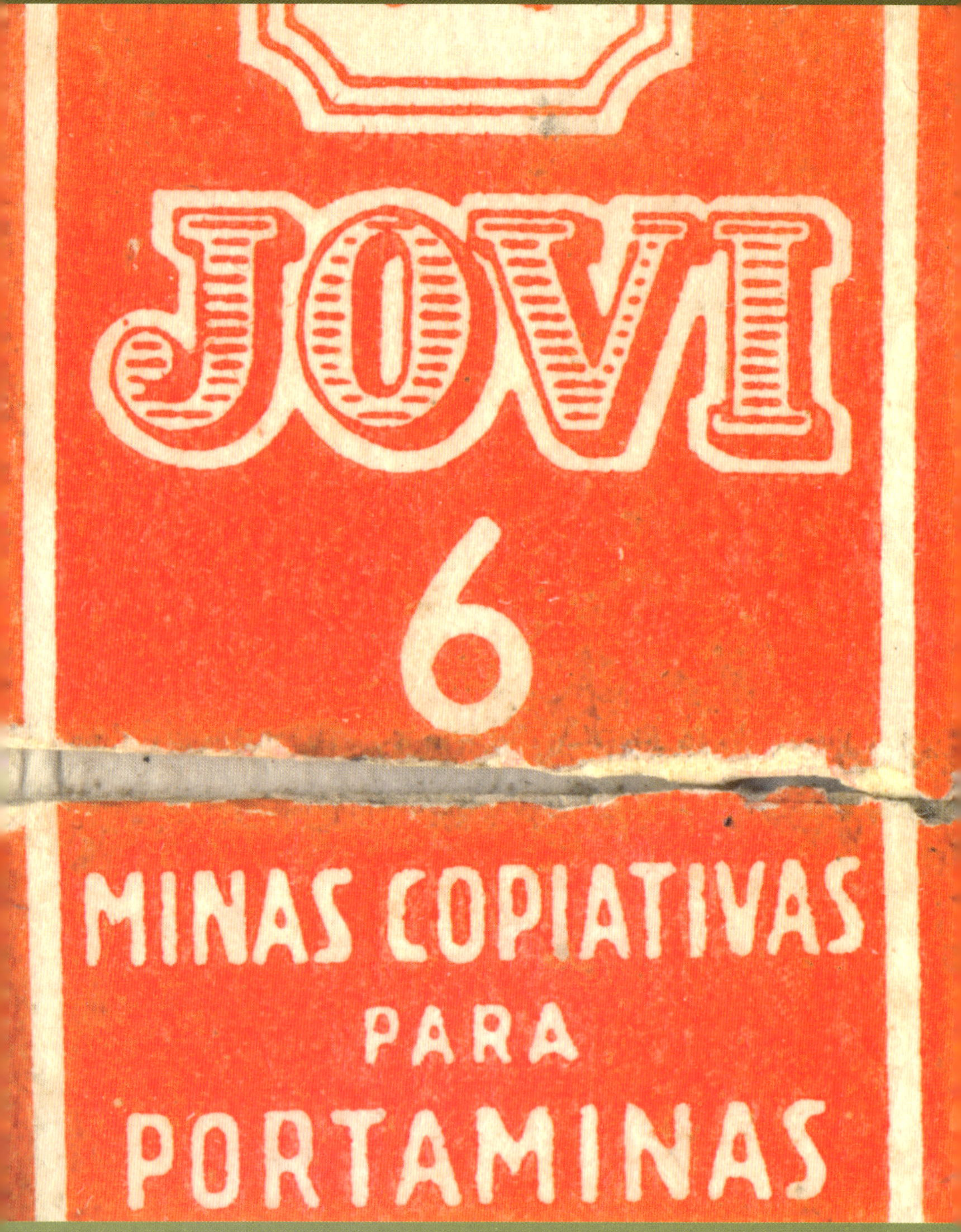
JOVI
6
MINAS COPIATIVAS
PARA
PORTAMINAS

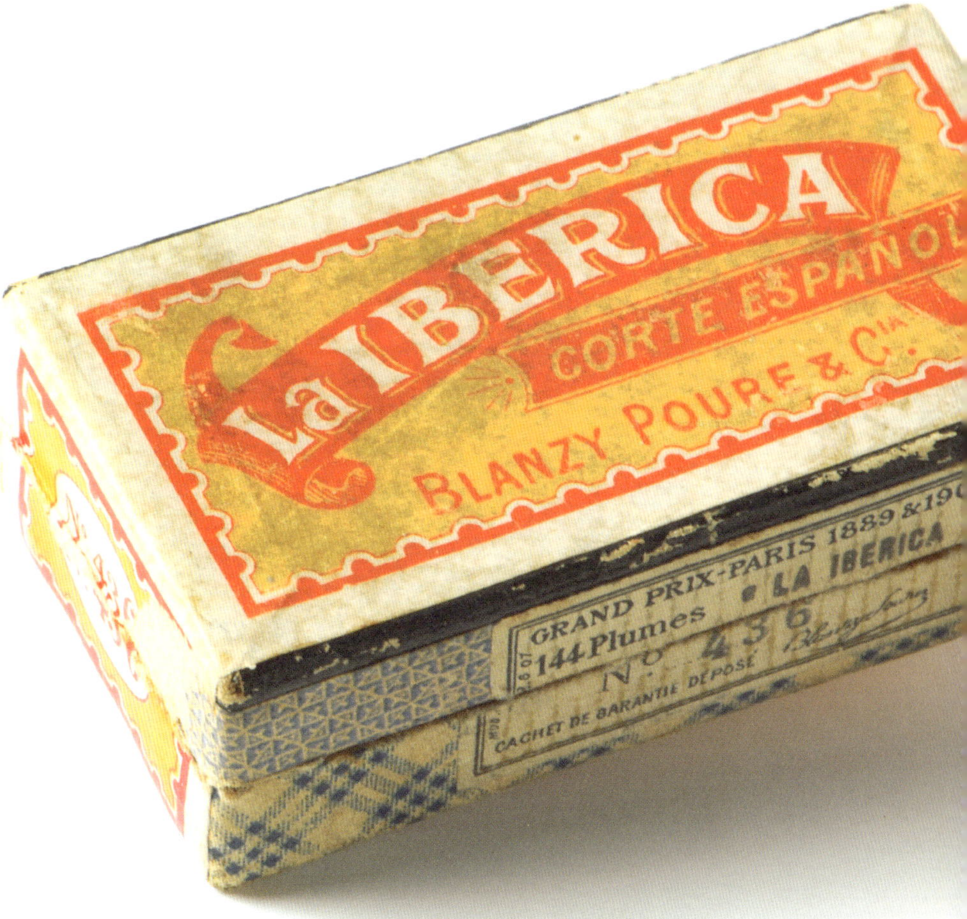
La IBERICA
CORTE ESPAÑOL
BLANZY POURE & Cia
GRAND PRIX-PARIS 1889 &
144 Plumes • LA IBERICA
No 436
CACHET DE GARANTIE DÉPOSÉ

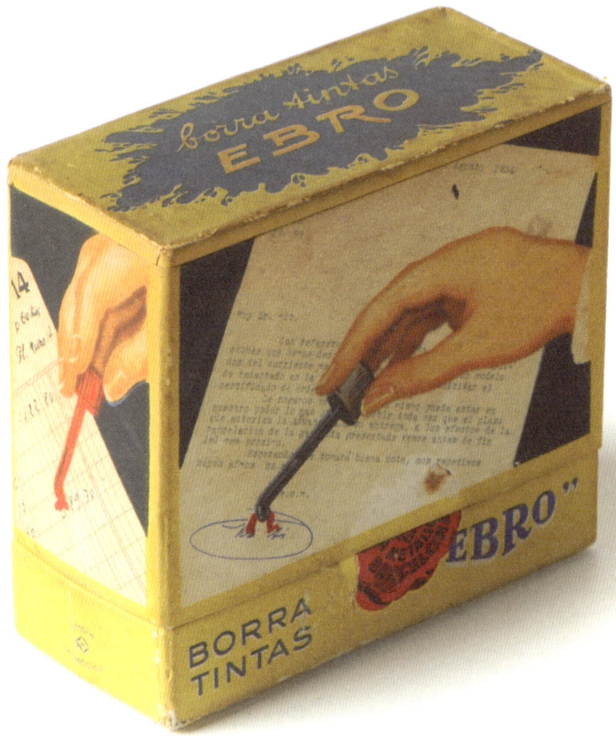
borra tintas
EBRO
"EBRO"
BORRA
TINTAS

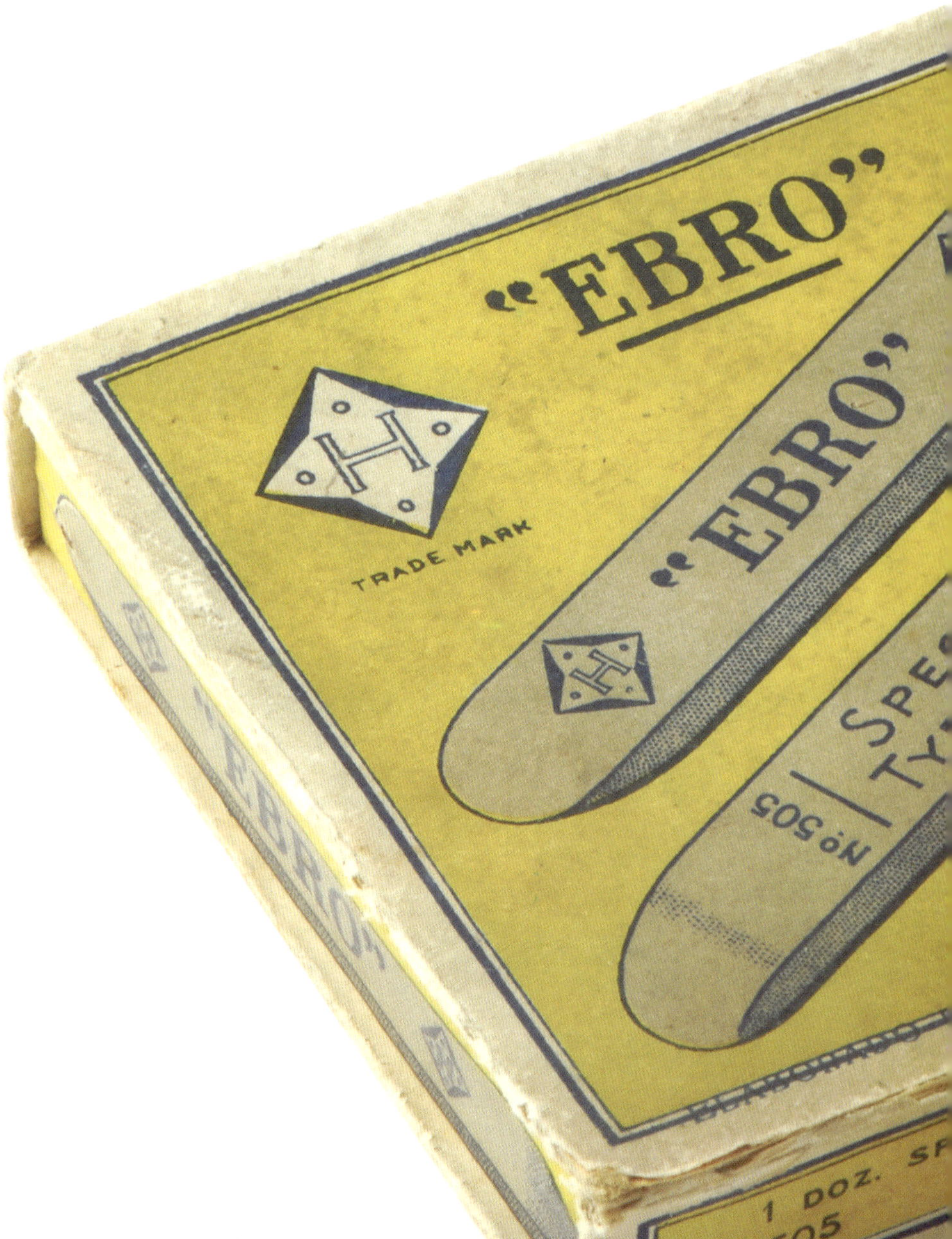
"EBRO"
TRADE MARK
"EBRO"
No 505

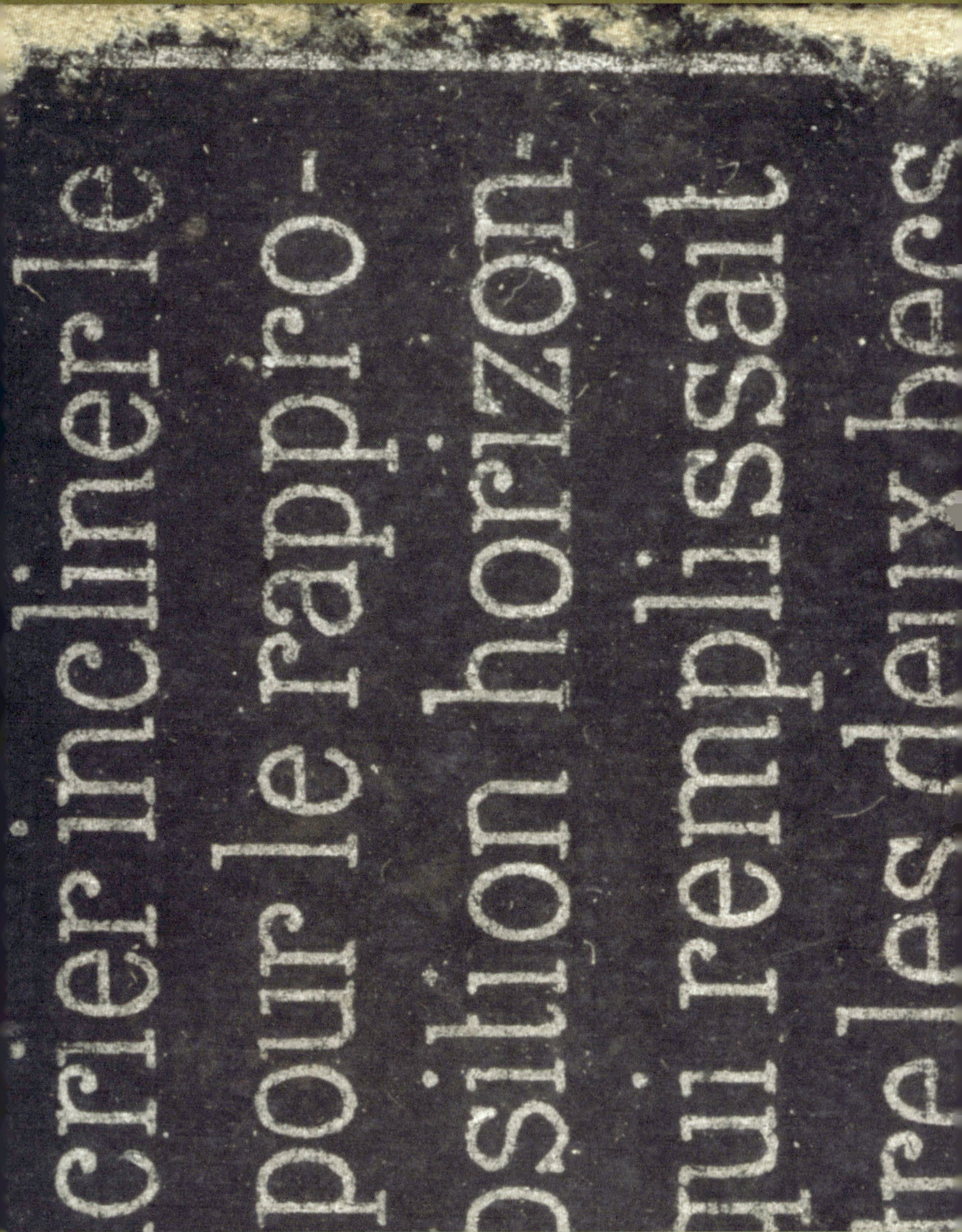

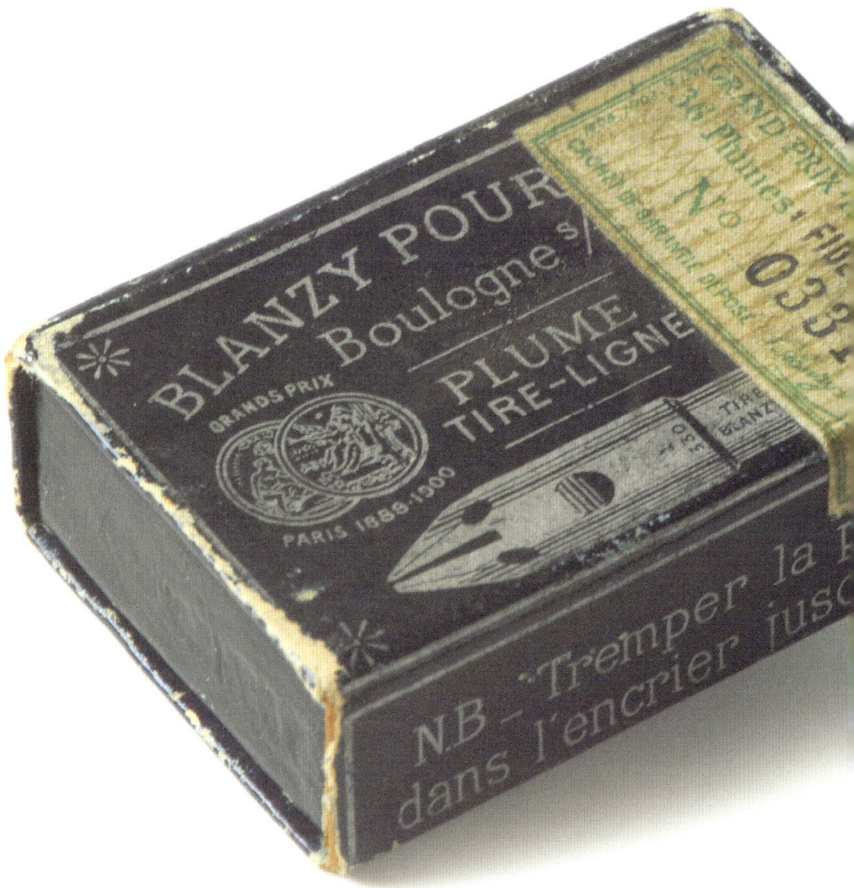
BLANZY POUR
Boulogne s/
PLUME
TIRE-LIGNE
GRANDS PRIX
PARIS 1889·1900
GRAND PRIX
Plumes
N°
N.B – Tremper la
dans l'encrier

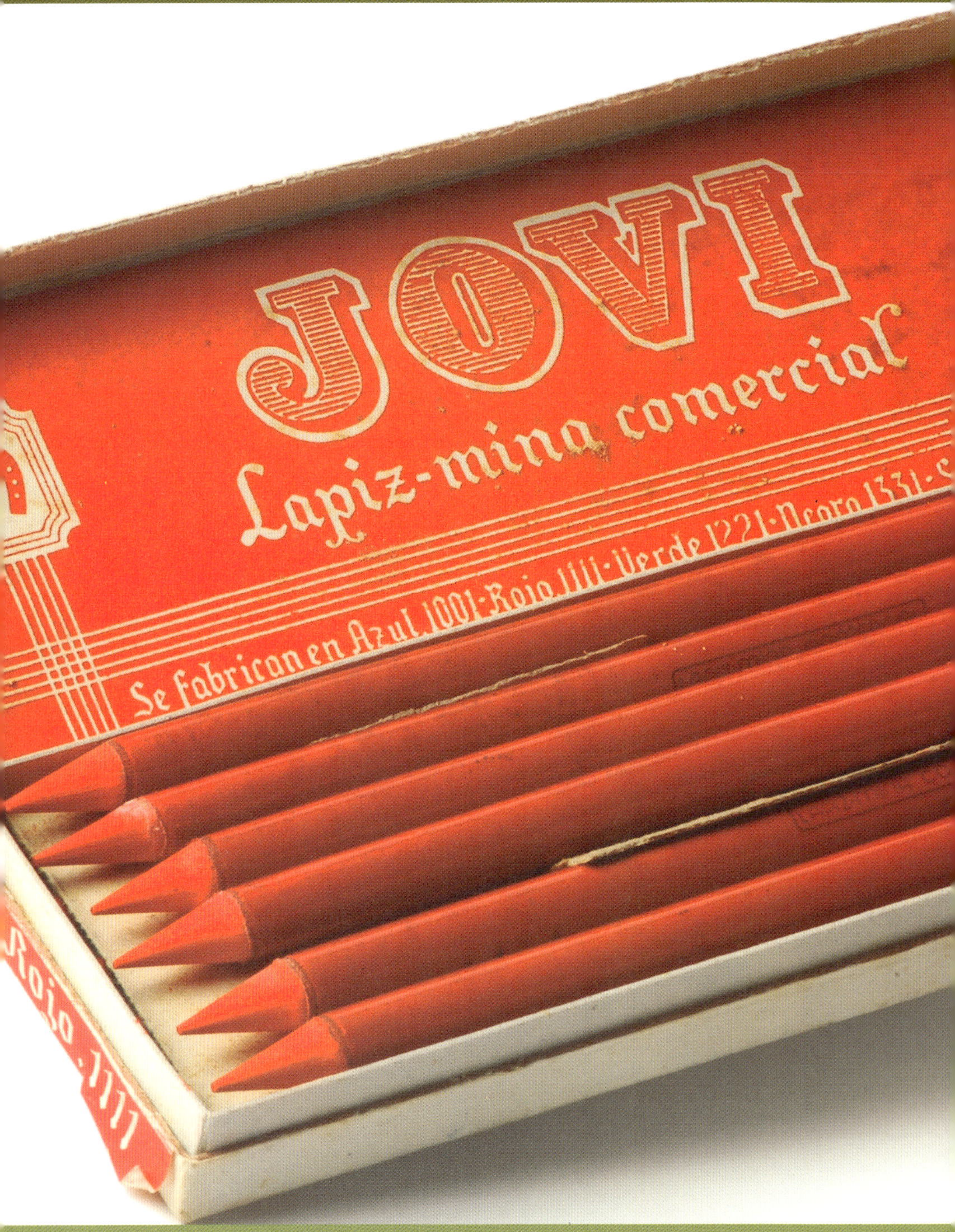
JOVI
Lapiz-mina comercial
Se fabrican en Azul 1001-Rojo 1111-Verde 1221-Negro 1331-
Rojo 1111

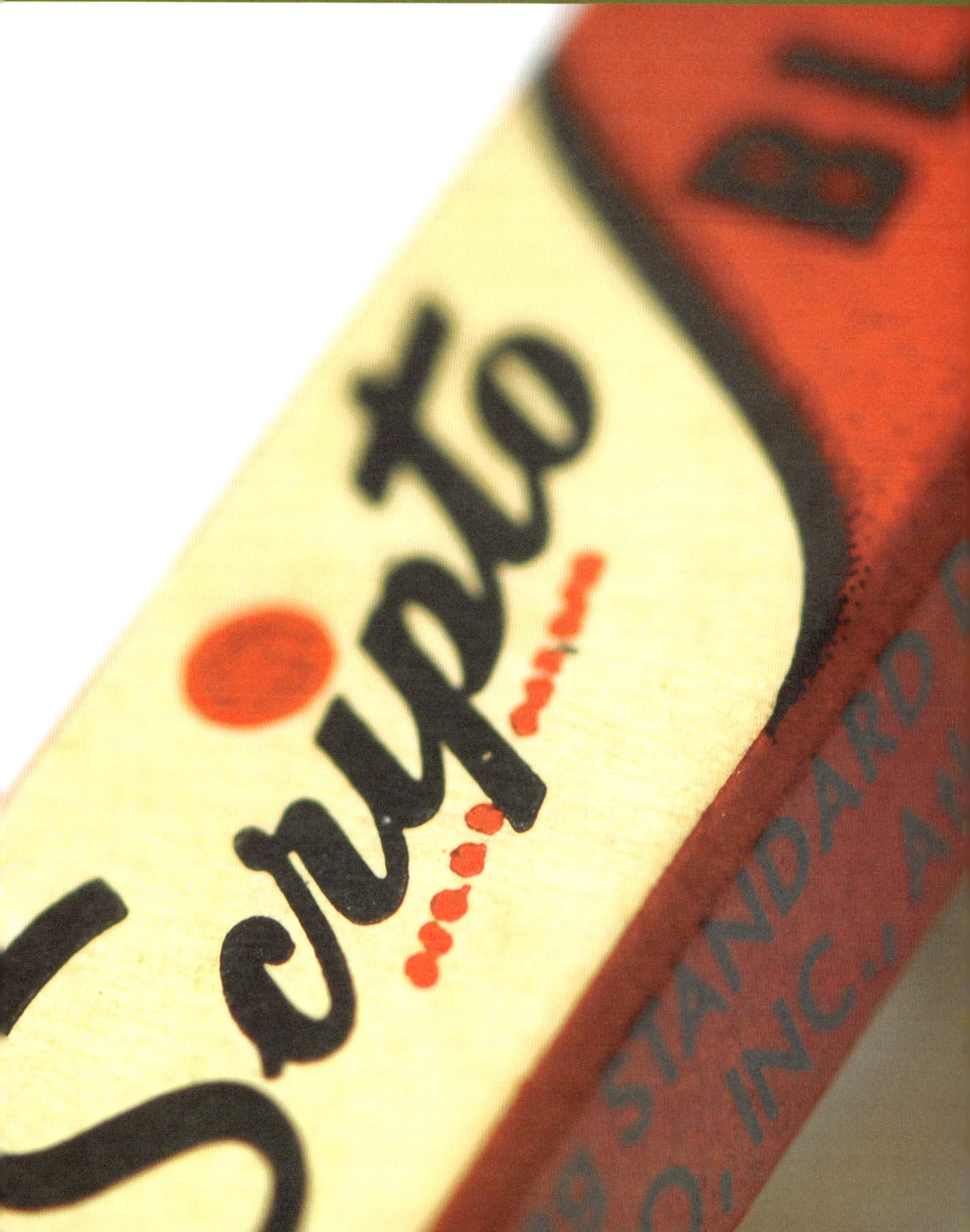
Scripto
STANDARD
INC.

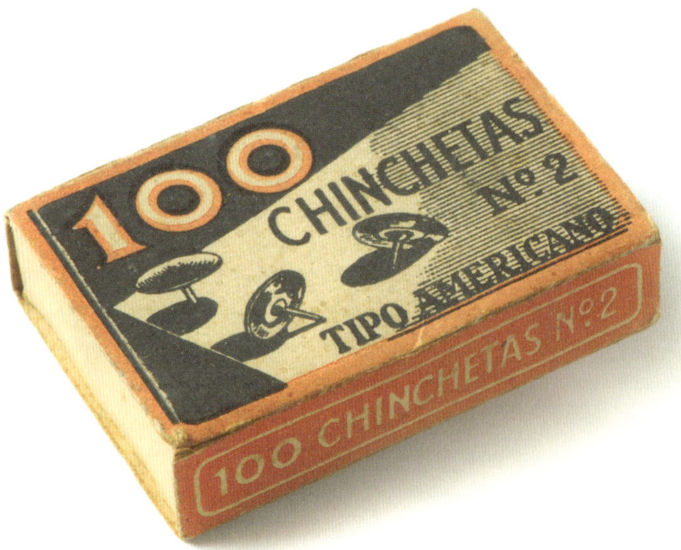
100
CHINCHETAS
Nº2
TIPO AMERICANO
100 CHINCHETAS Nº2

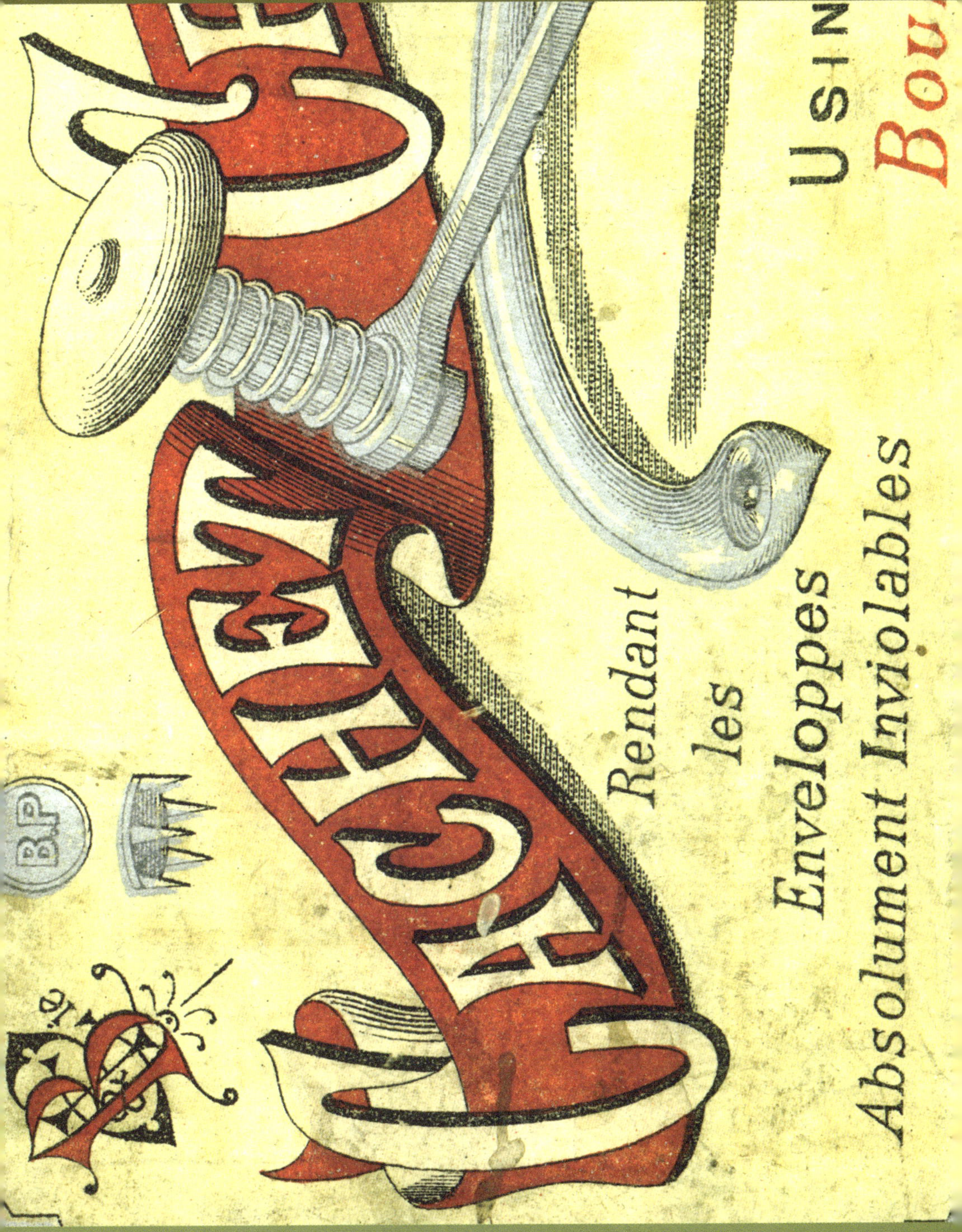
B.P
Rendant
les
Enveloppes
Absolument Inviolables

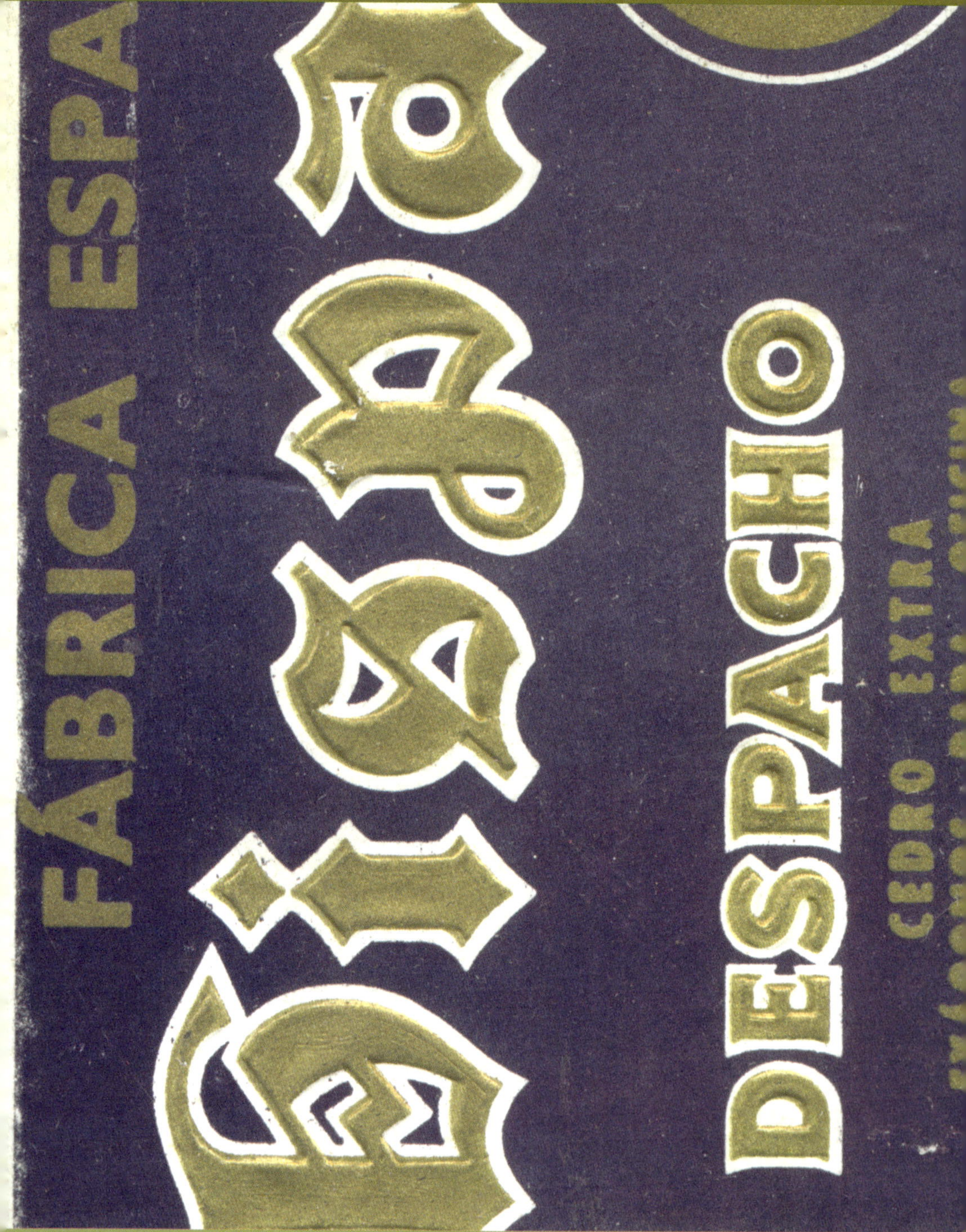
FÁBRICA ESPA
DESPACHO
CEDRO EXTRA

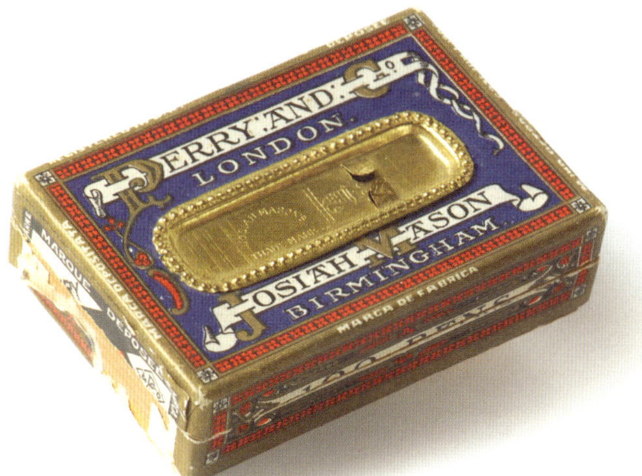
PERRY AND Co.
LONDON.
JOSIAH MASON
BIRMINGHAM.
MARCA DE FABRICA
MARQUE DEPOSÉE

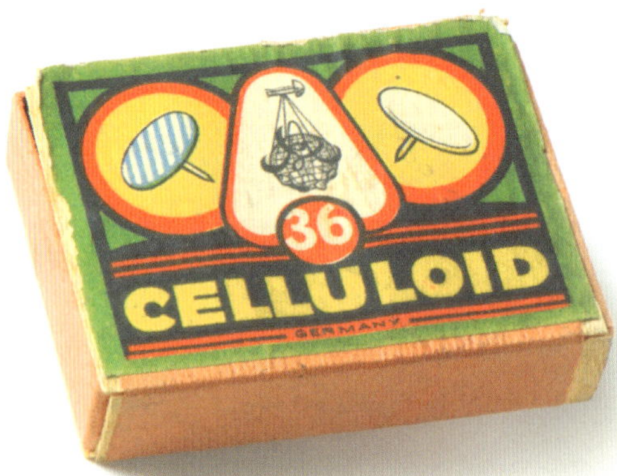
36
CELLULOID
GERMANY

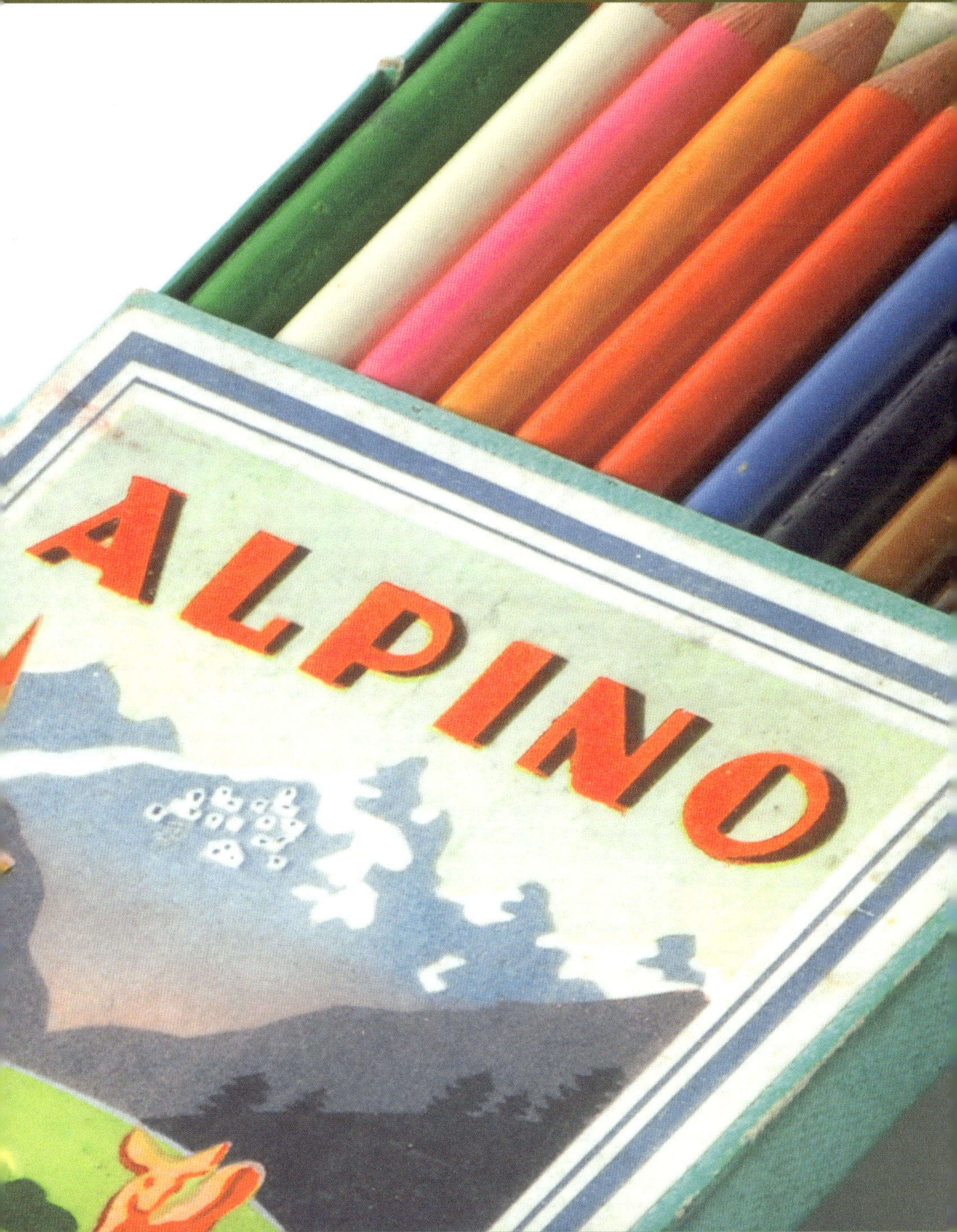
ALPINO

NTERTA

TERTAINMENTEN

ENTERTAINMENT

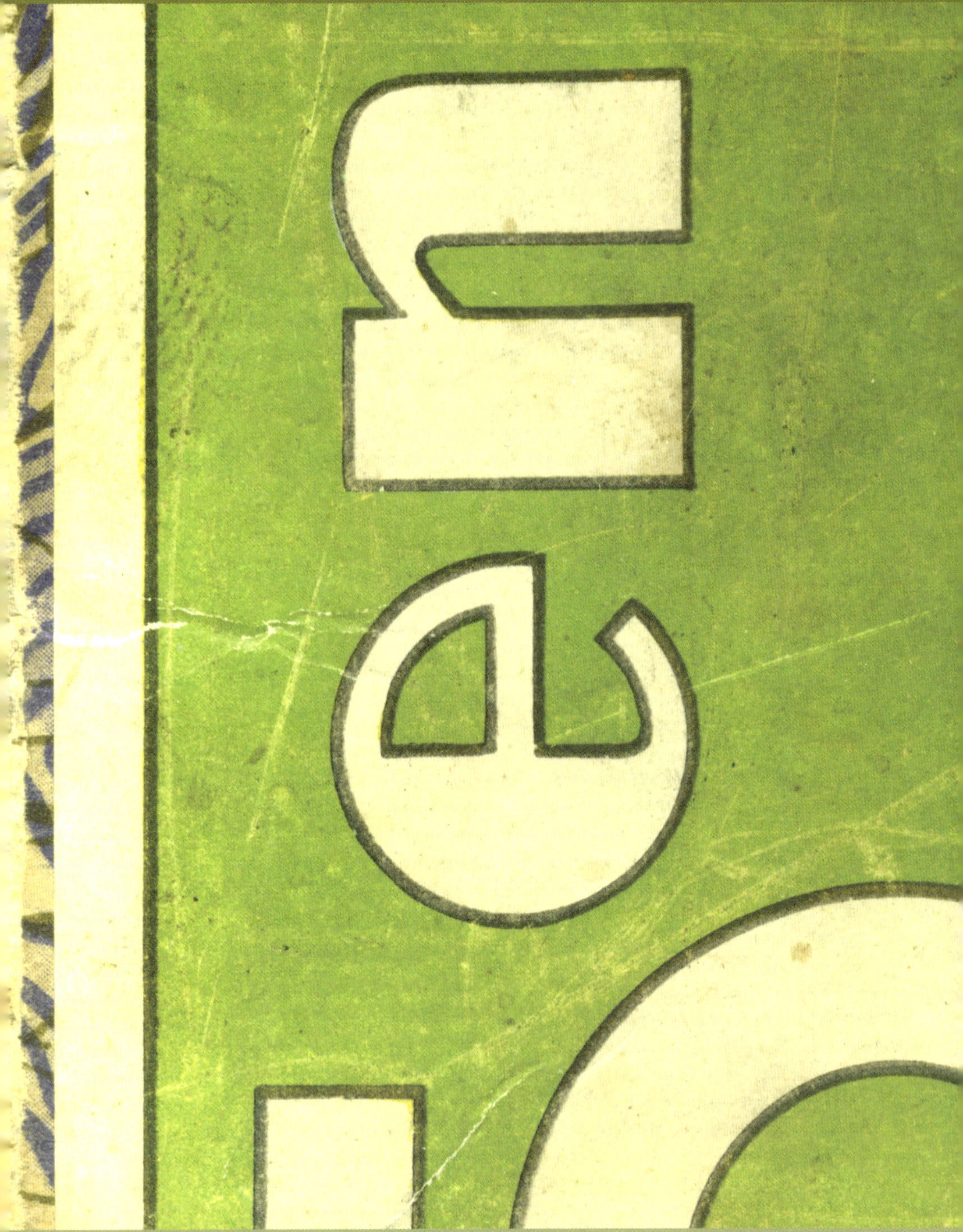

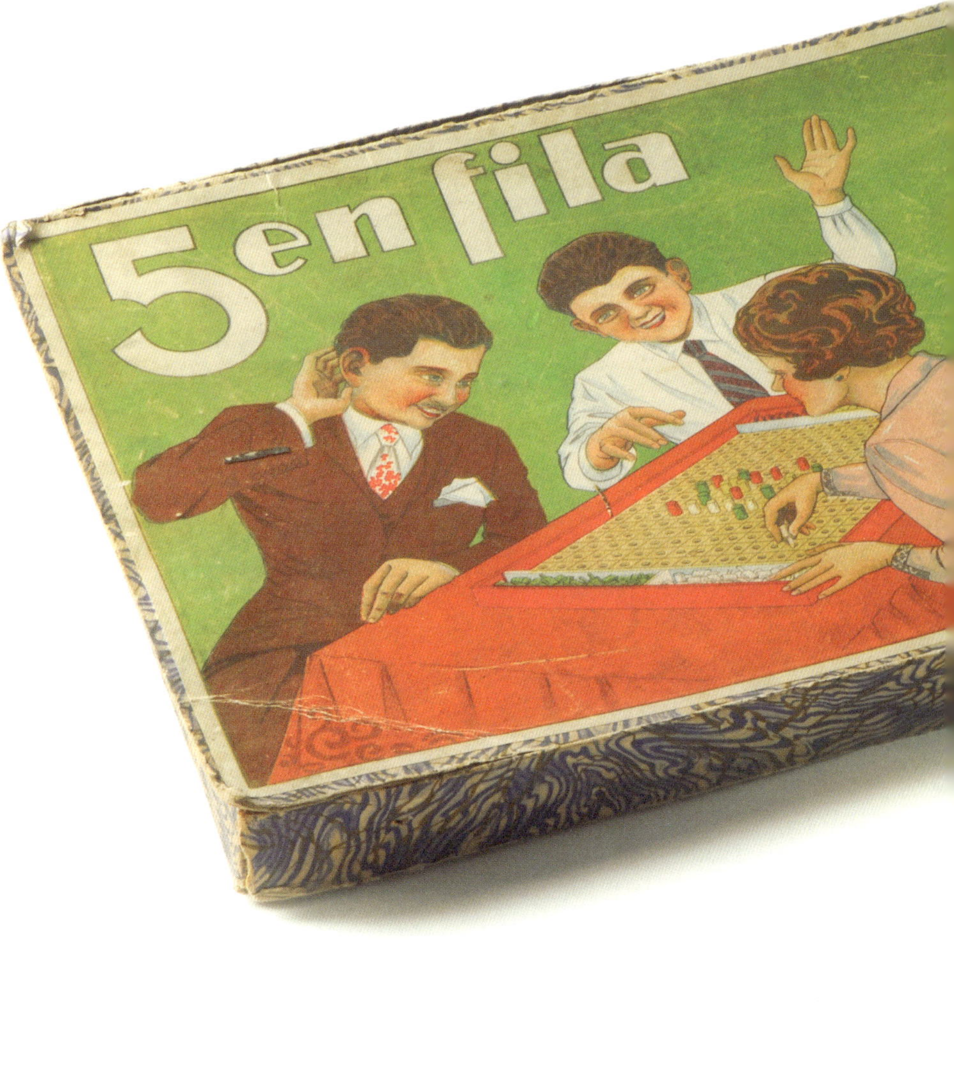
5 en fila

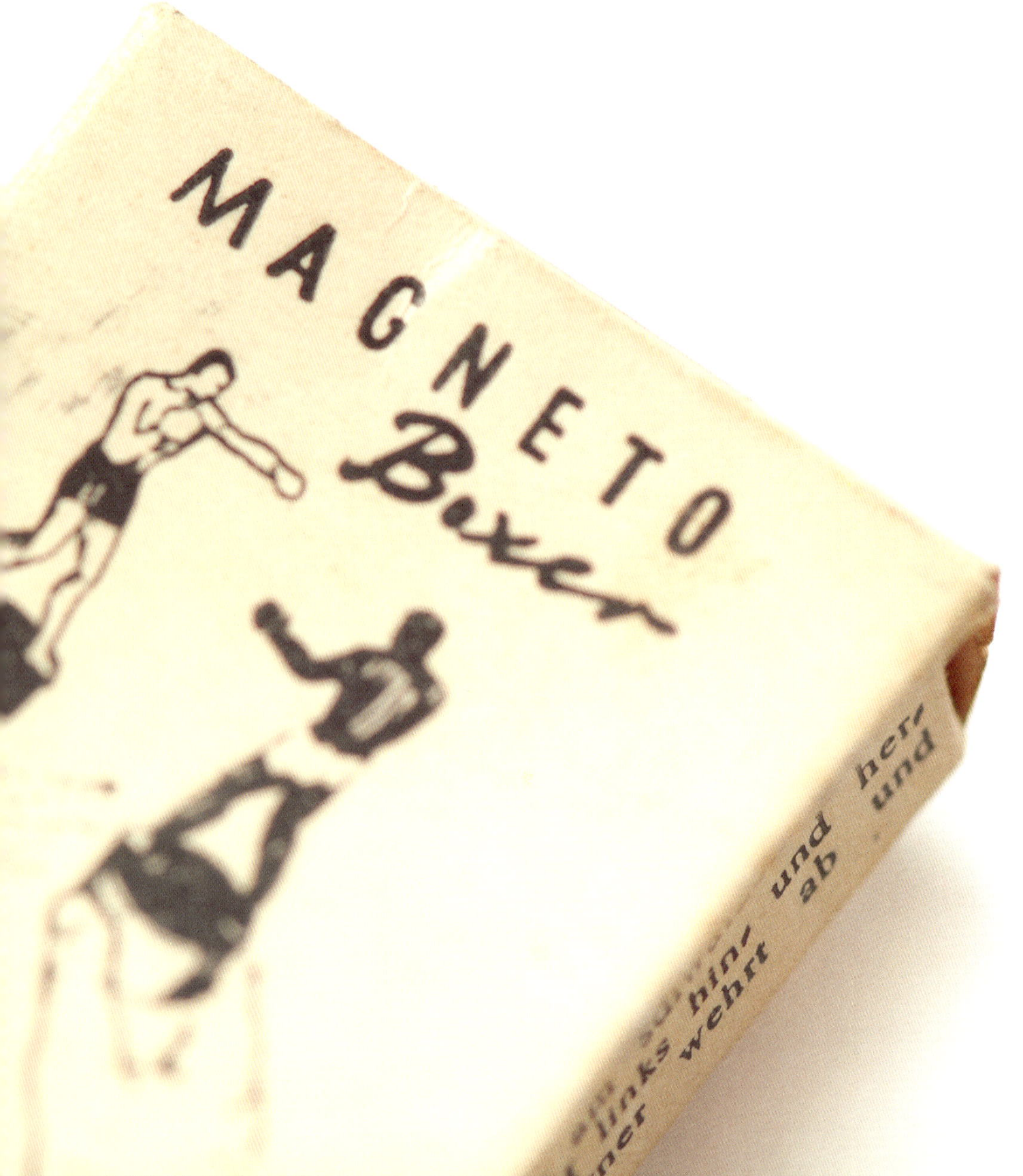
MAGNETO
Boxer
links hin- und her-
wehrt ab und

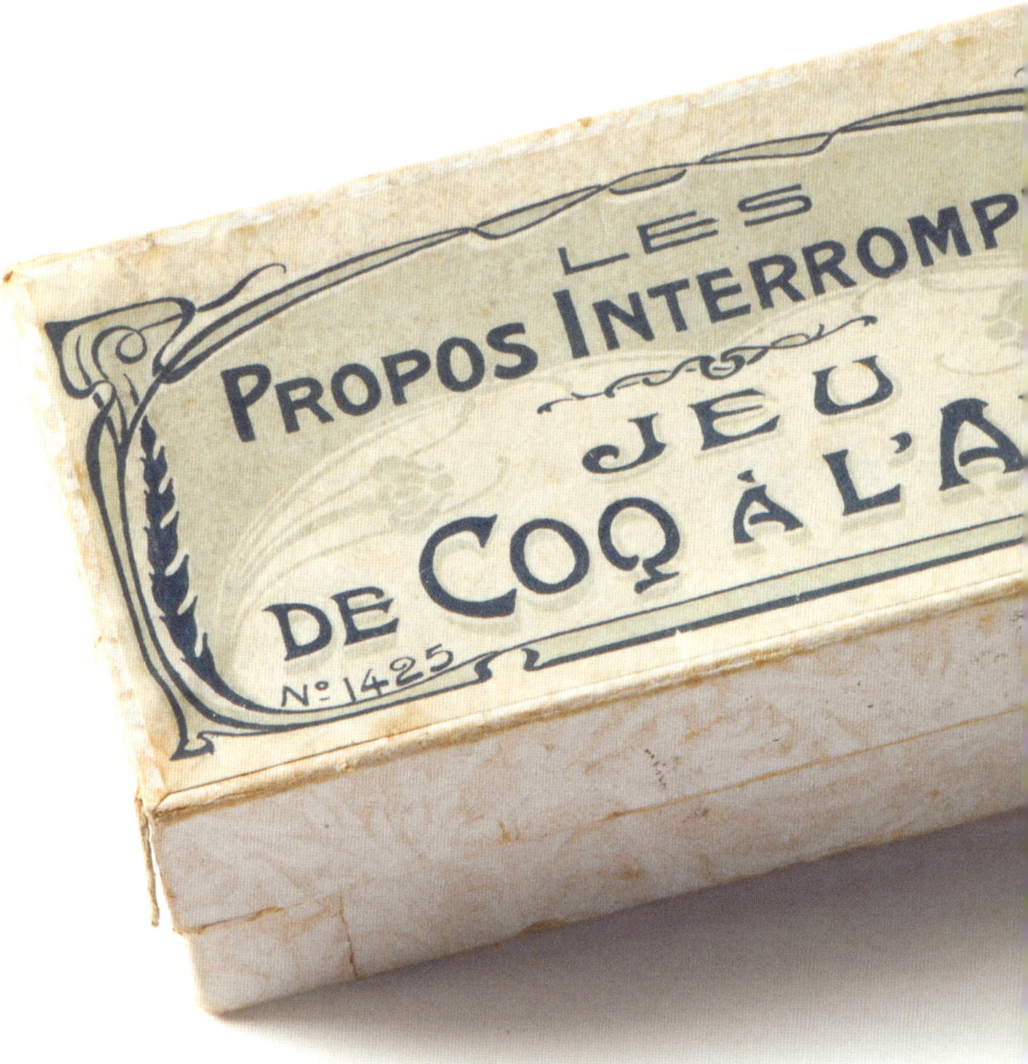
LES
PROPOS INTERROMP
JEU
DE COQ À L'A
No 1425

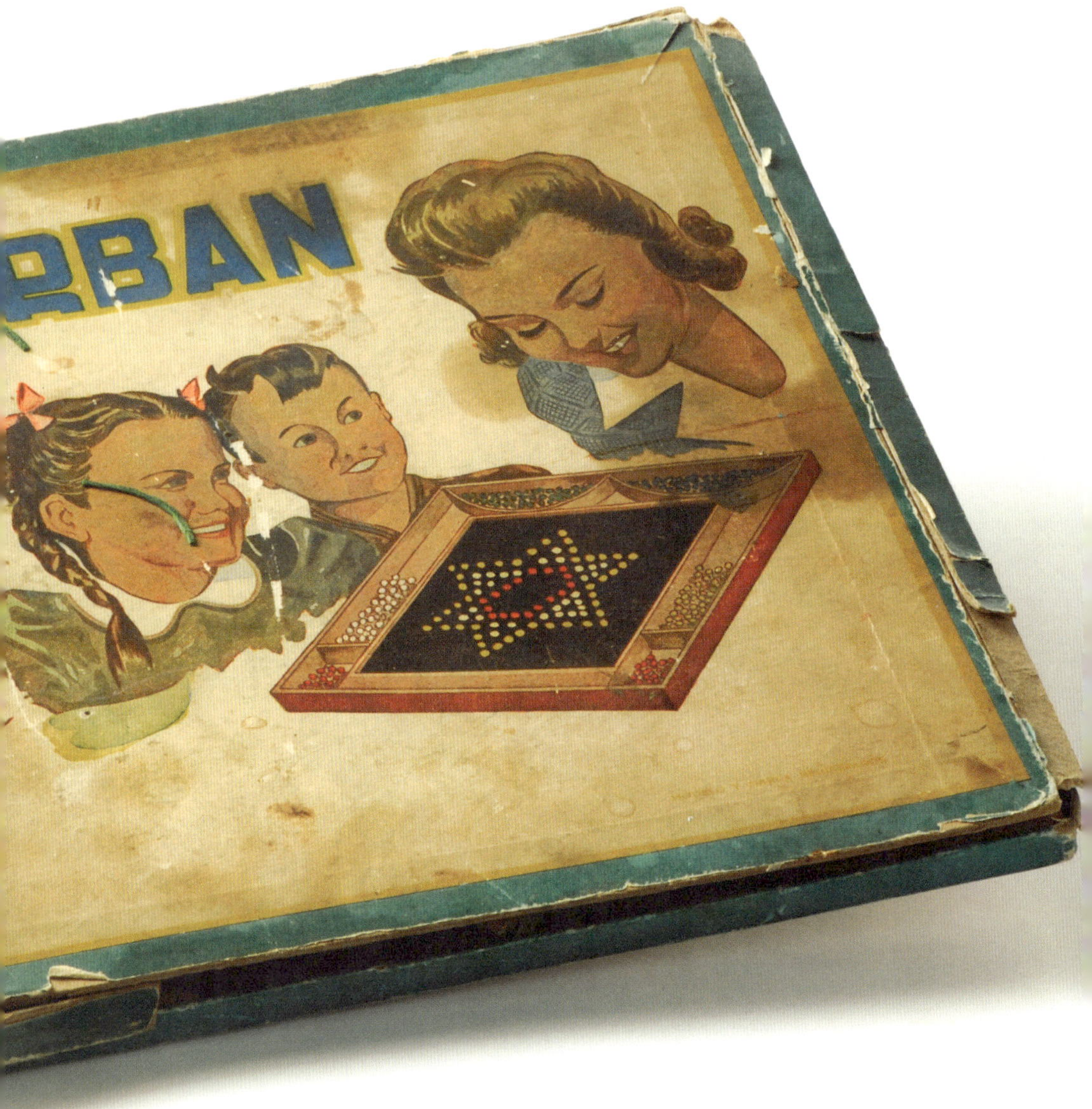
RBAN

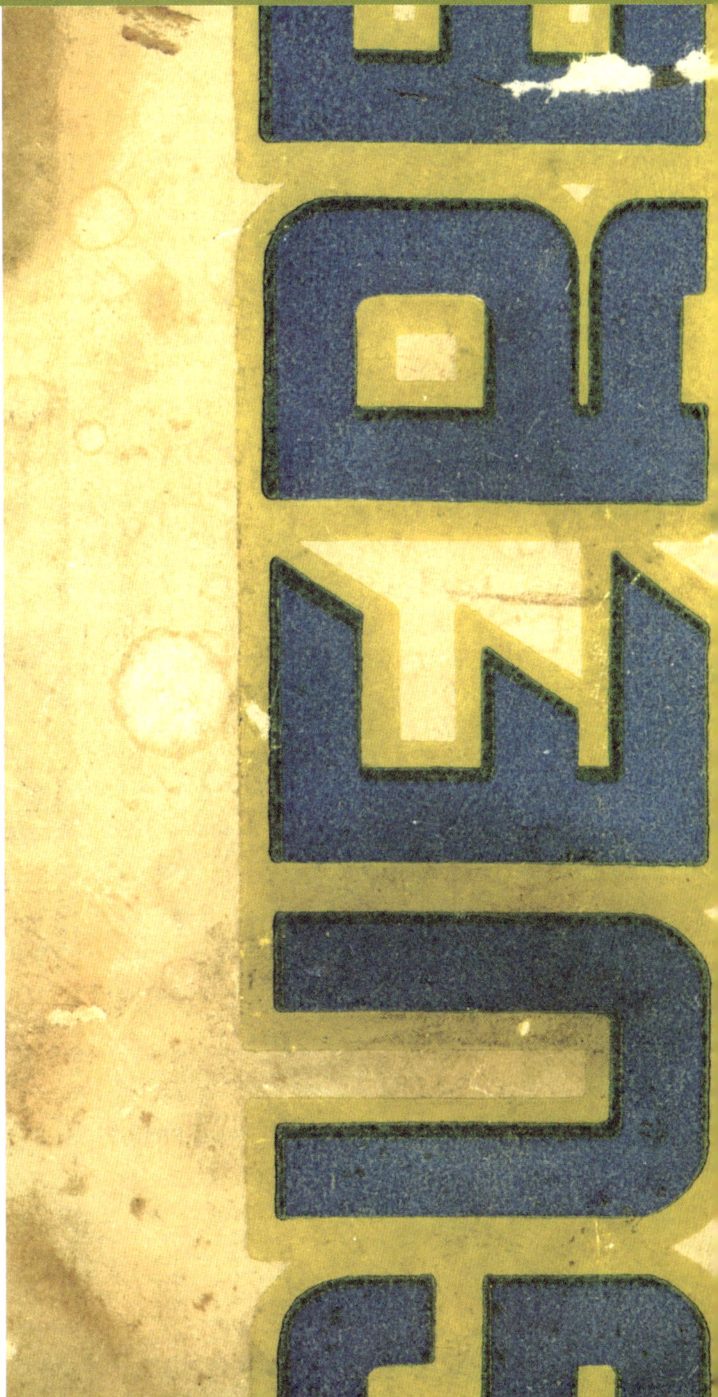

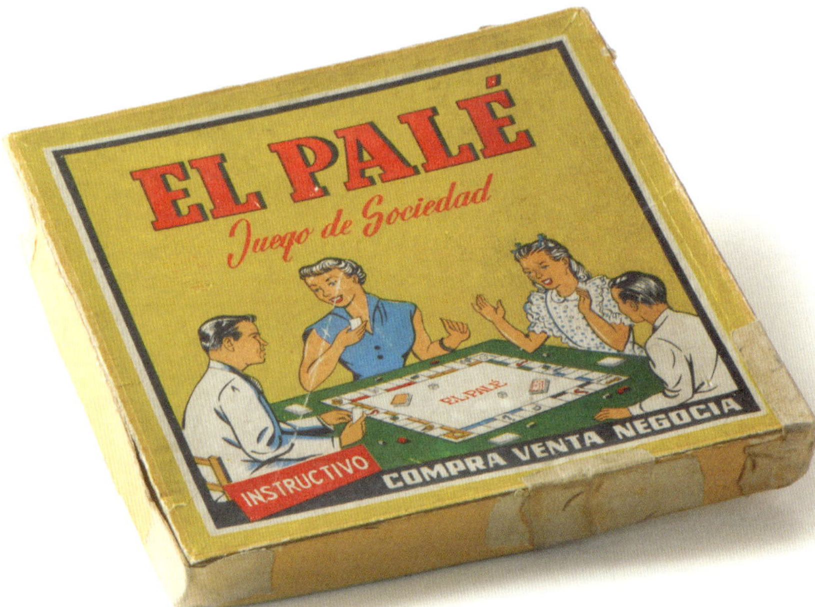
EL PALÉ
Juego de Sociedad
EL PALÉ
INSTRUCTIVO
COMPRA VENTA NEGOCIA

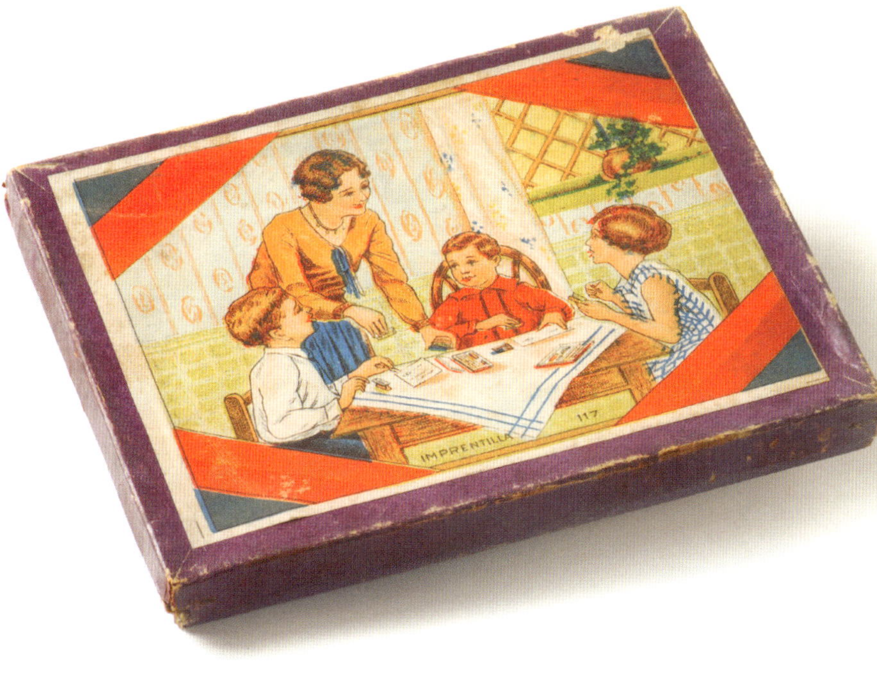
IMPRENTILLA
117

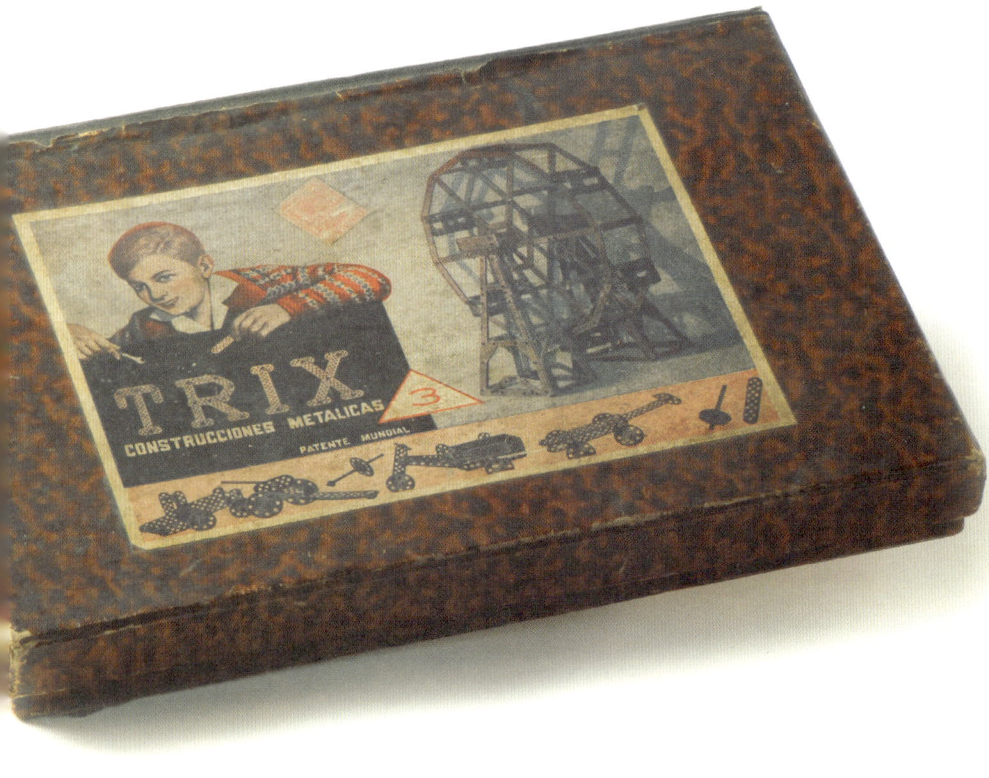
TRIX
3
CONSTRUCCIONES METALICAS
PATENTE MUNDIAL

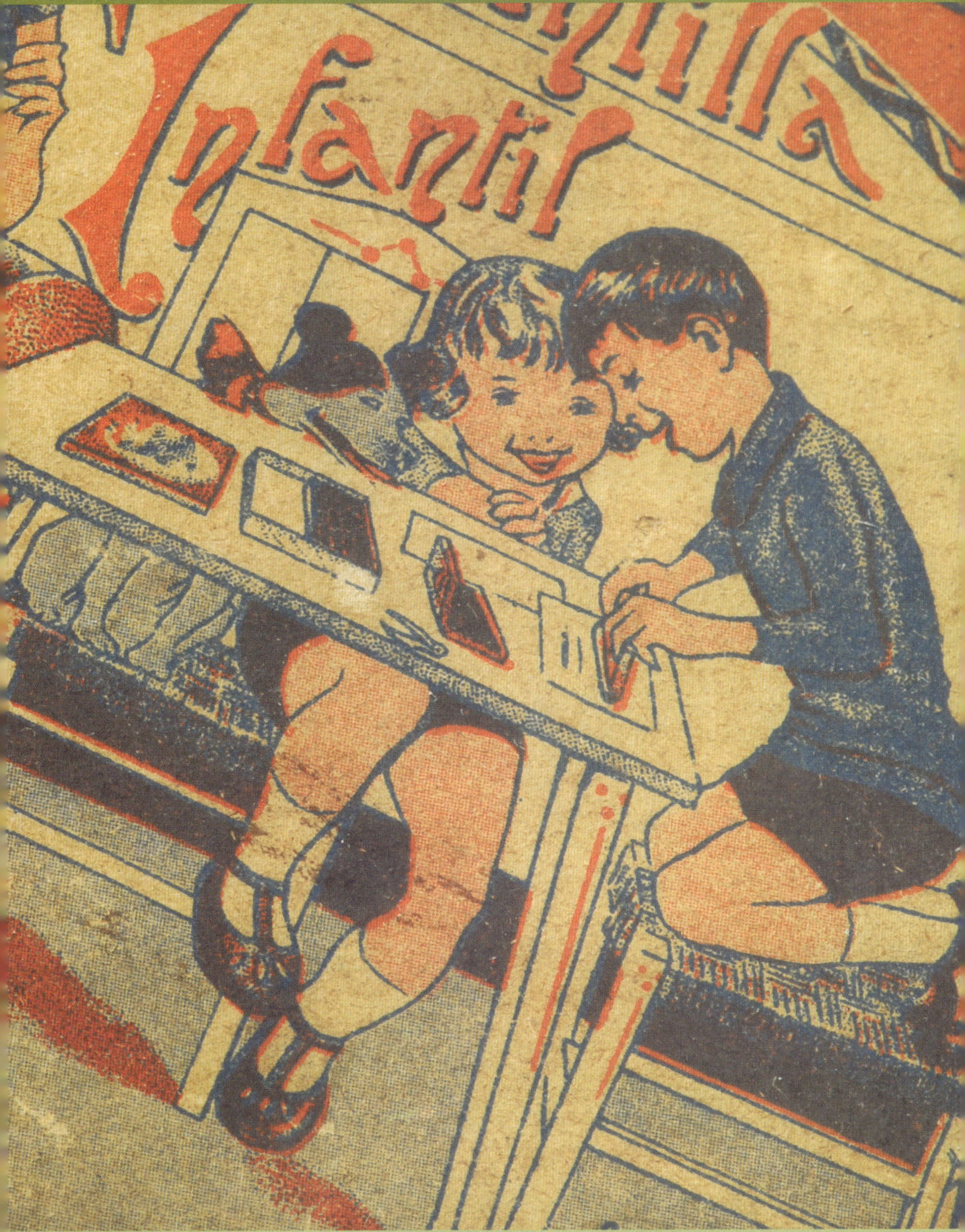
Infantil

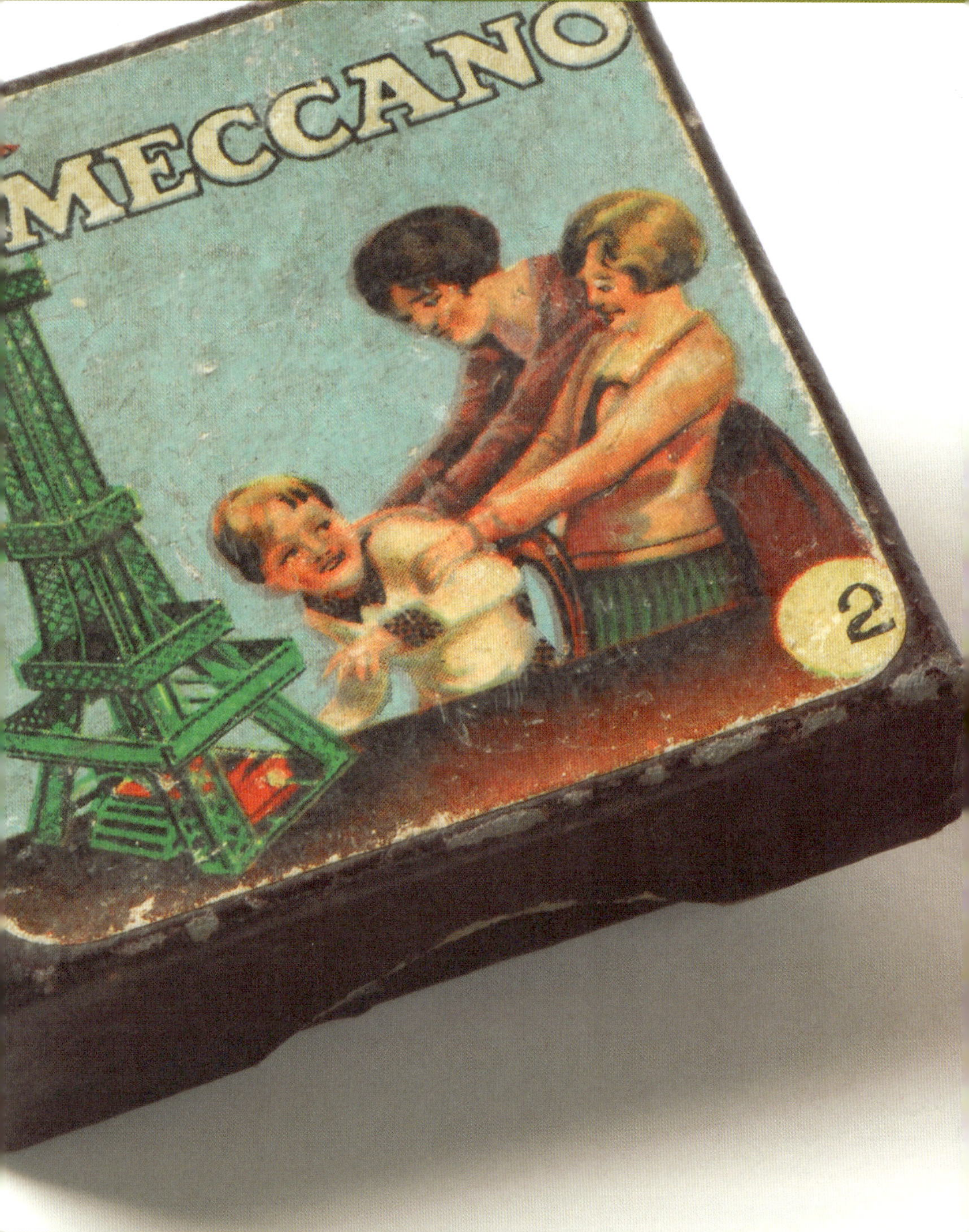
MECCANO
2

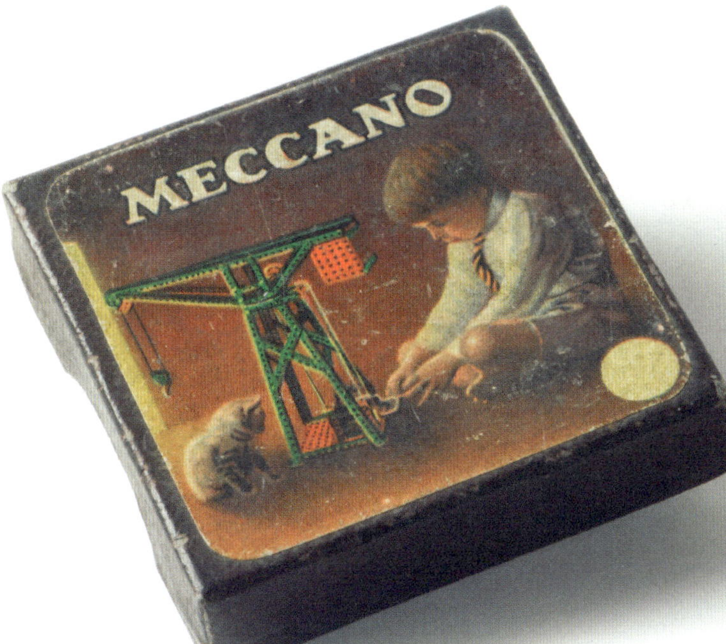
MECCANO

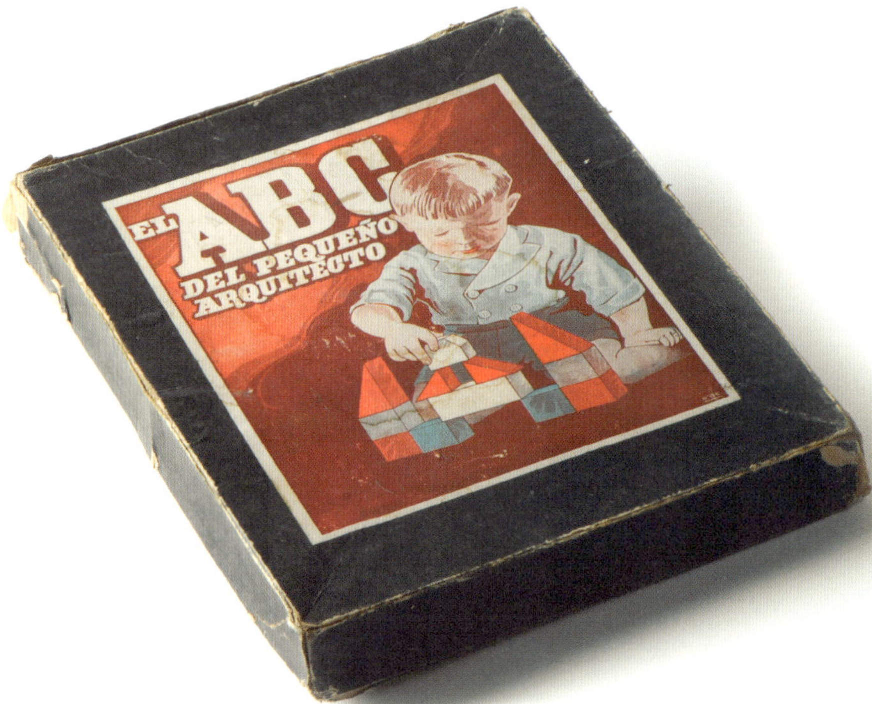
EL ABC
DEL PEQUEÑO ARQUITECTO

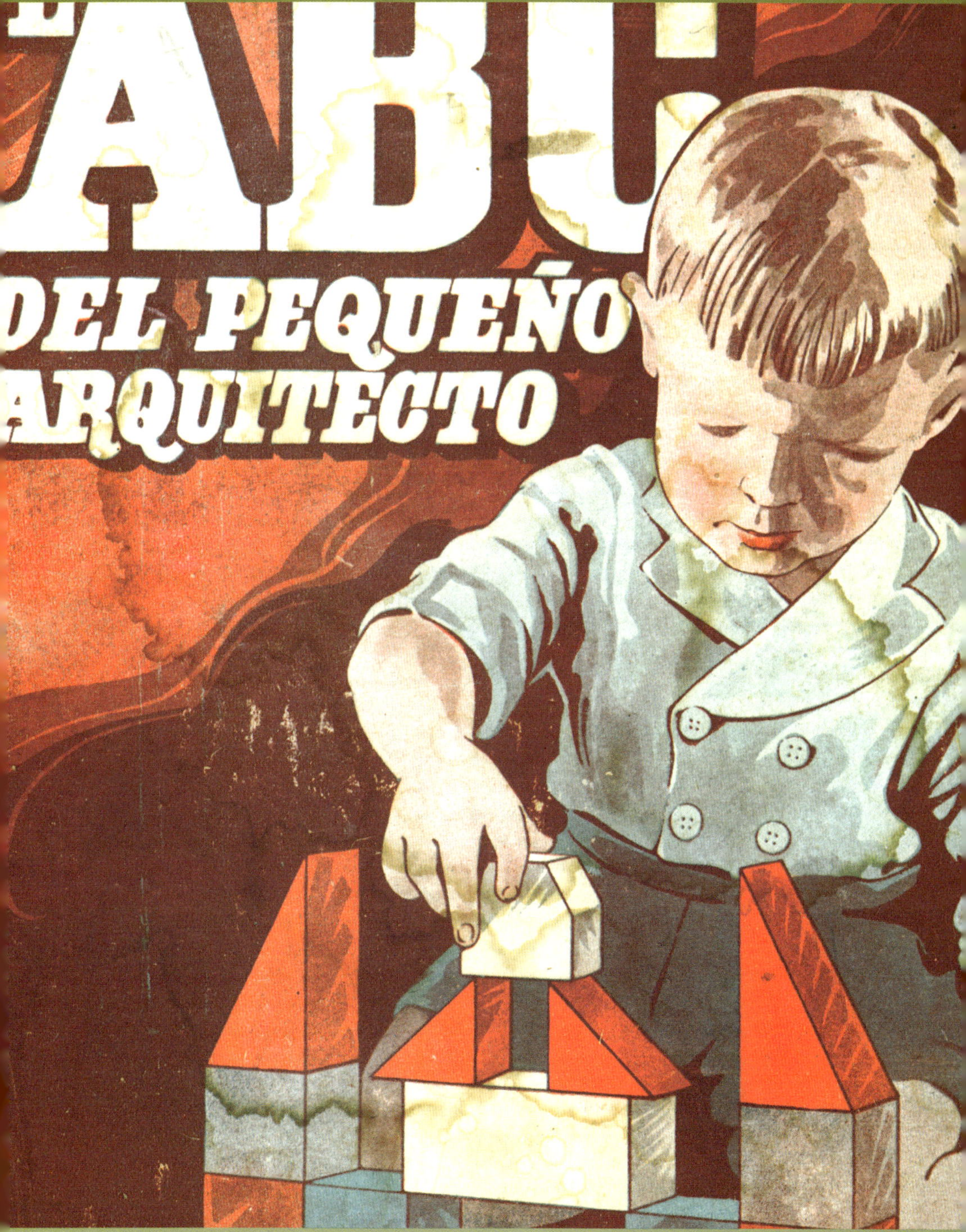
ABC
DEL PEQUEÑO
ARQUITECTO

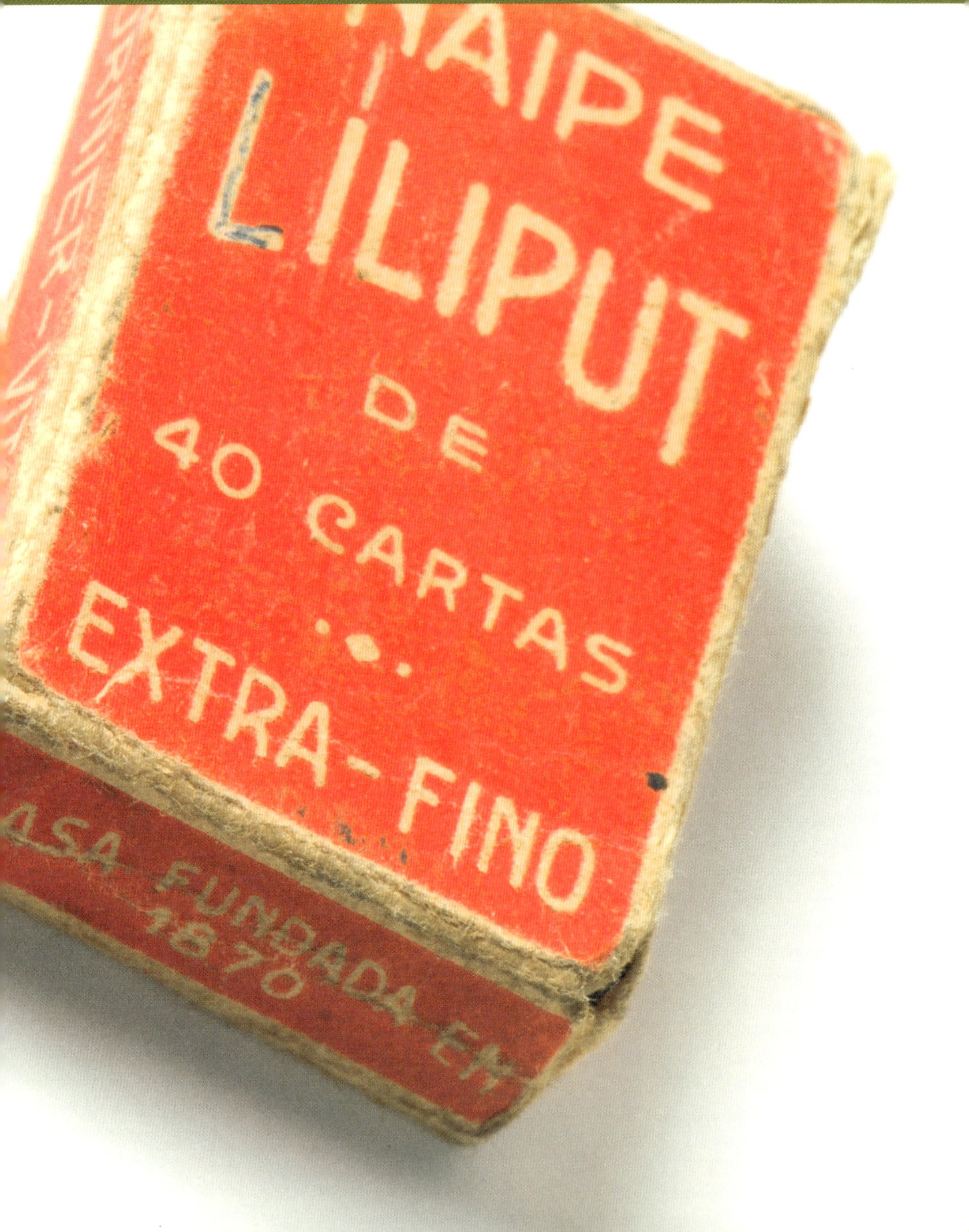
NAIPE
LILIPUT
DE
40 CARTAS
EXTRA-FINO
ASA FUNDADA EM
1870

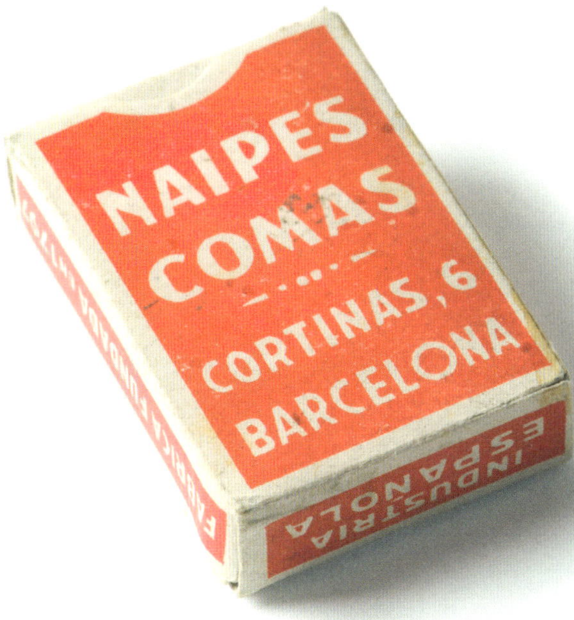
NAIPES
COMAS
CORTINAS, 6
BARCELONA
INDUSTRIA
ESPAÑOLA

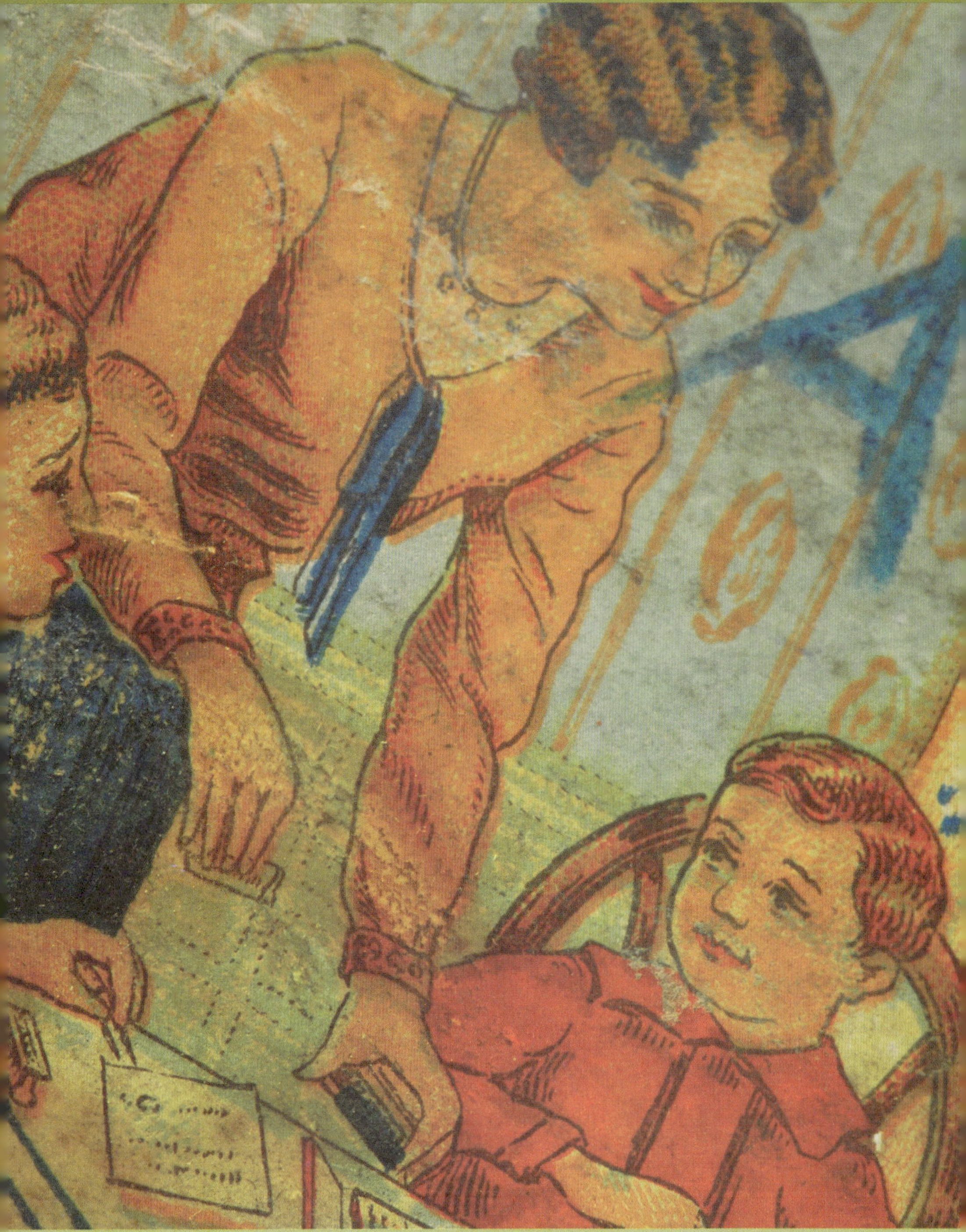

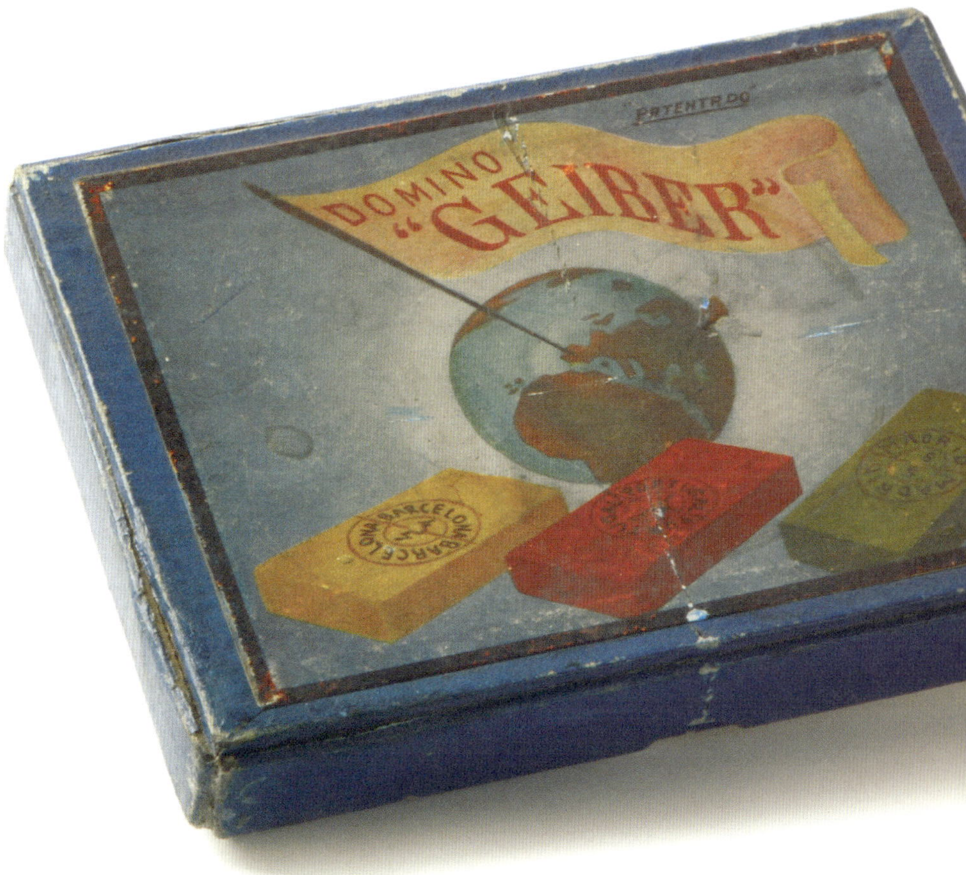
DOMINO
"GEIBER"

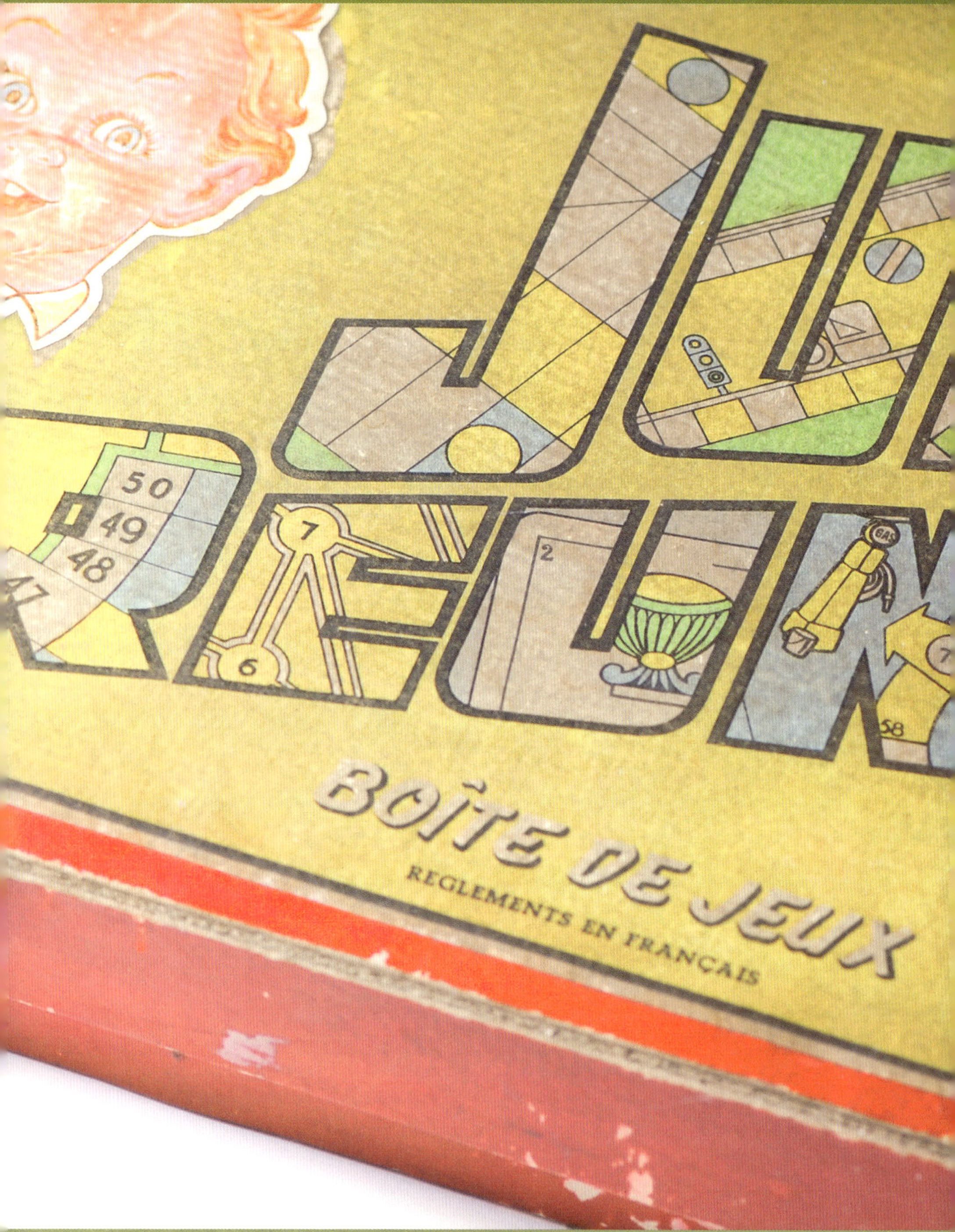
BOÎTE DE JEUX
REGLEMENTS EN FRANÇAIS

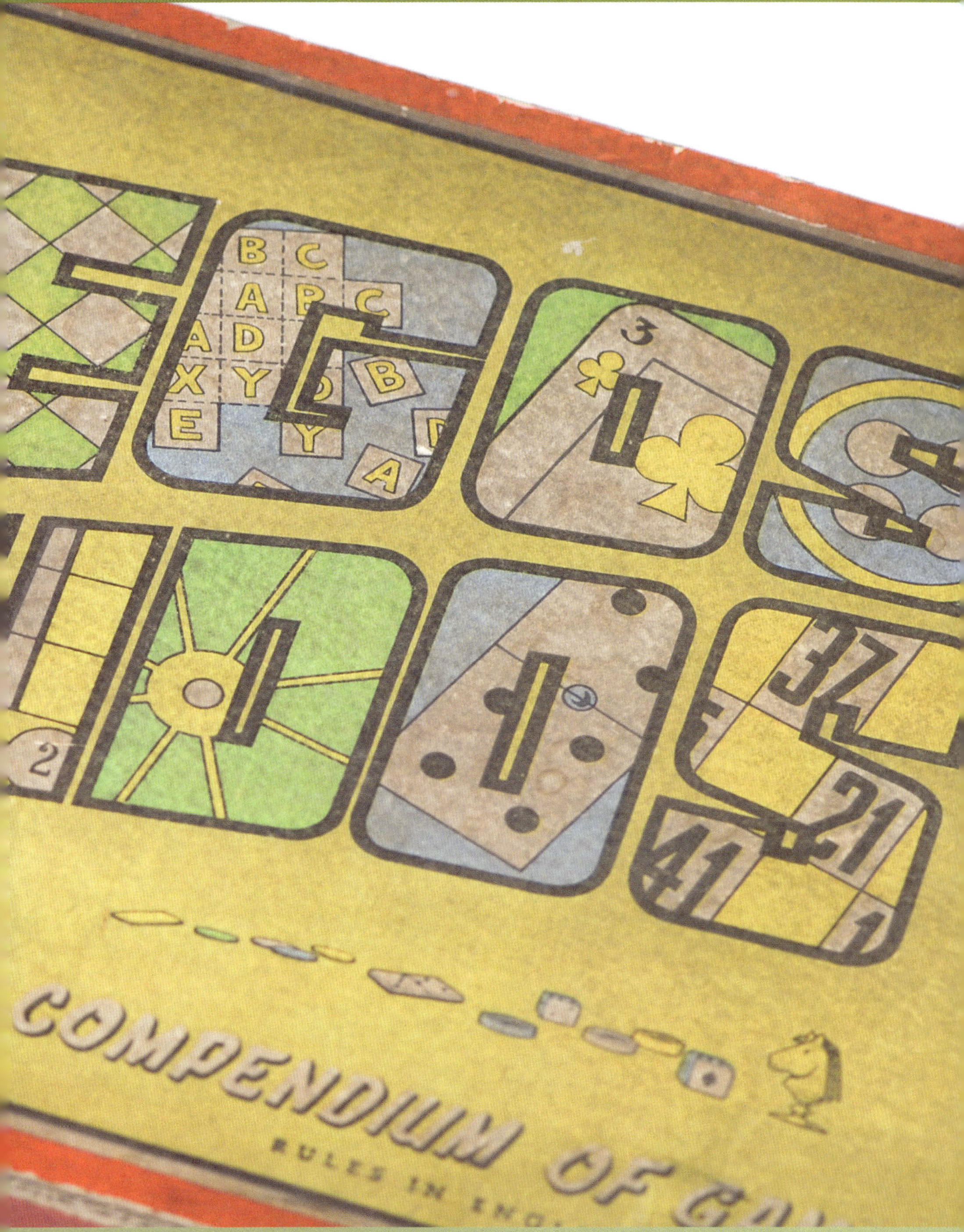
COMPENDIUM OF GA
RULES IN ENG

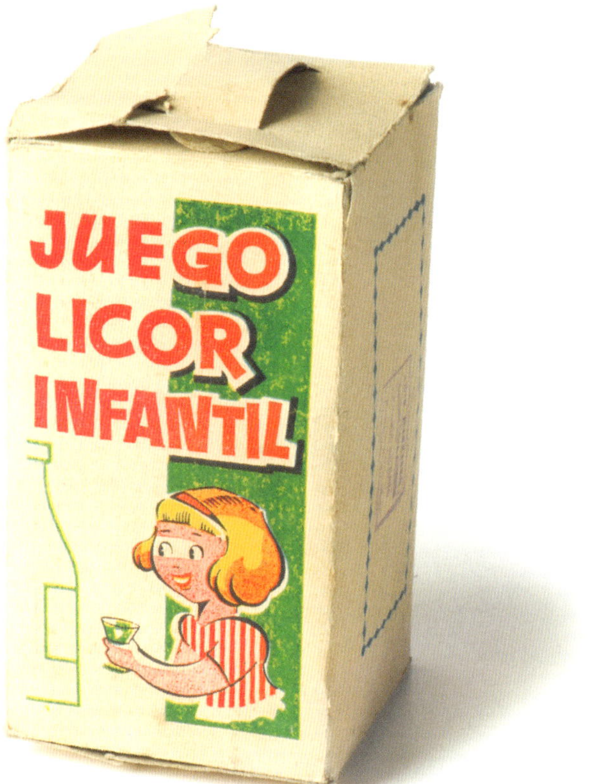
JUEGO
LICOR
INFANTIL

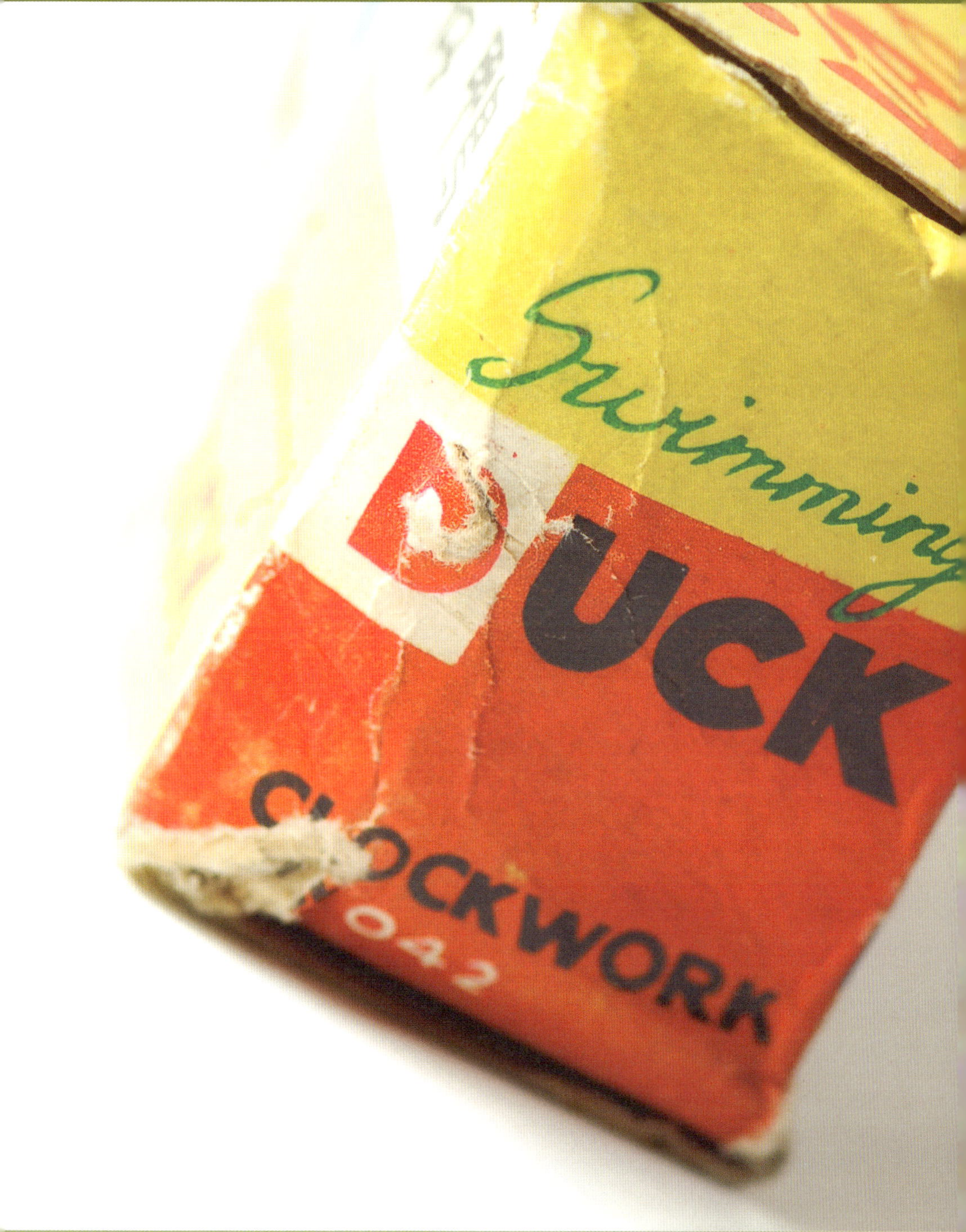
UCK
OCKWORK
042

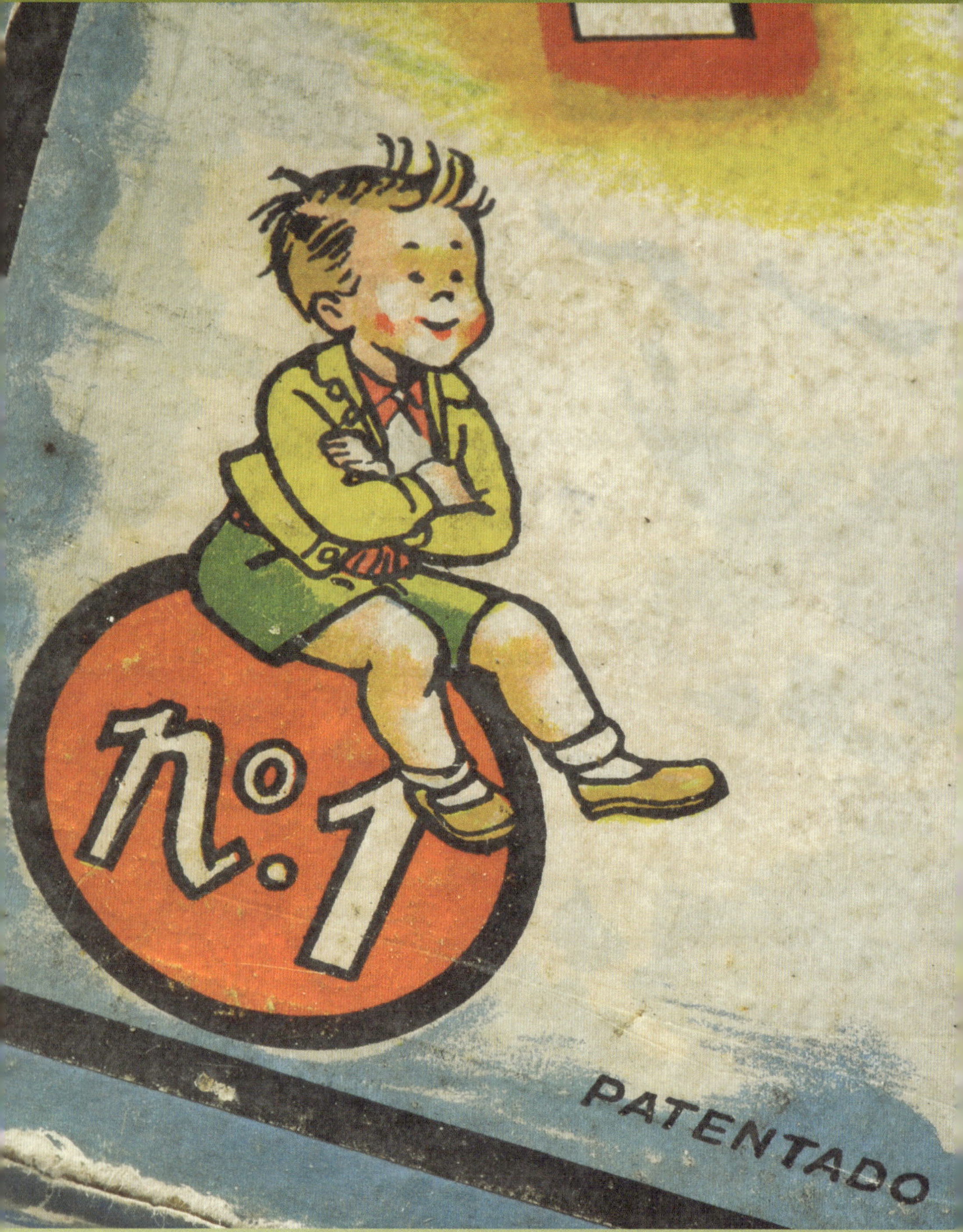

nº1
PATENTADO

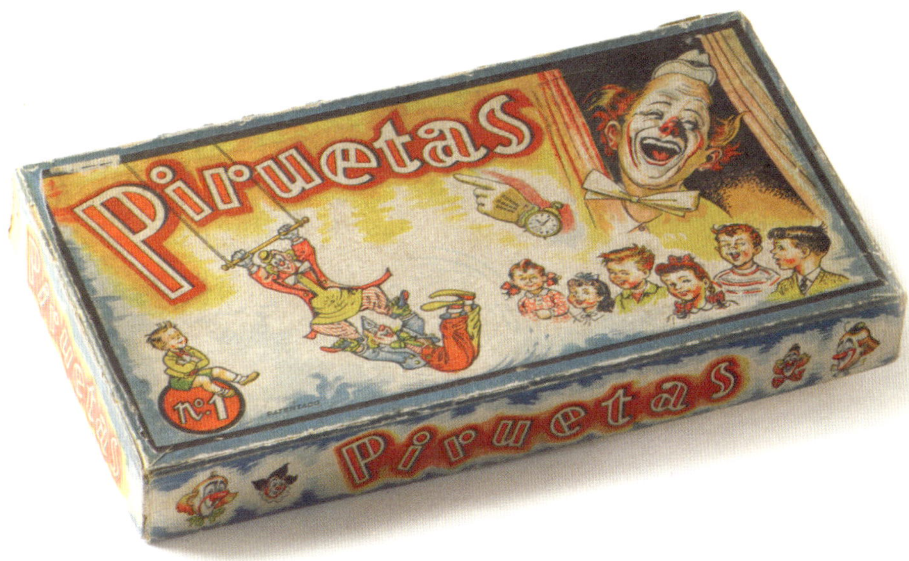
Piruetas
n.º 1
Piruetas
Piruetas

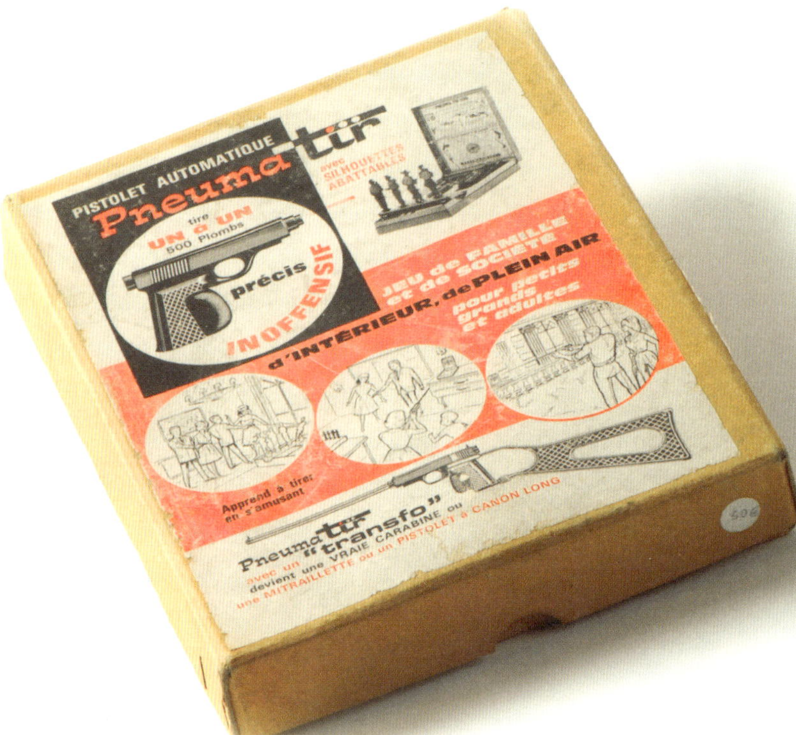
PISTOLET AUTOMATIQUE
Pneuma tir
tire
UN à UN
500 Plombs
précis
INOFFENSIF
avec SILHOUETTES ABATTABLES
JEU de FAMILLE et de SOCIÉTÉ
d'INTÉRIEUR, de PLEIN AIR
pour petits grands et adultes
Apprend à tirer en s'amusant
Pneuma tir
avec un "transfo"
devient une VRAIE CARABINE ou
une MITRAILLETTE ou un PISTOLET à CANON LONG

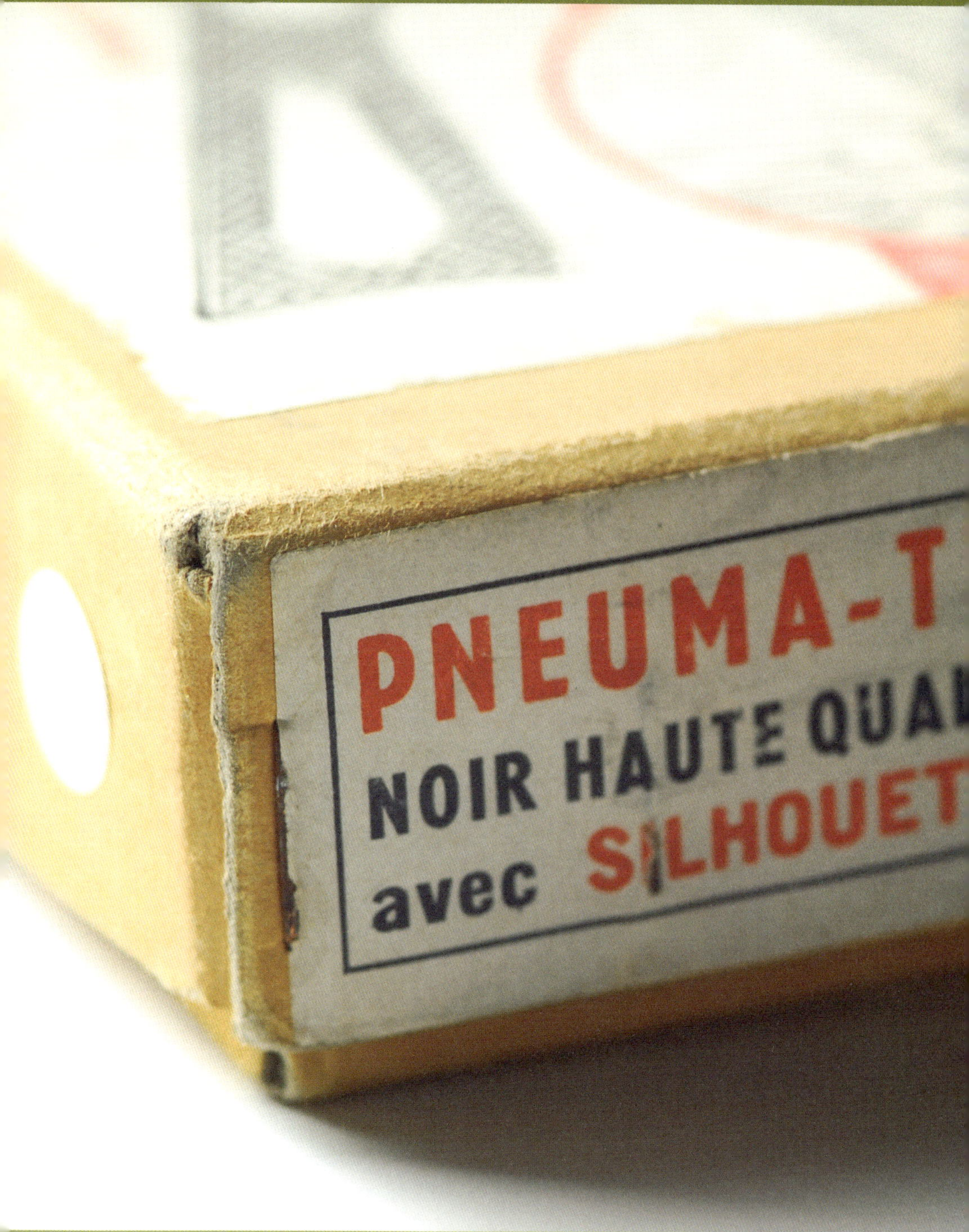
PNEUMA-T
NOIR HAUTE QUA
avec SILHOUET

ELEMEN

RELEMENTSCAR

TSCARE
ELEMENTSCARELE

CAR ELEMENTS

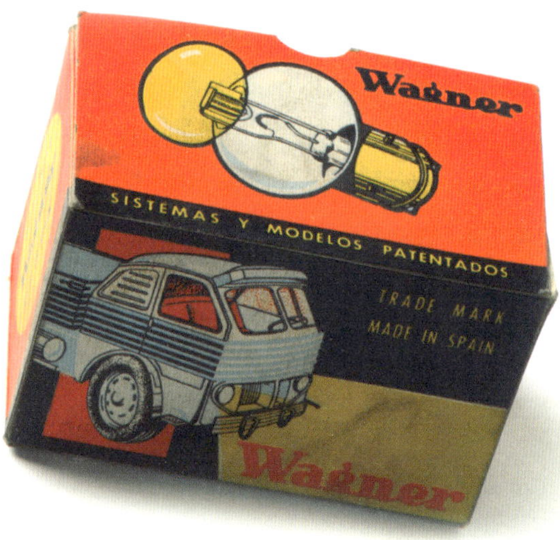
Wagner
SISTEMAS Y MODELOS PATENTADOS
TRADE MARK
MADE IN SPAIN
Wagner

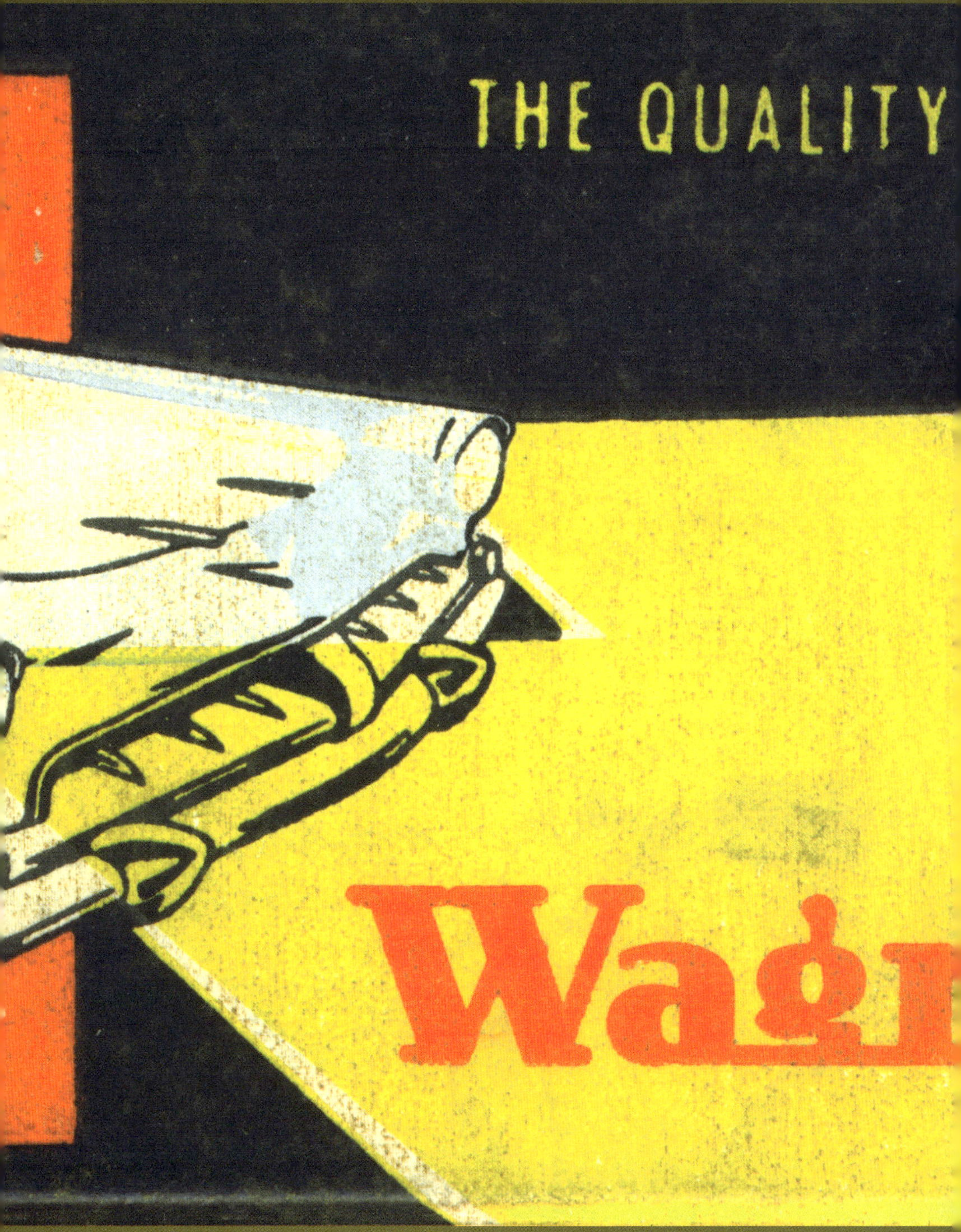
THE QUALITY

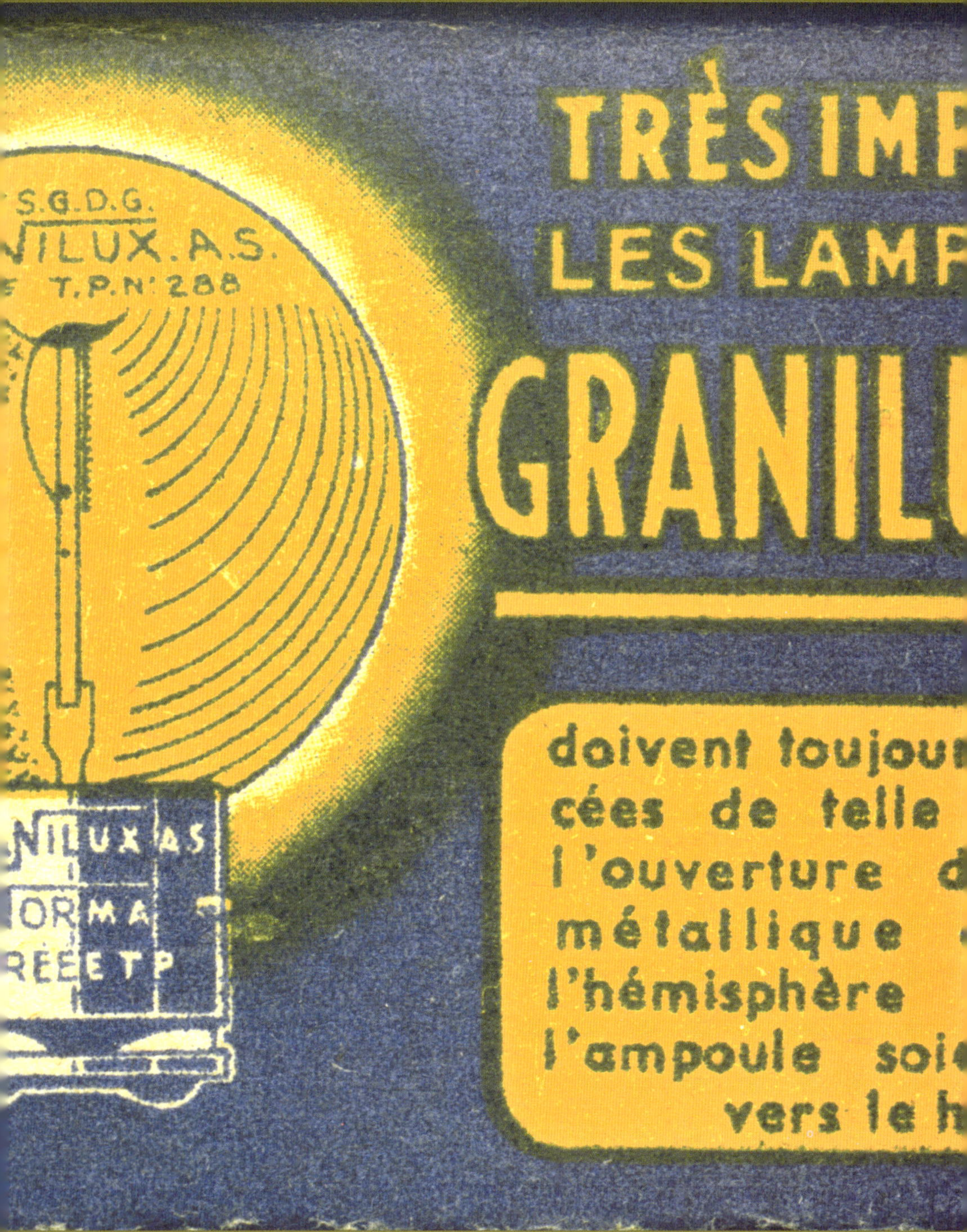
TRÈS IMP
LES LAMP
GRANILU
S.G.D.G.
NILUX.A.S.
T.P.N°288
NILUX AS
ORMA
RÉEETP
doivent toujour
cées de telle
l'ouverture d
métallique
l'hémisphère
l'ampoule soi
vers le h

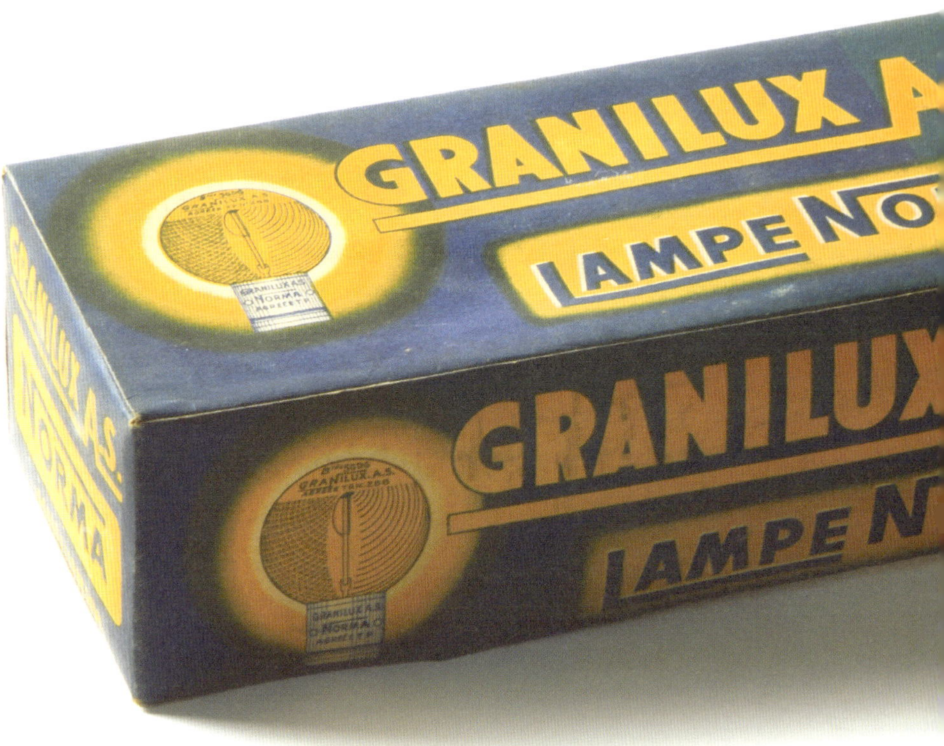
GRANILUX
LAMPE
GRANILUX A.S.
NORMA
GRANILUX
LAMPE

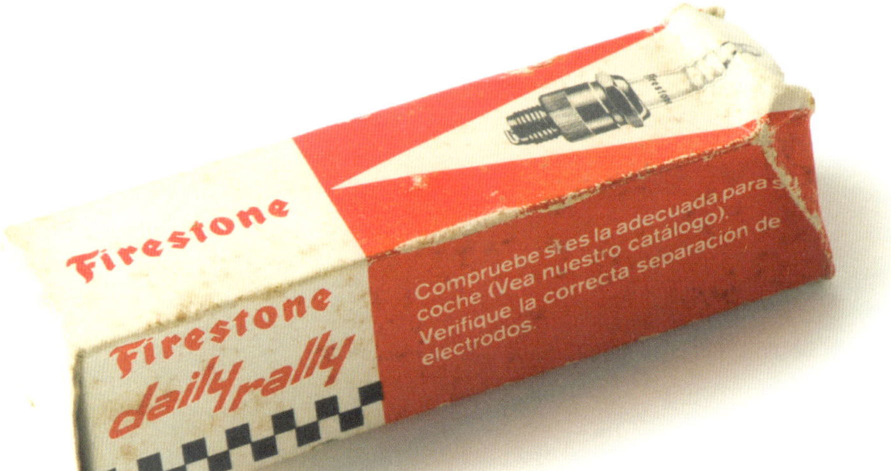
Firestone
Firestone
daily rally
Compruebe si es la adecuada para su
coche (Vea nuestro catálogo).
Verifique la correcta separación de
electrodos.

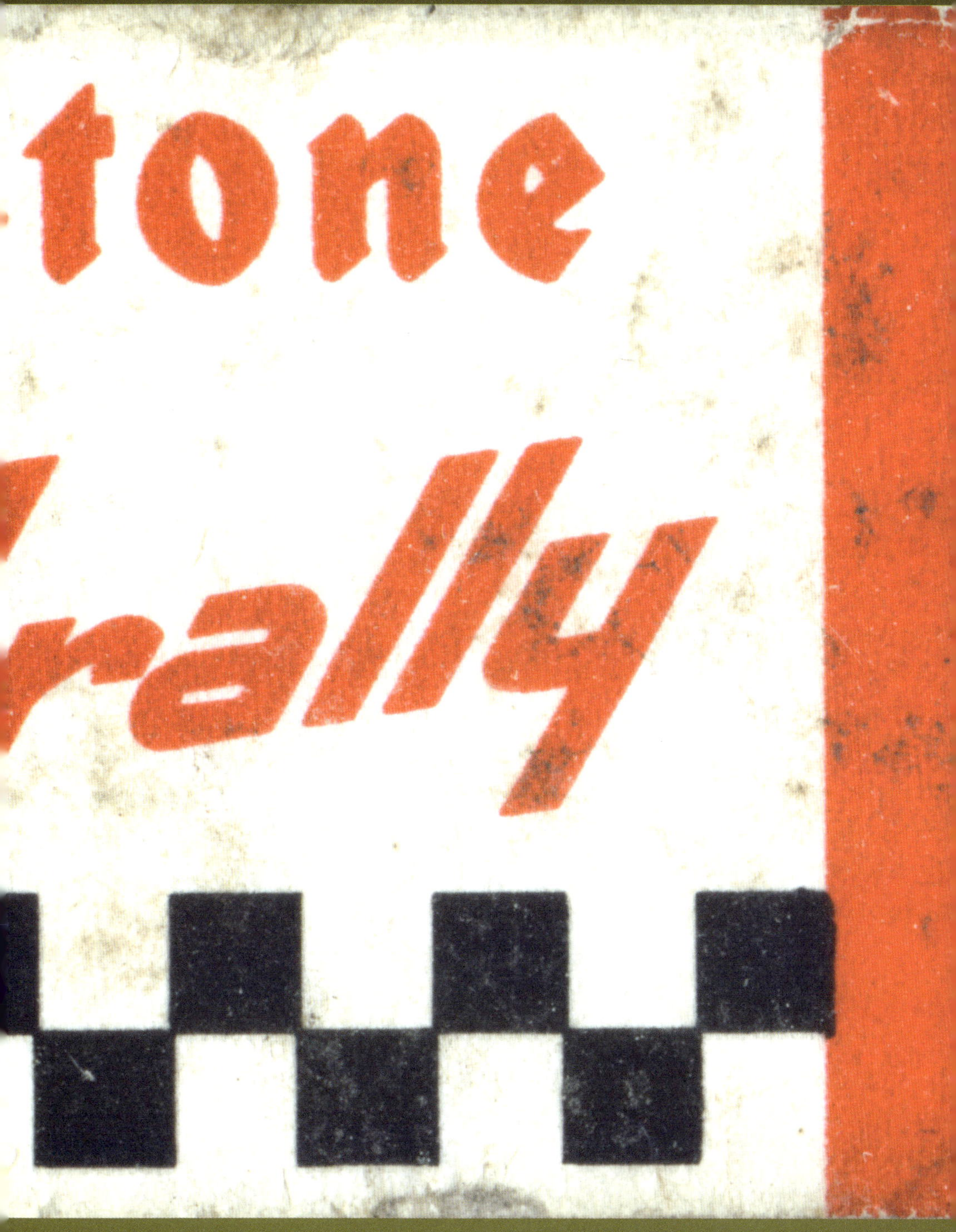
tone
rally

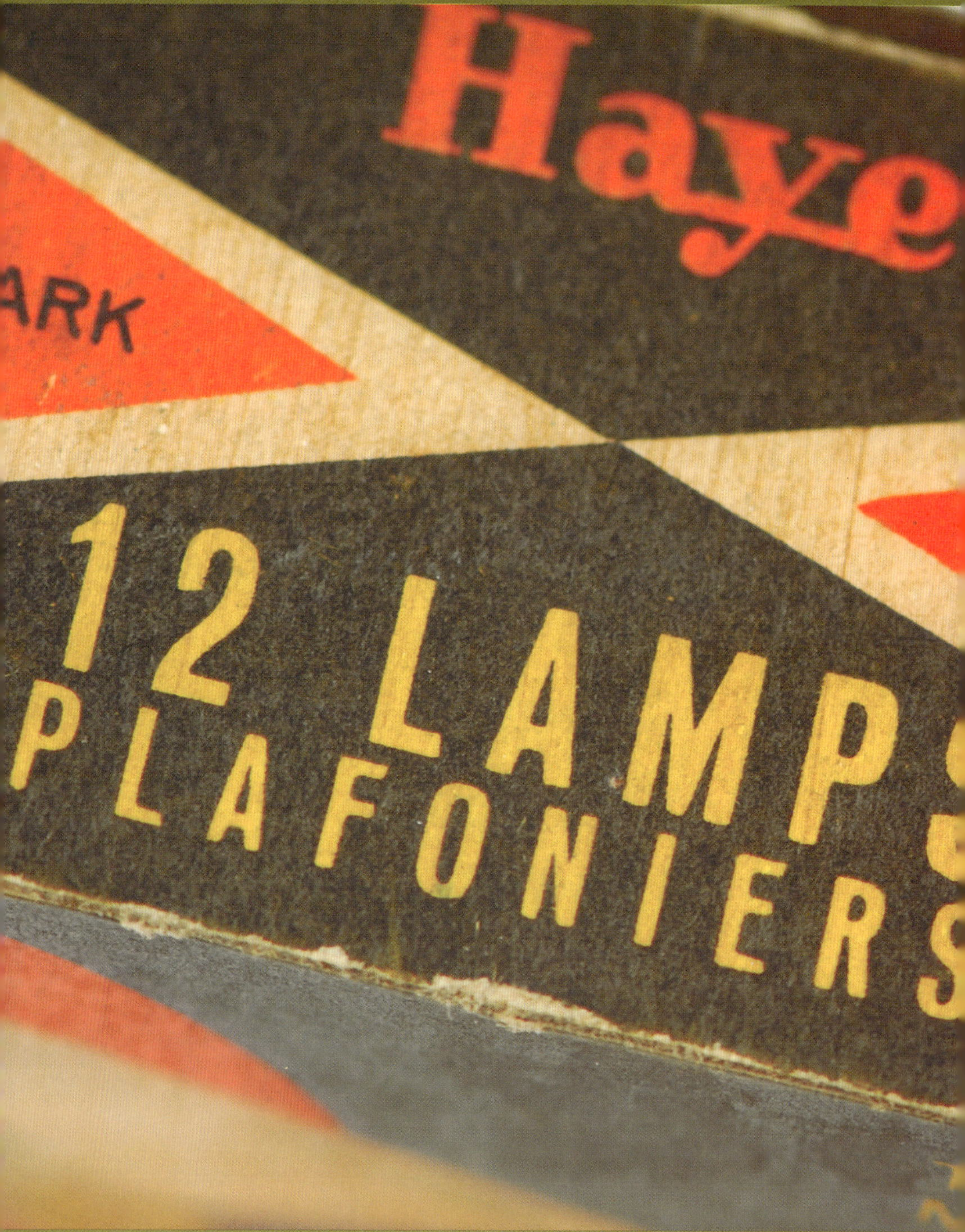
Haye
ARK
12 LAMP
PLAFONIERS

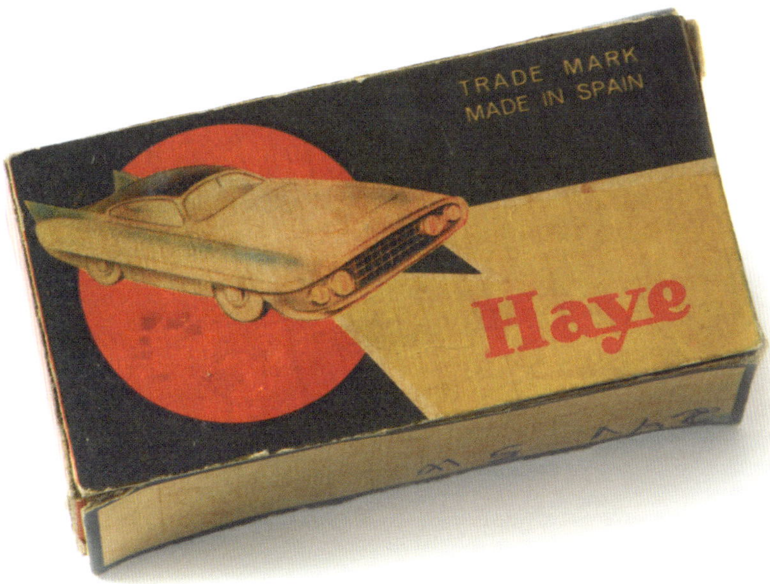
TRADE MARK
MADE IN SPAIN
Haye

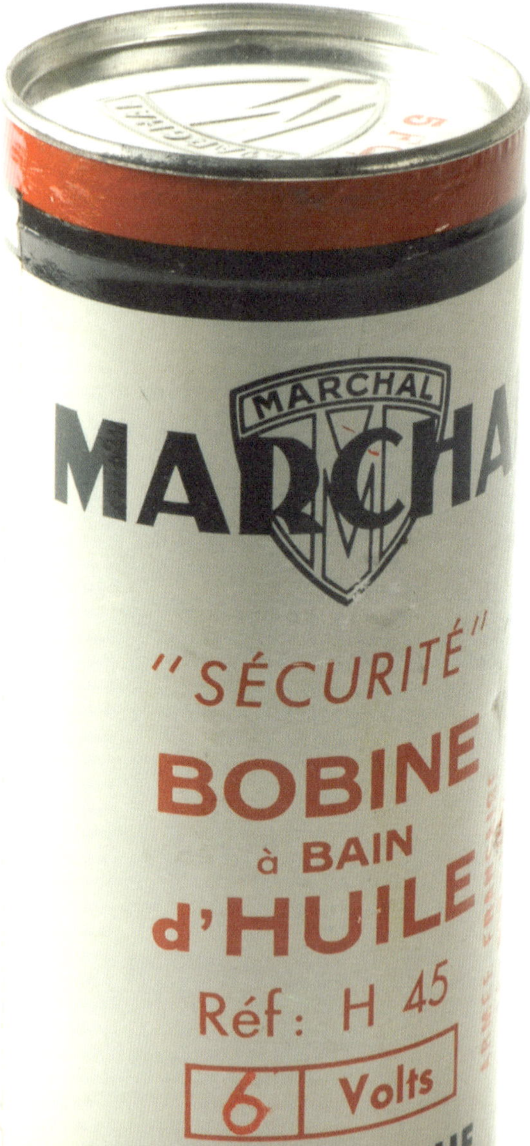
MARCHAL
MARCHAL
"SÉCURITÉ"
BOBINE
à BAIN
d'HUILE
Réf: H 45
6 Volts

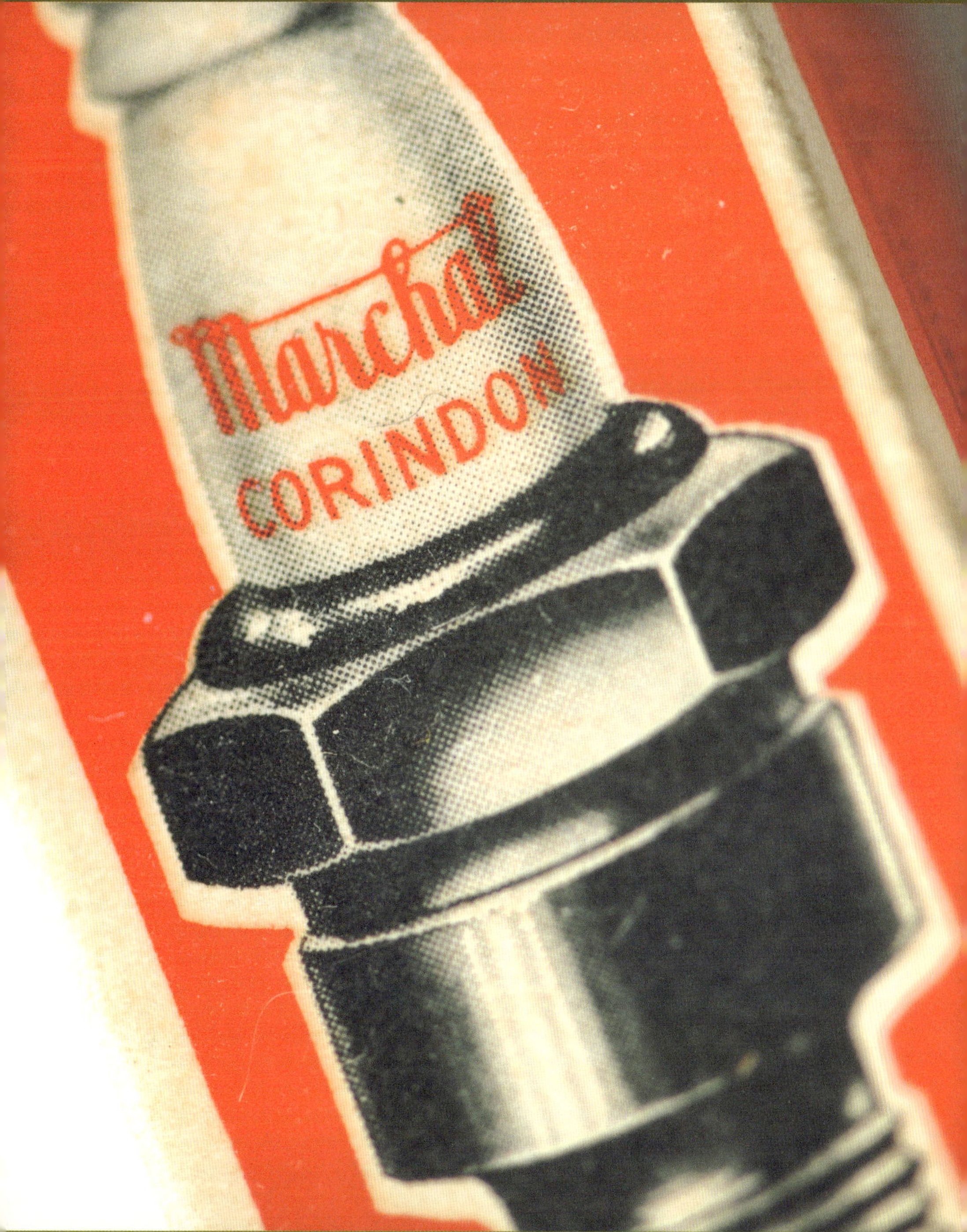
Marchal
CORINDON

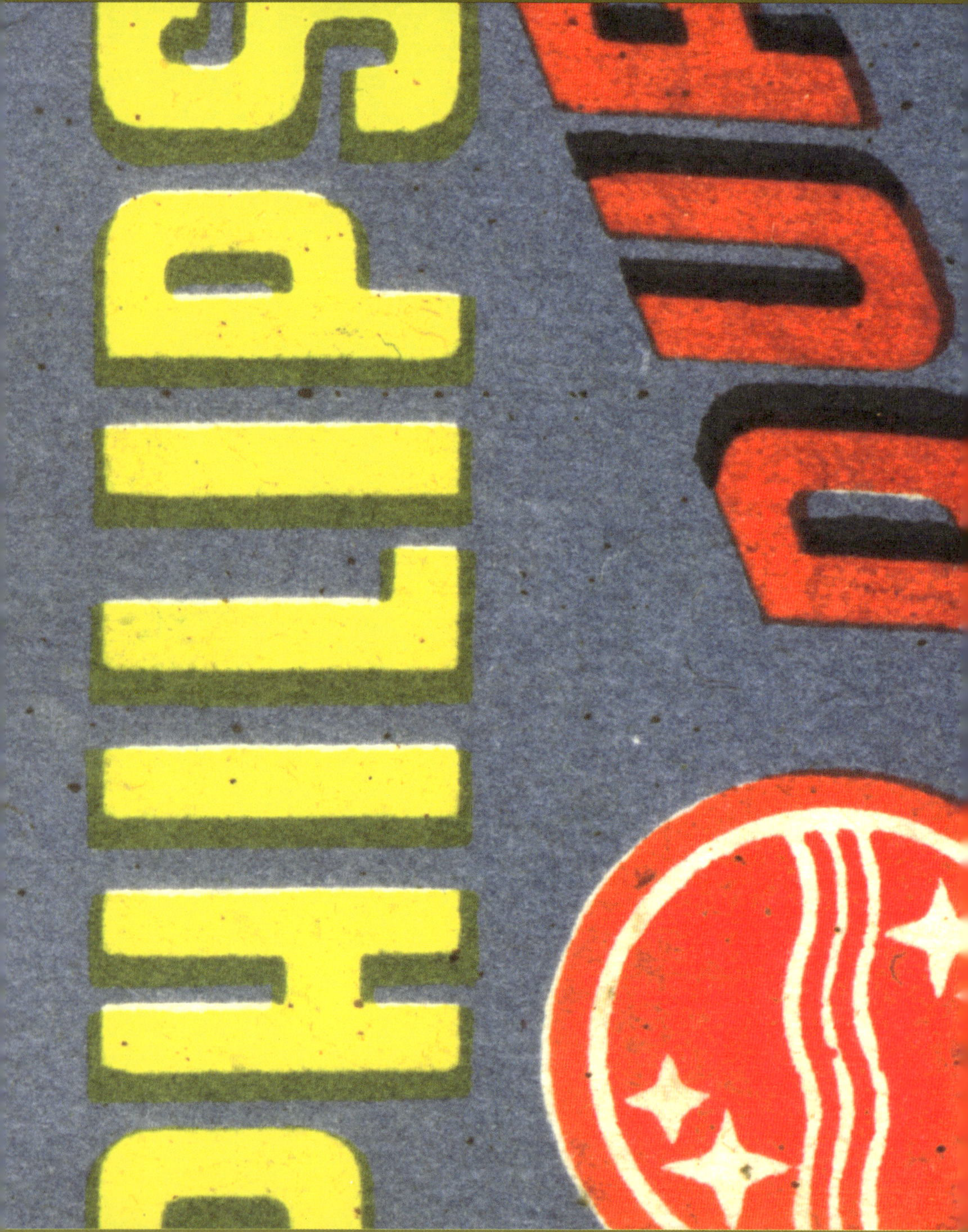

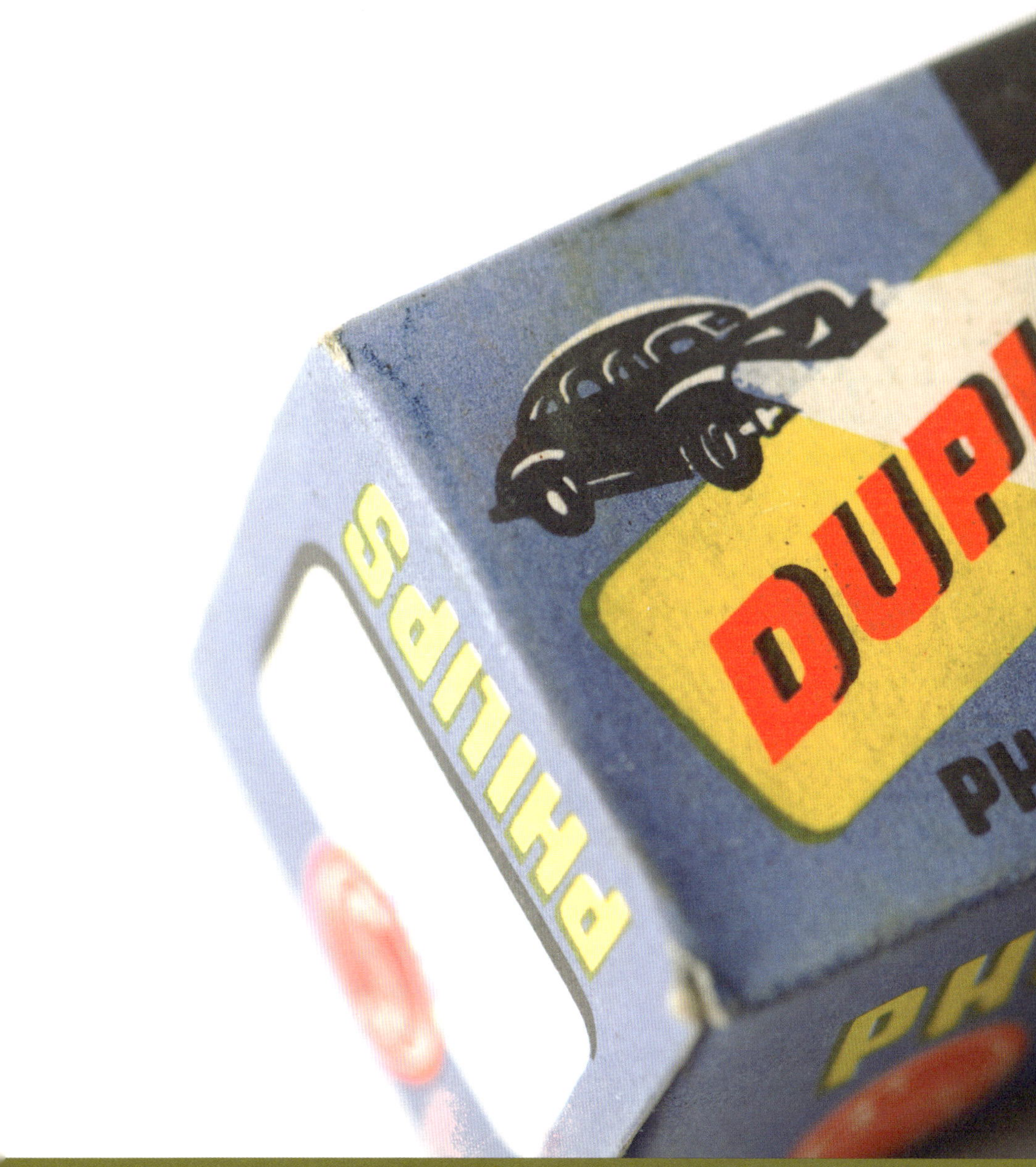
PHILIPS

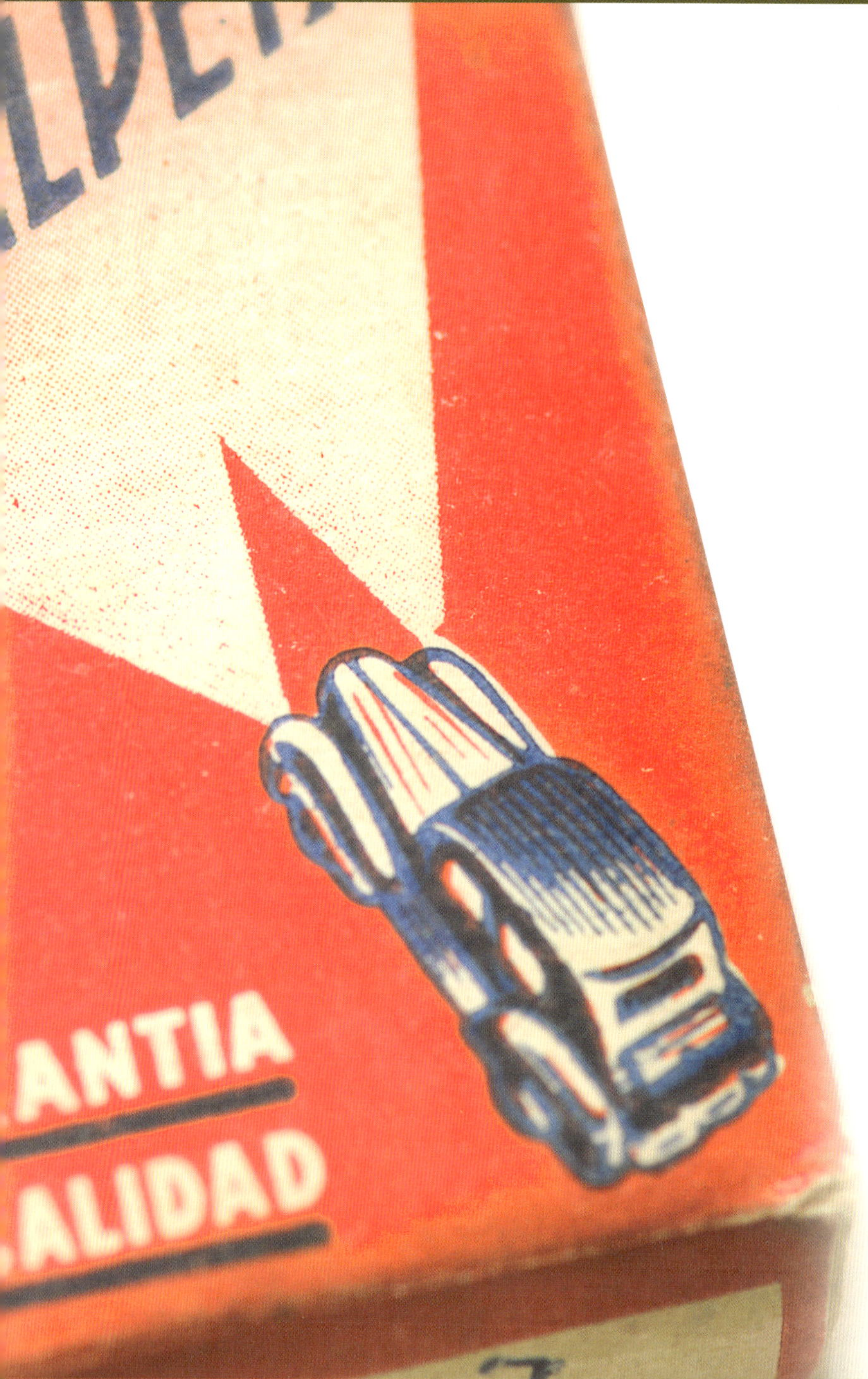
ANTIA
ALIDAD

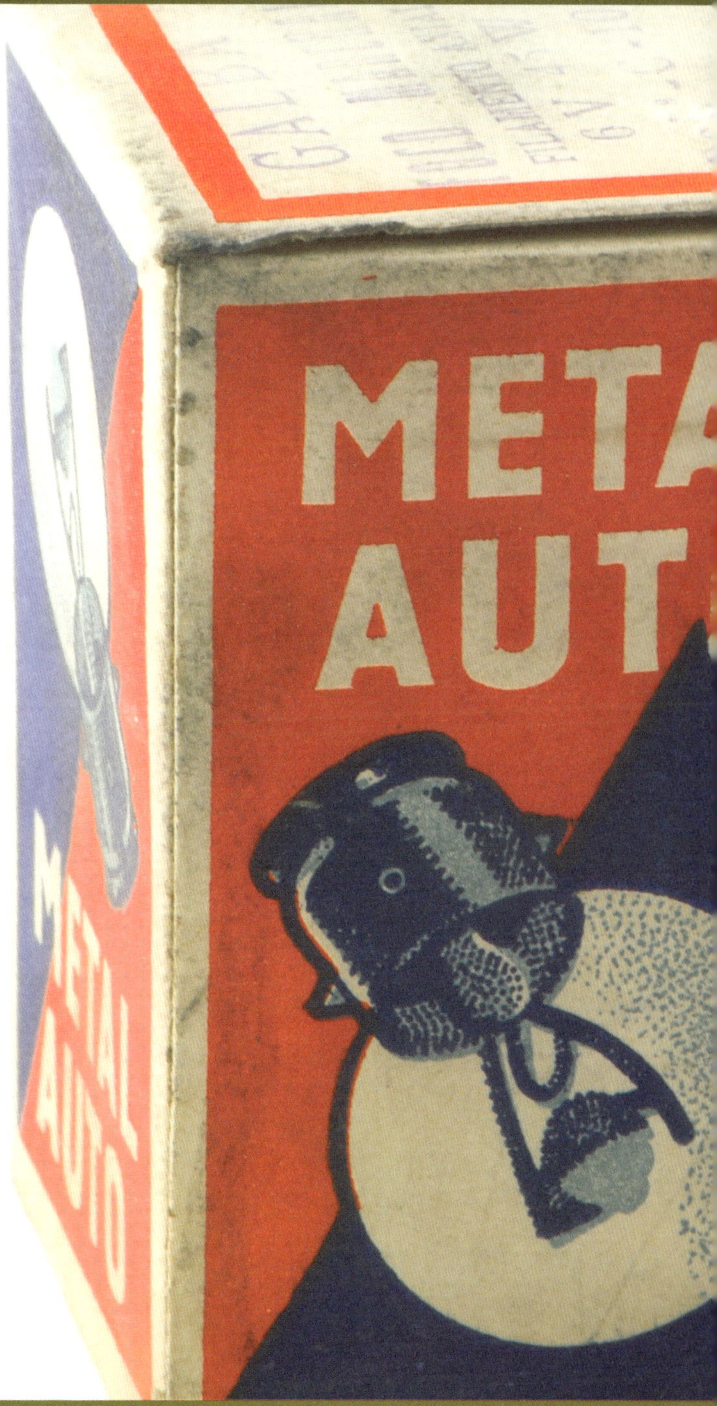
METAL
AUTO

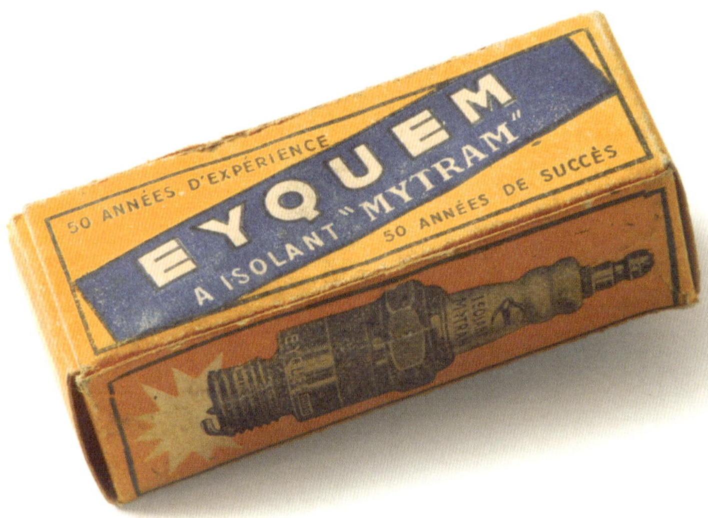
50 ANNÉES D'EXPÉRIENCE
EYQUEM
A ISOLANT "MYTRAM"
50 ANNÉES DE SUCCÈS

"WICKERS"

is made of raw materials of the very best qualities.

All the pieces are interchangeable. The absolutely parallel and concentric rolling, rectified after tempering, and the perfect fitting, render it incomparably easy running.

Clean with petrol without dismouting and oil with fine oil.

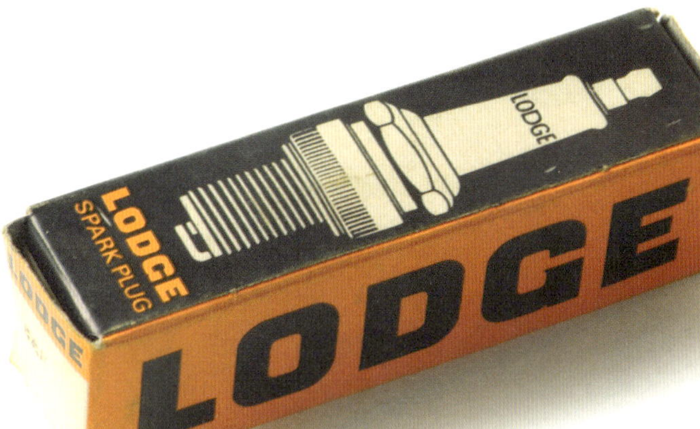
LODGE
SPARK PLUG
LODGE
LODGE

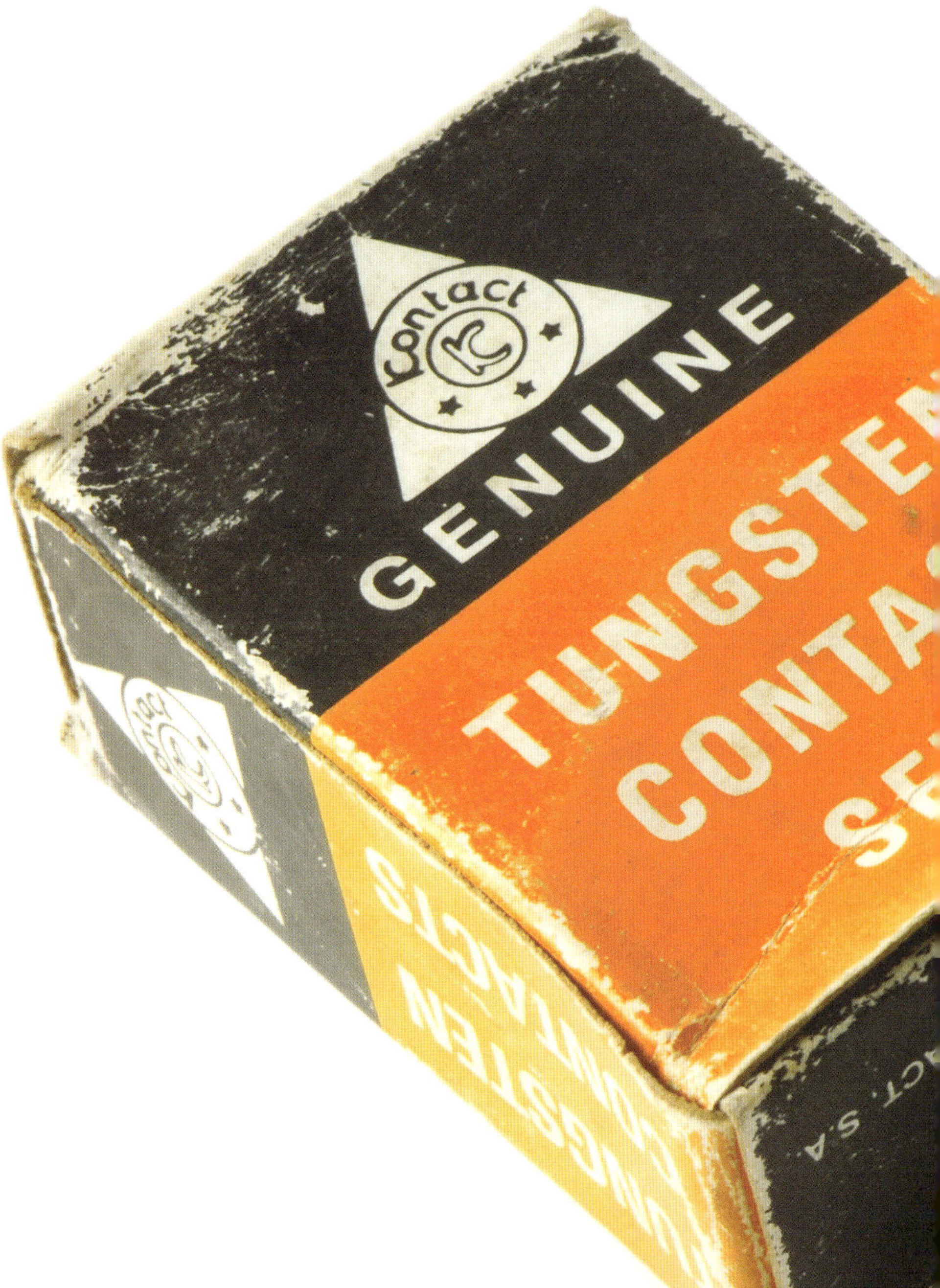
kontact
GENUINE
TUNGSTEN
CONTACTS

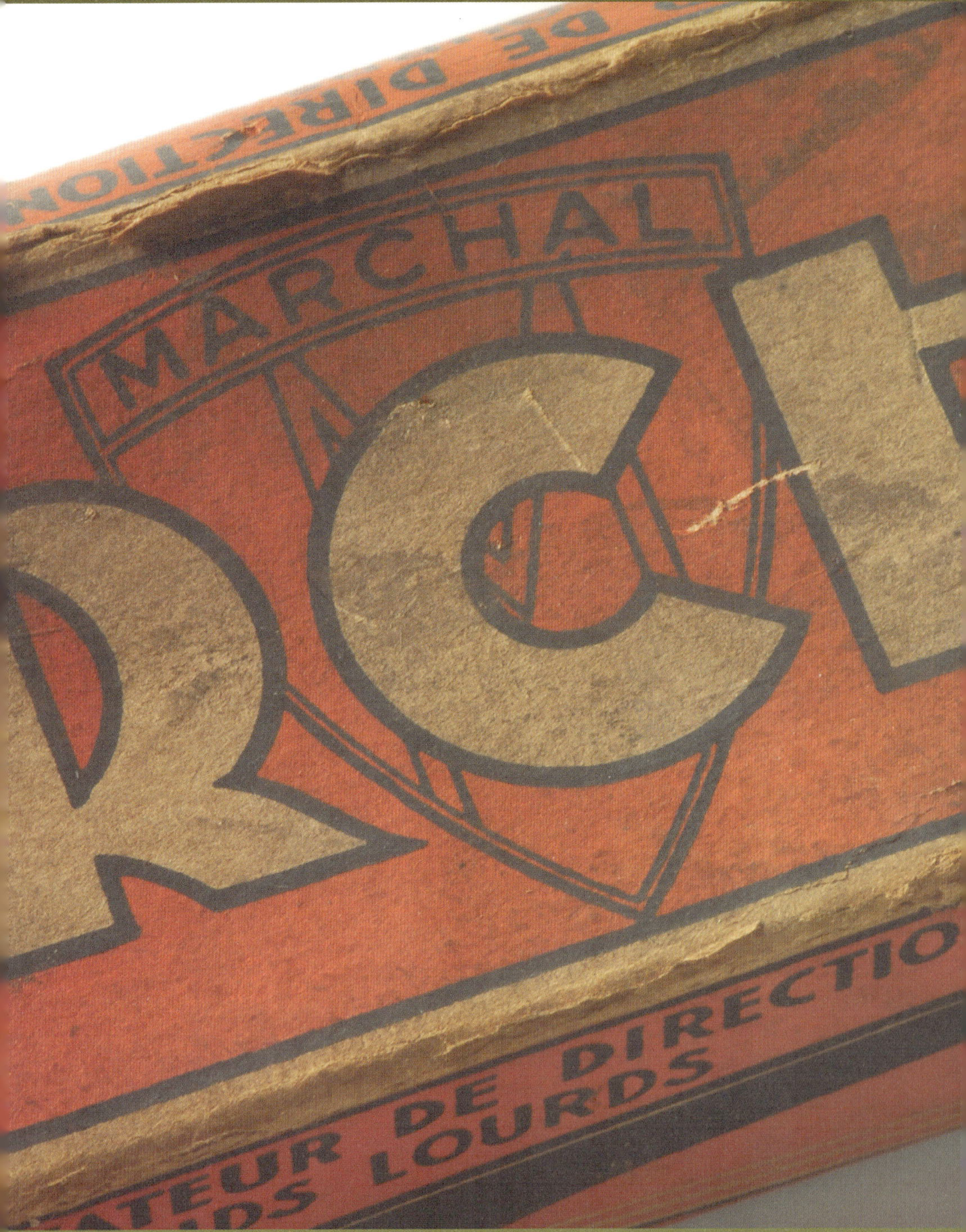
MARCHAL
DE
LOURDS

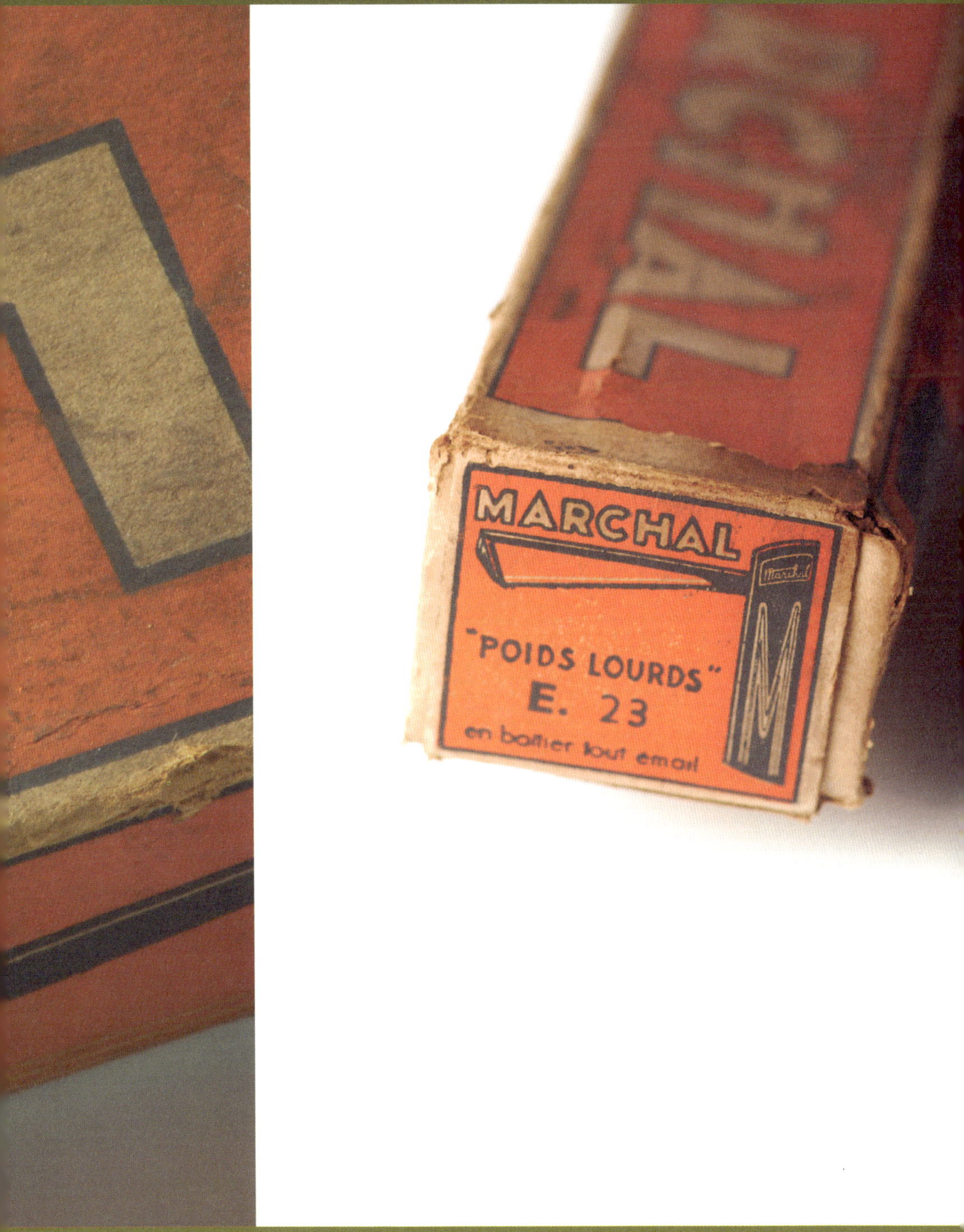
MARCHAL
"POIDS LOURDS"
E. 23

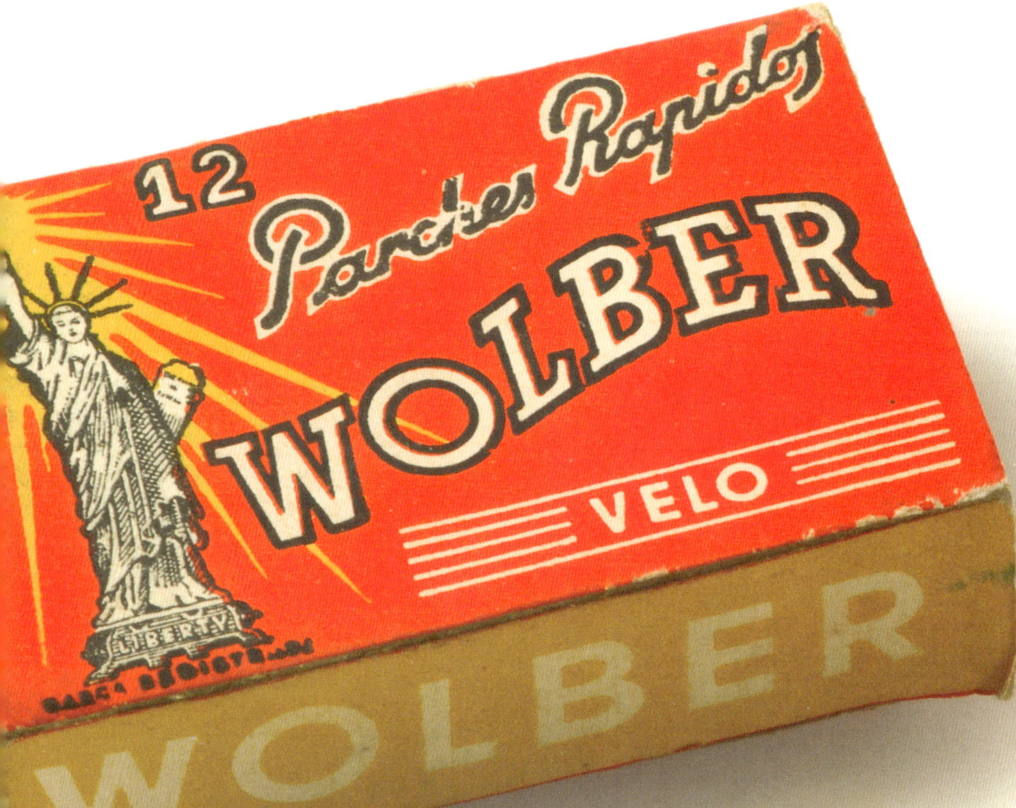
12
Parches Rapidos
WOLBER
VELO
LIBERTY
WOLBER

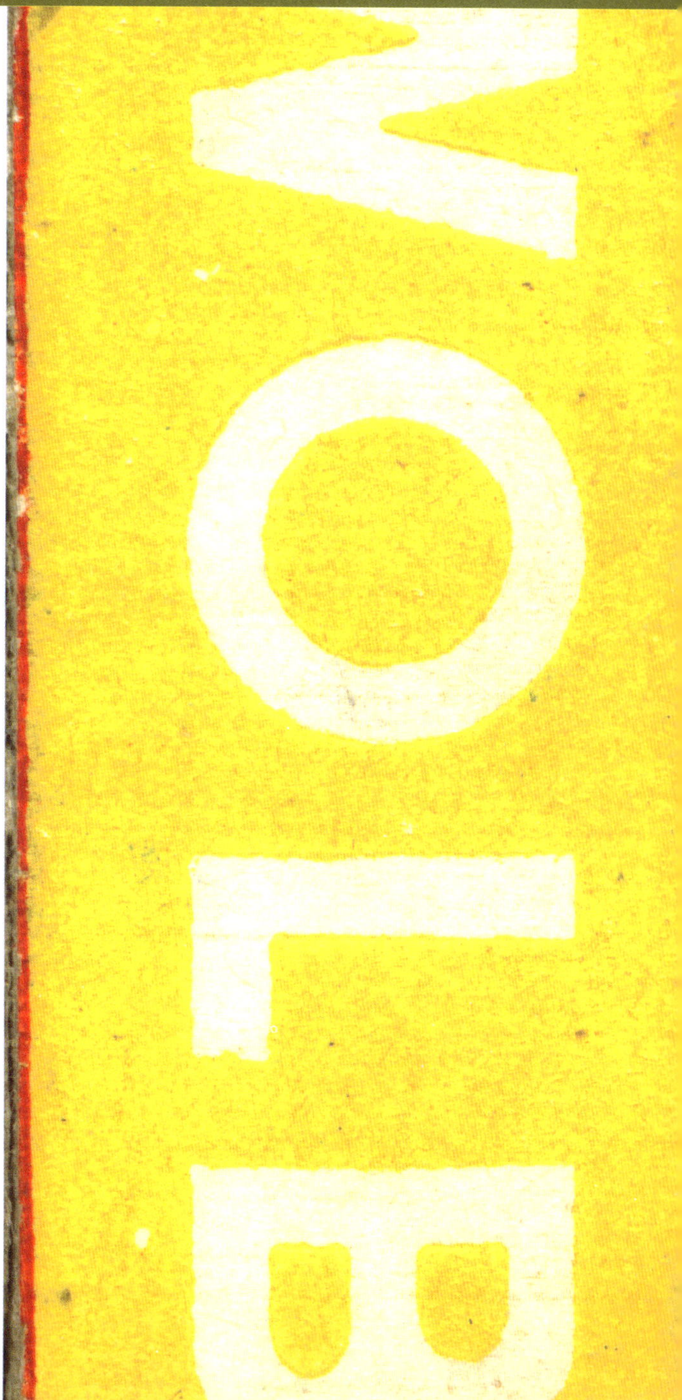

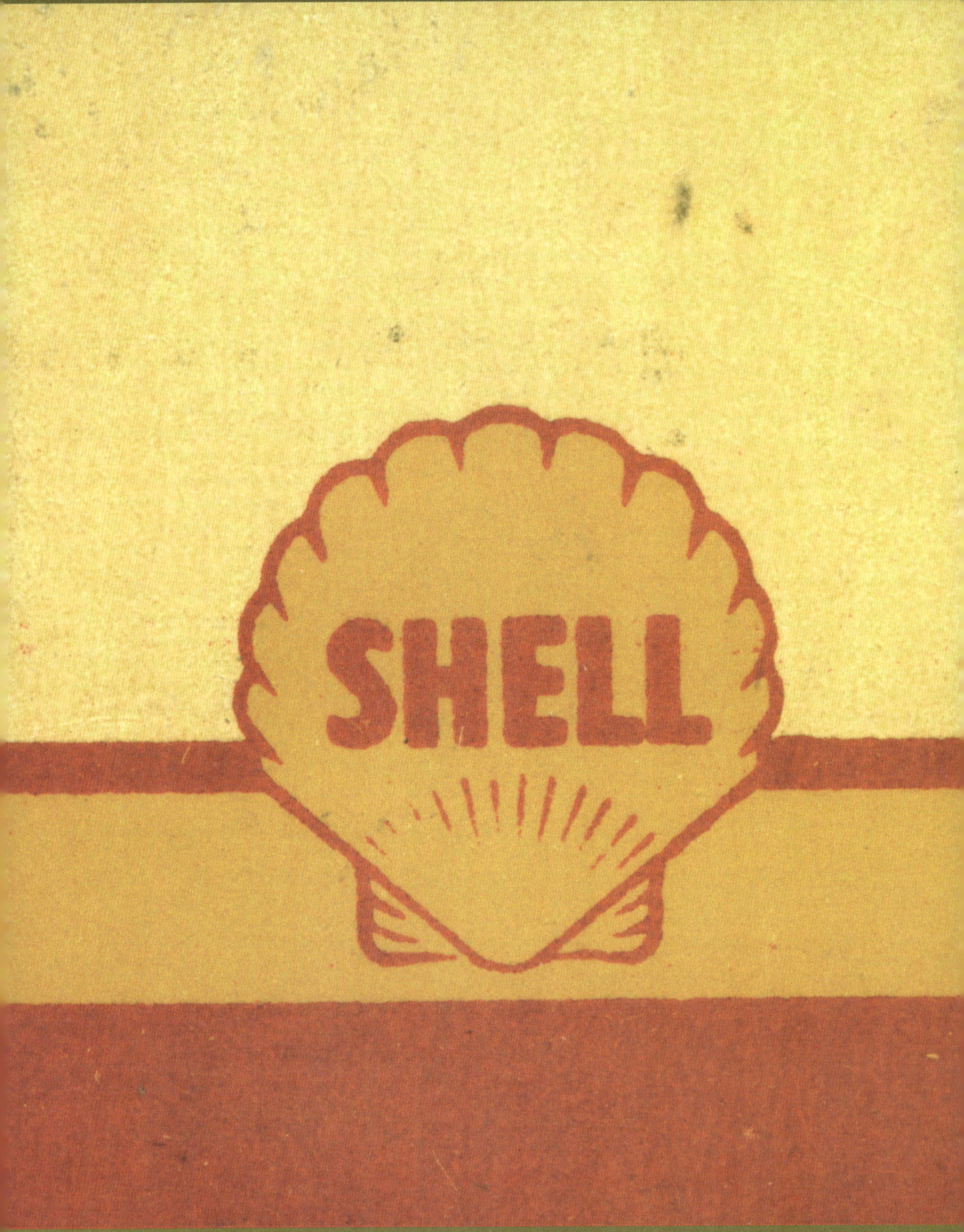
SHELL

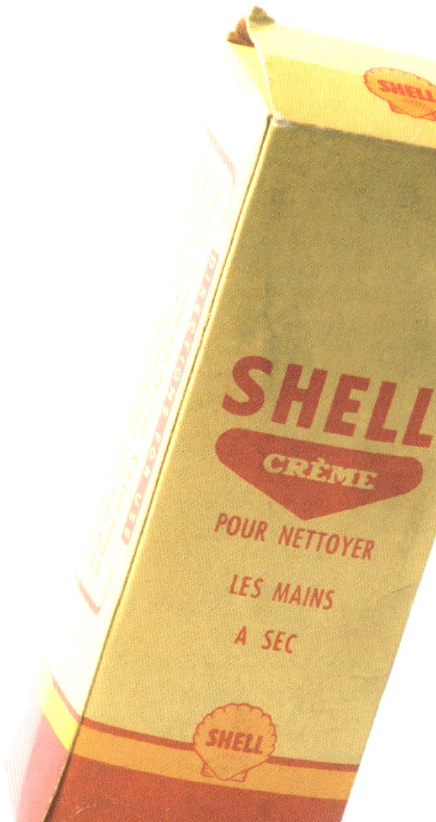
SHELL
SHELL
CRÈME
POUR NETTOYER
LES MAINS
A SEC
SHELL

RFUMEI

RFUMEINDUSTRY

PERFUME INDUSTRY

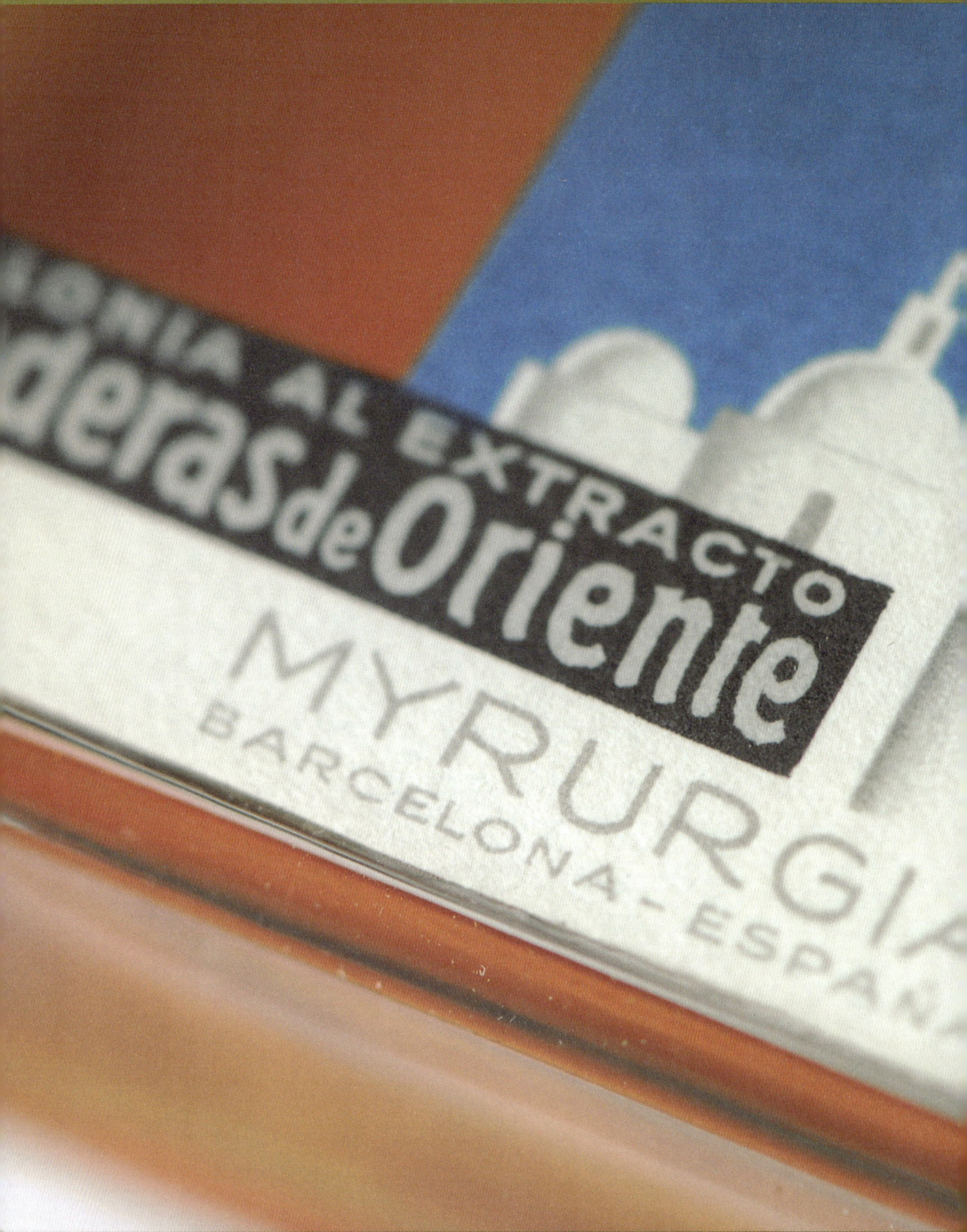
AL EXTRACTO
de Oriente
BARCELONA -

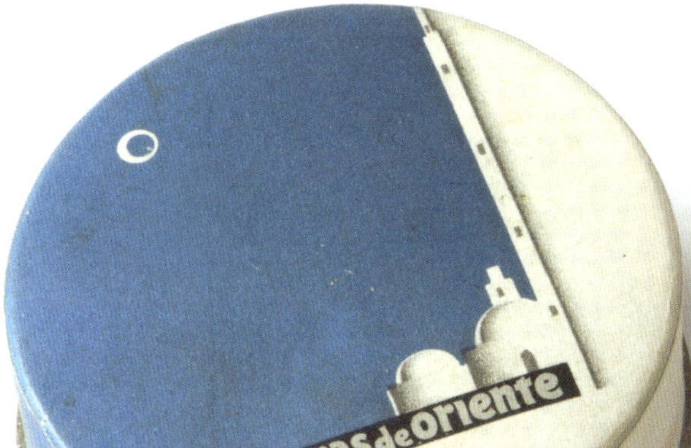
maderas de Oriente
MYRURGIA

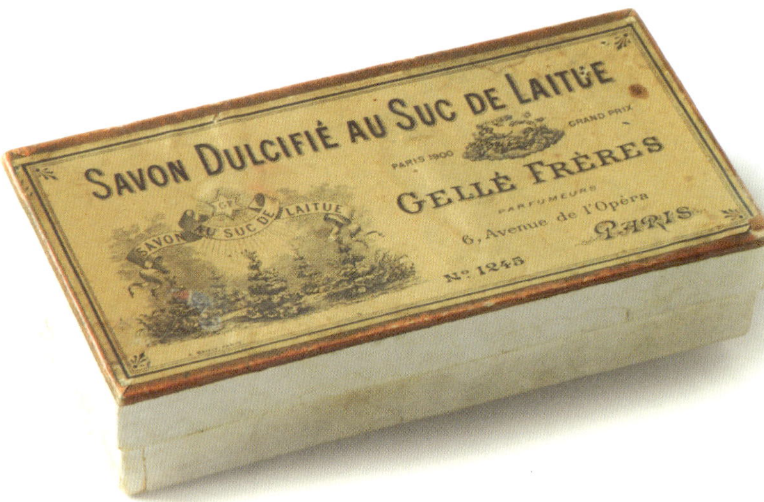
SAVON DULCIFIÉ AU SUC DE LAITUE
SAVON AU SUC DE LAITUE
PARIS 1900
GRAND PRIX
GELLÉ FRÈRES
PARFUMEURS
6, Avenue de l'Opéra
PARIS
Nº 1245

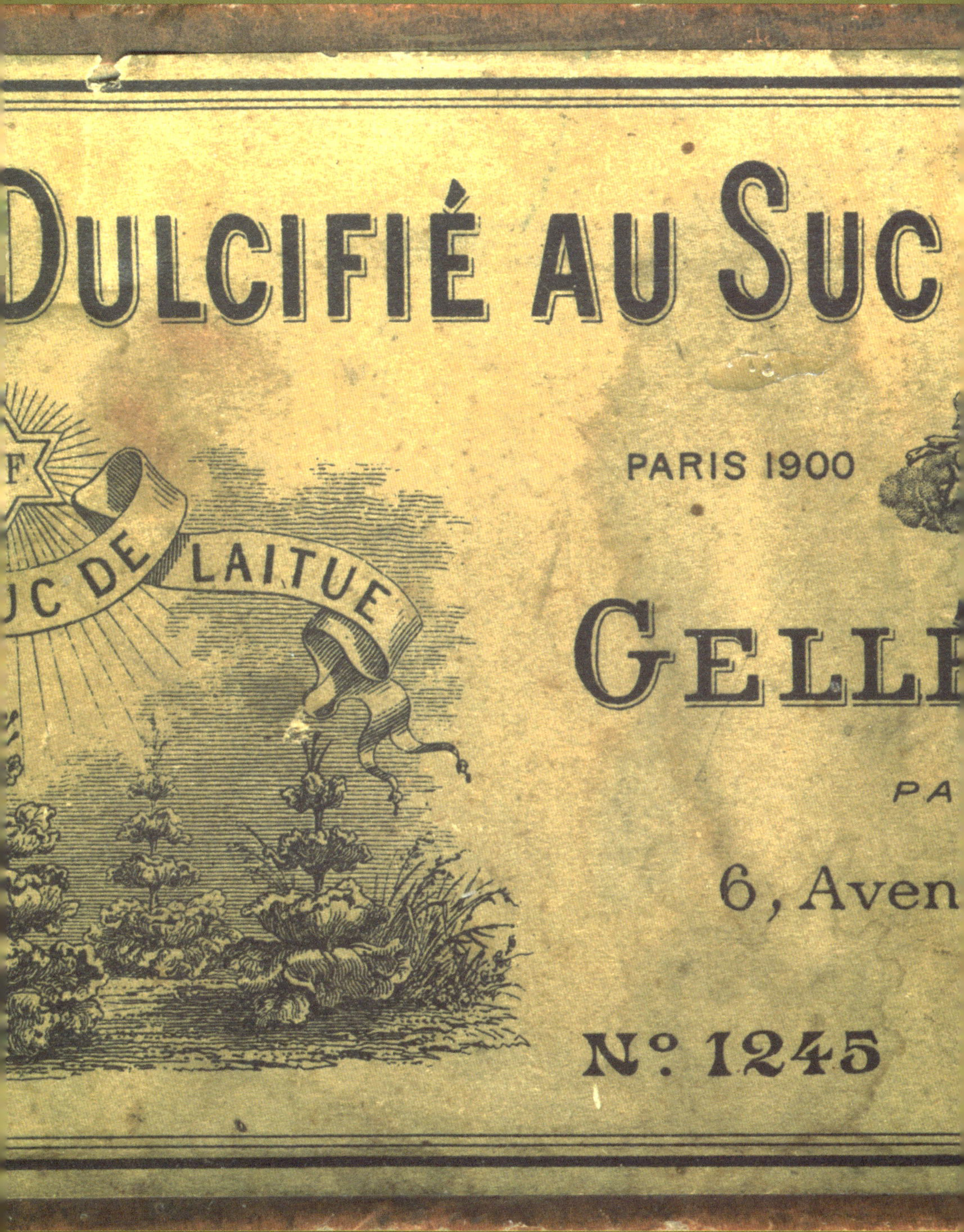
DULCIFIÉ AU SUC
UC DE LAITUE
PARIS 1900
GELLÉ
PA
6, Aven
N° 1245

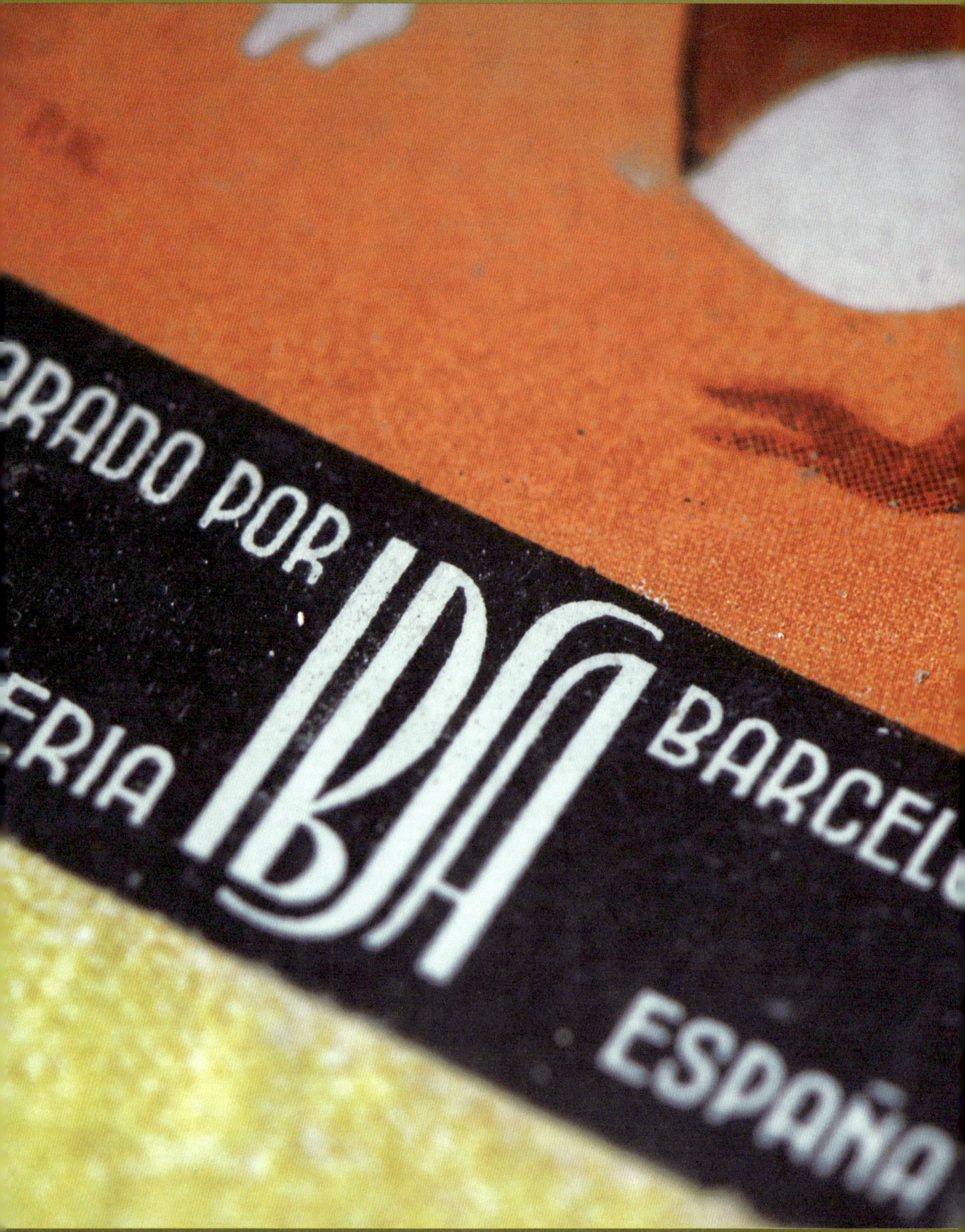
RADO POR
FRIA
BARCEL
ESPAÑA

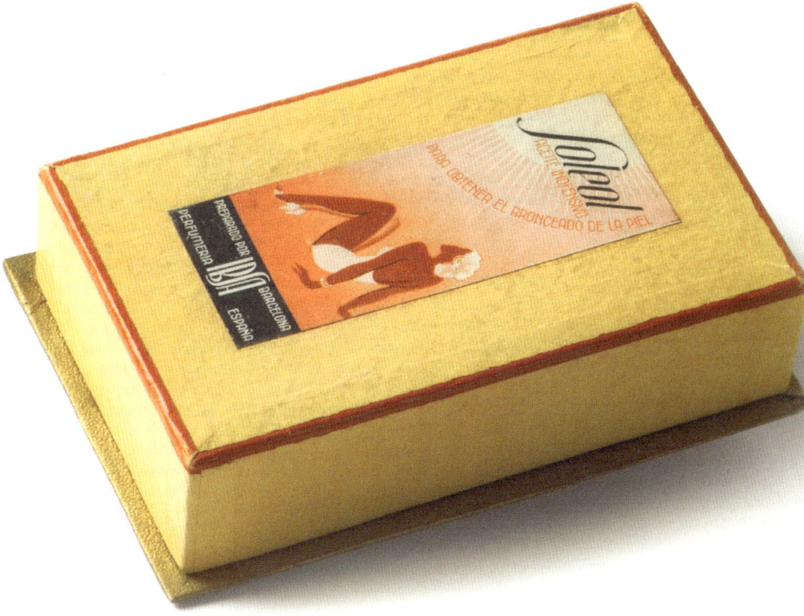
ACEITE INOFENSIVO
PARA OBTENER EL BRONCEADO DE LA PIEL
PREPARADO POR
PERFUMERIA
BARCELONA
ESPAÑA

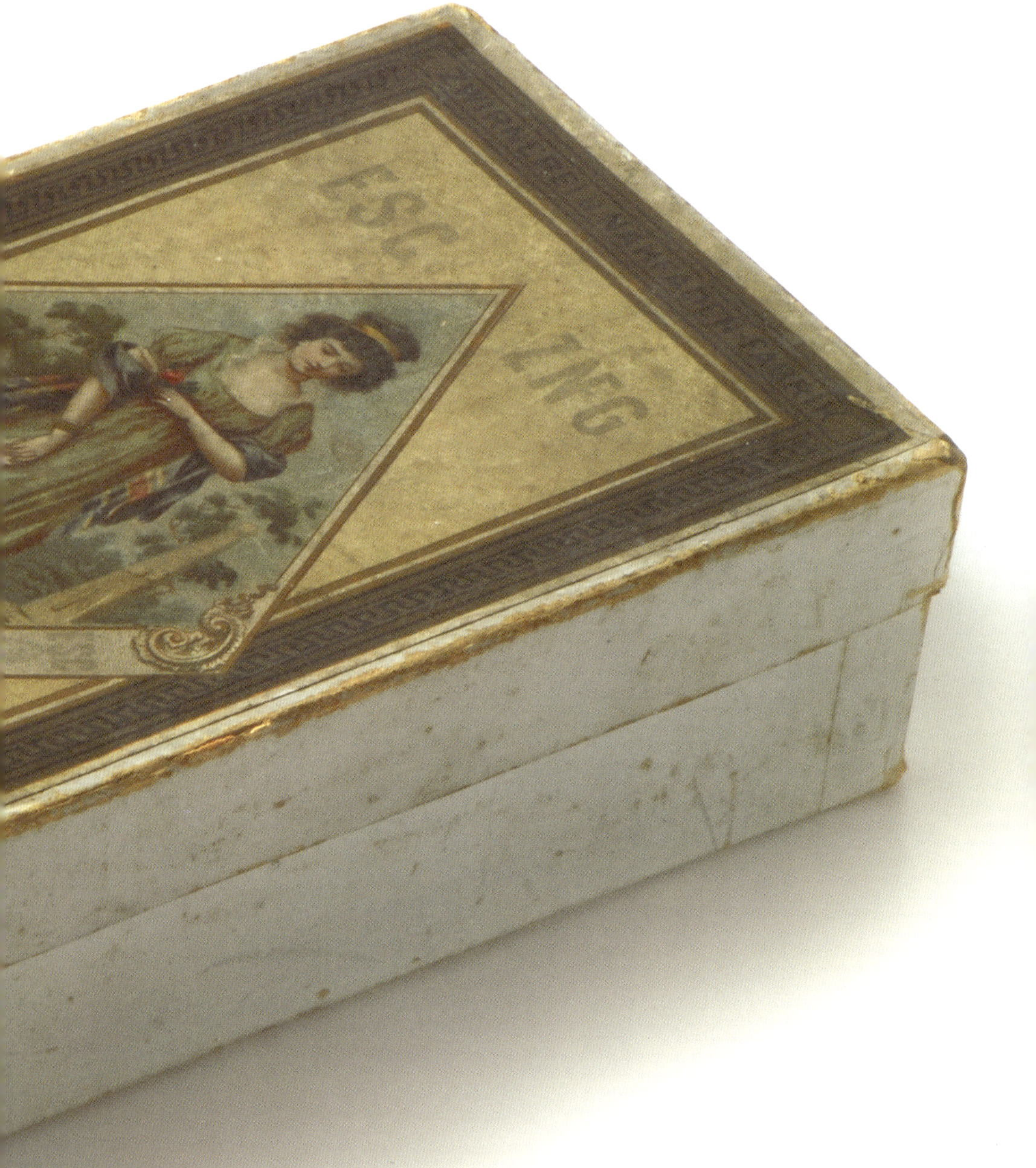

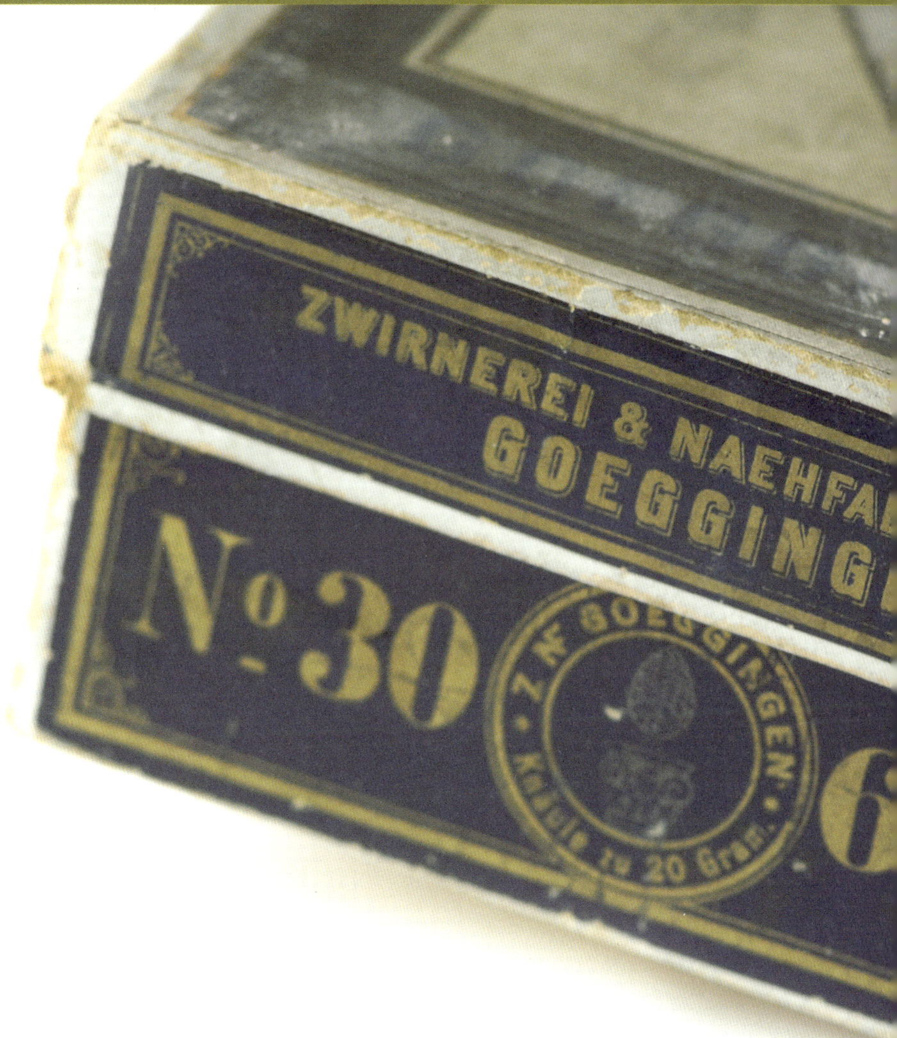
ZWIRNEREI &
No. 30
GOEGGINGEN
Knäule zu 20 Gram.

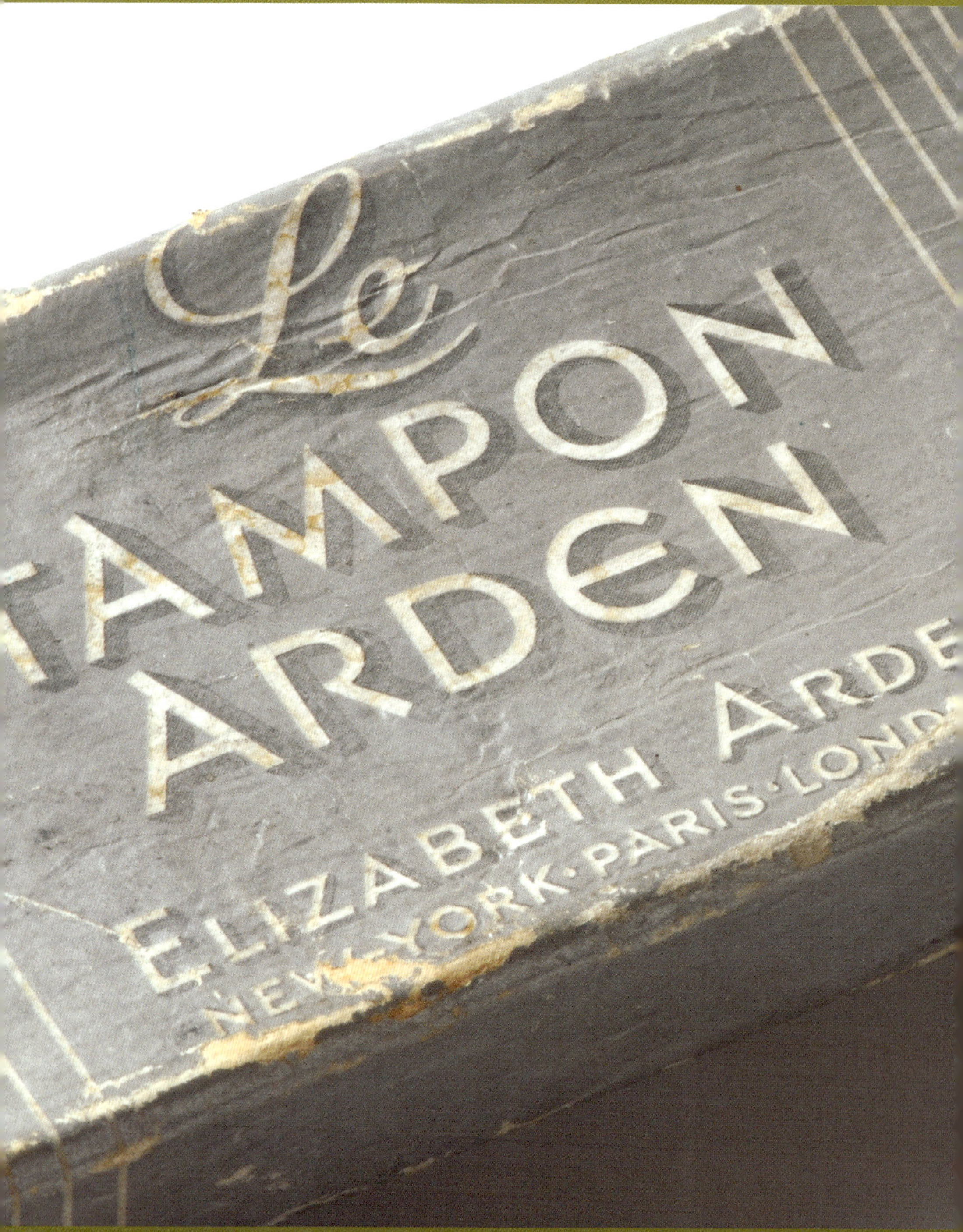
Le
TAMPON
ARDEN
ELIZABETH ARDE
NEW YORK · PARIS · LOND

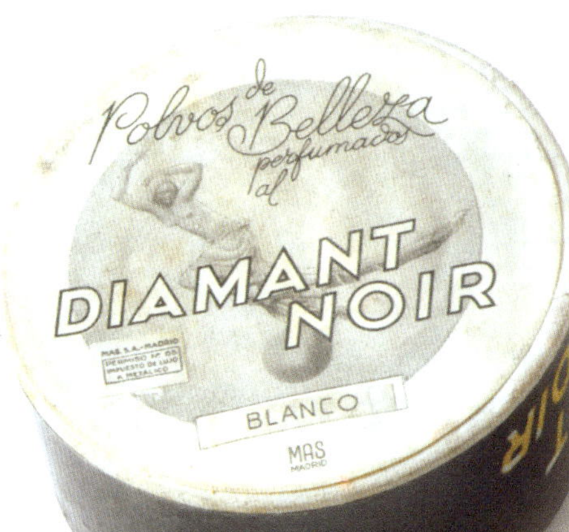
Polvos de Belleza perfumados al
DIAMANT NOIR
BLANCO
MAS
MADRID

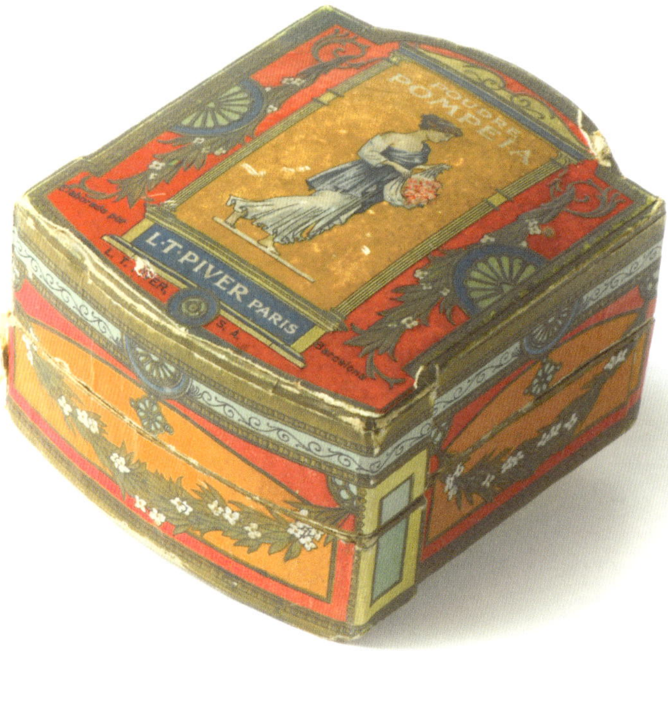
POUDRE
POMPEIA
L·T·PIVER PARIS

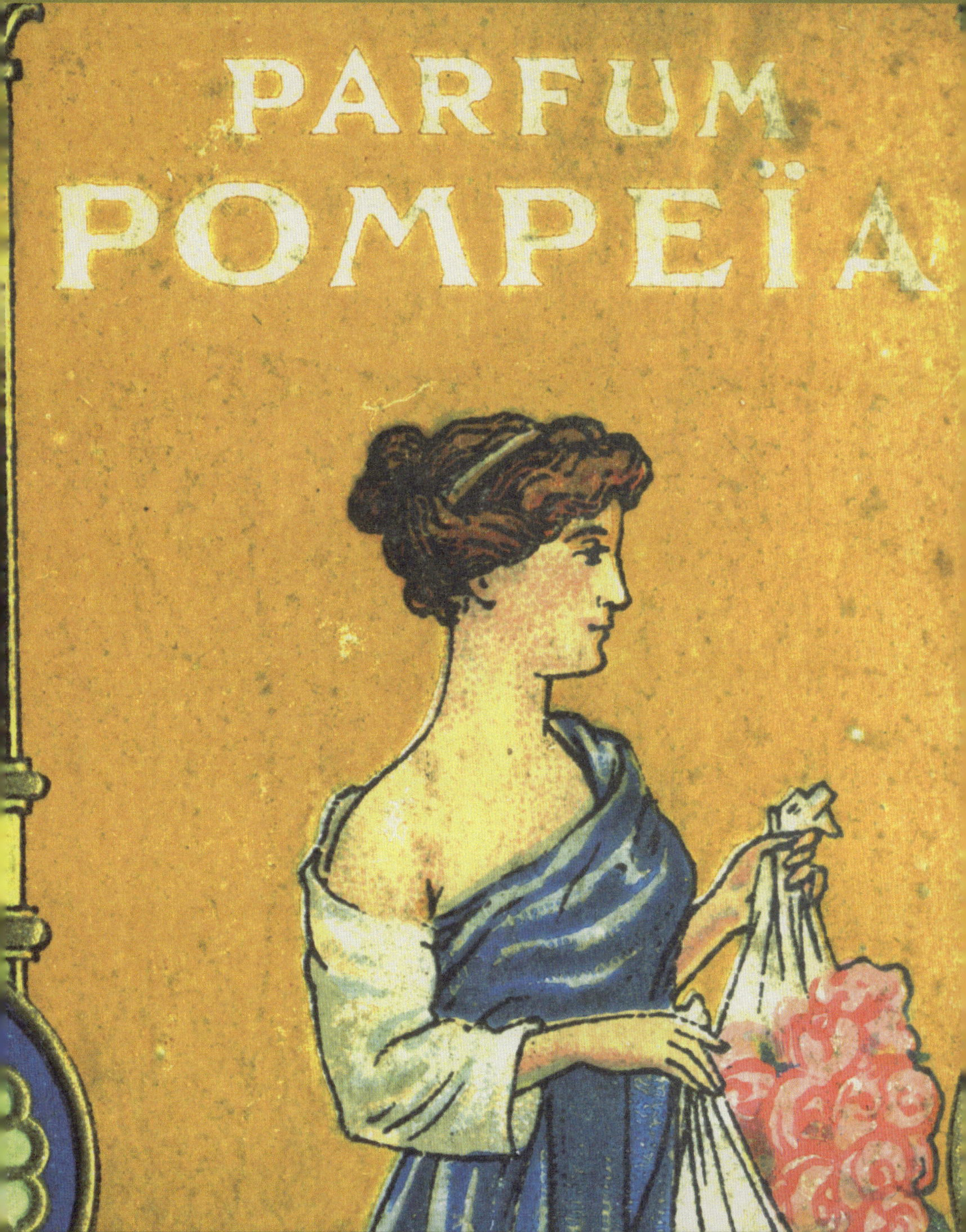
PARFUM
POMPEÏA

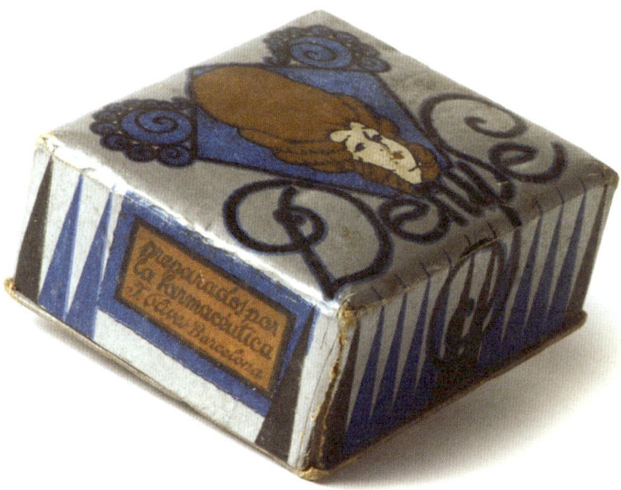
Preparados por
Barcelona

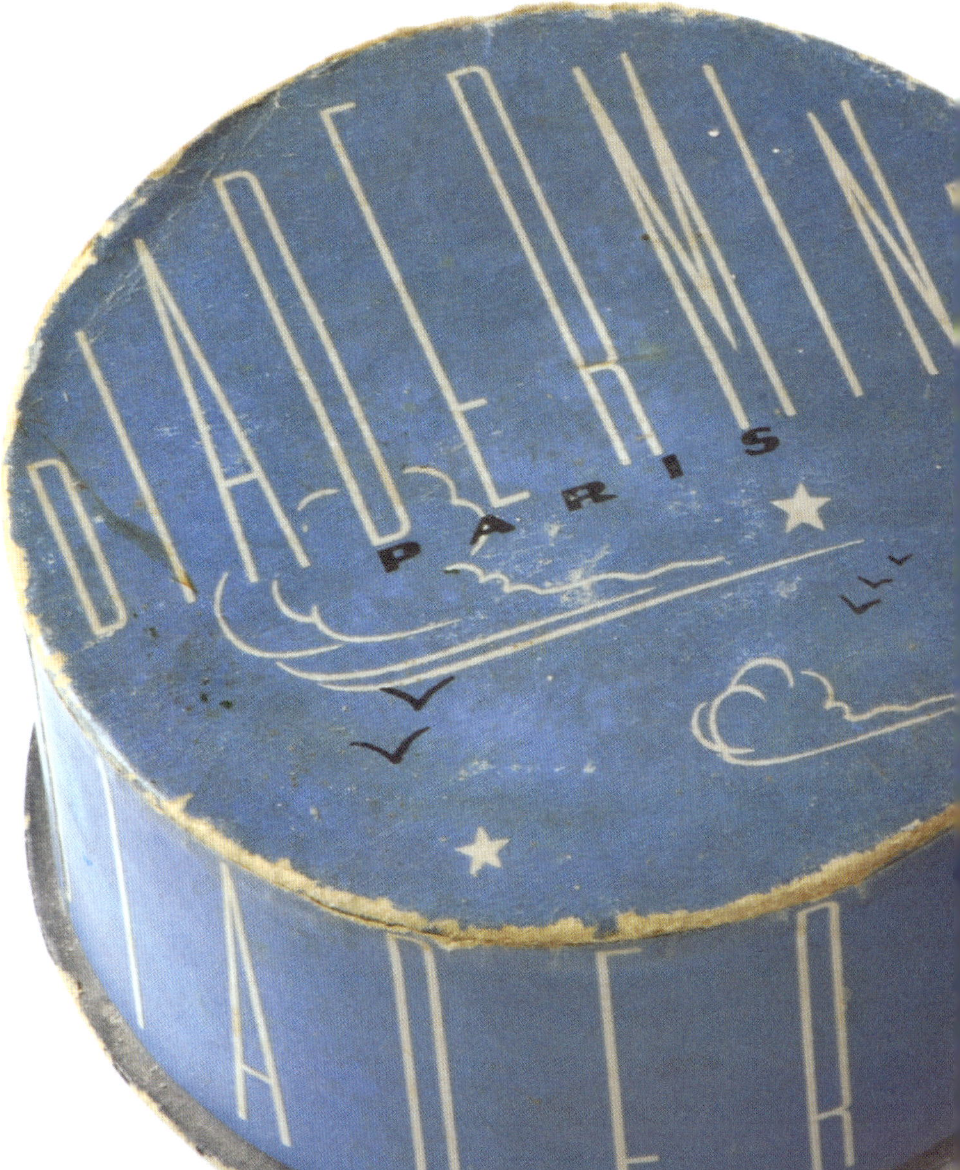
PARIS

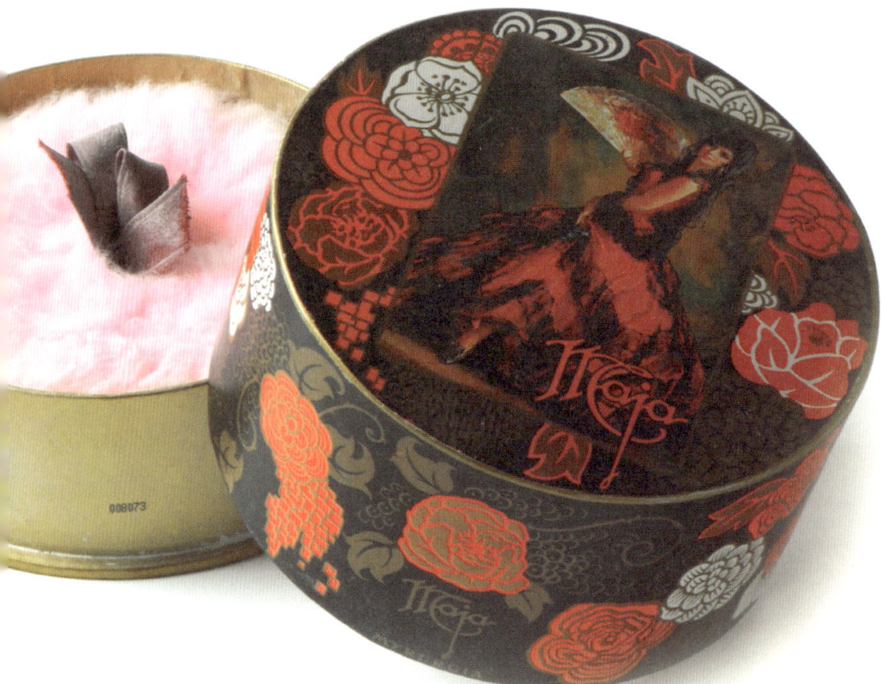
Maja
Maja
008073

267

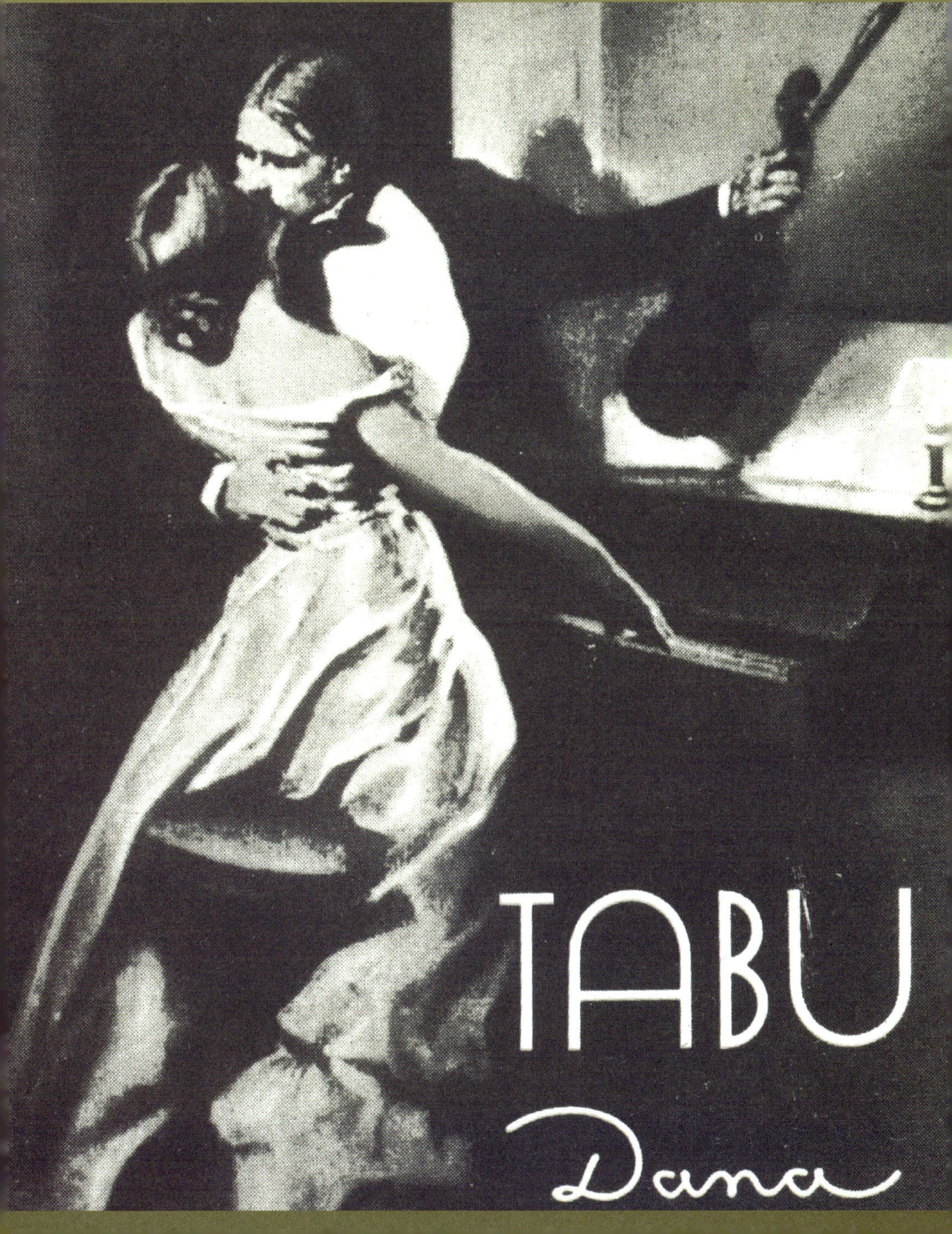

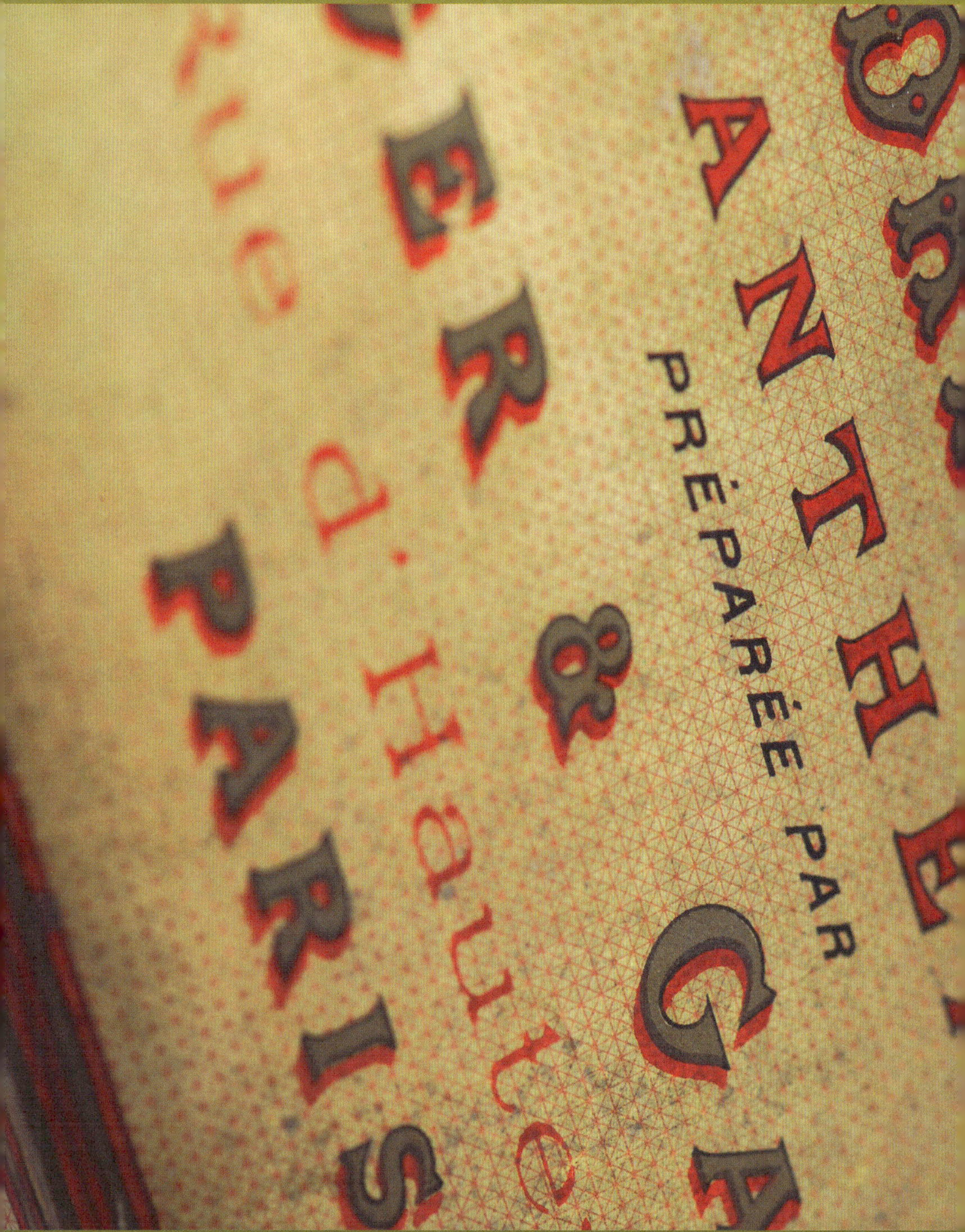
PRÉPARÉE PAR
PARIS

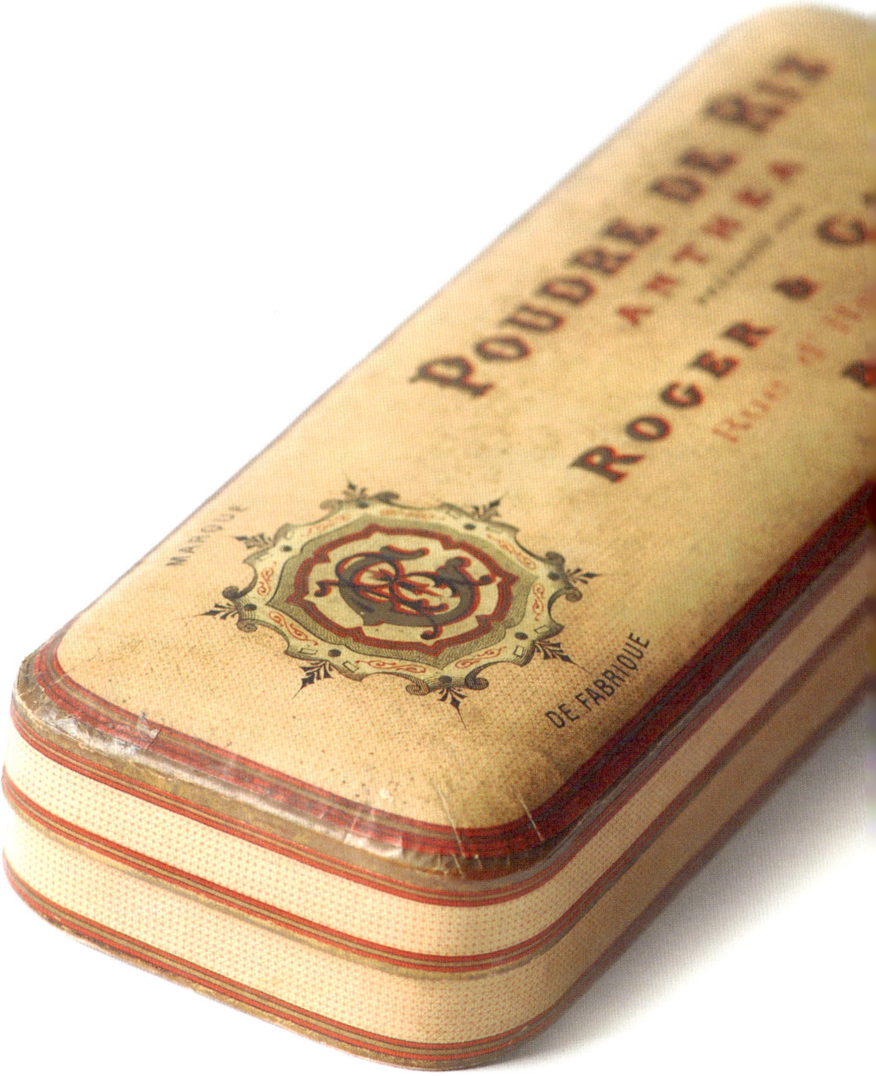
POUDRE DE RIZ
ANTHEA
ROGER &
MARQUE
DE FABRIQUE

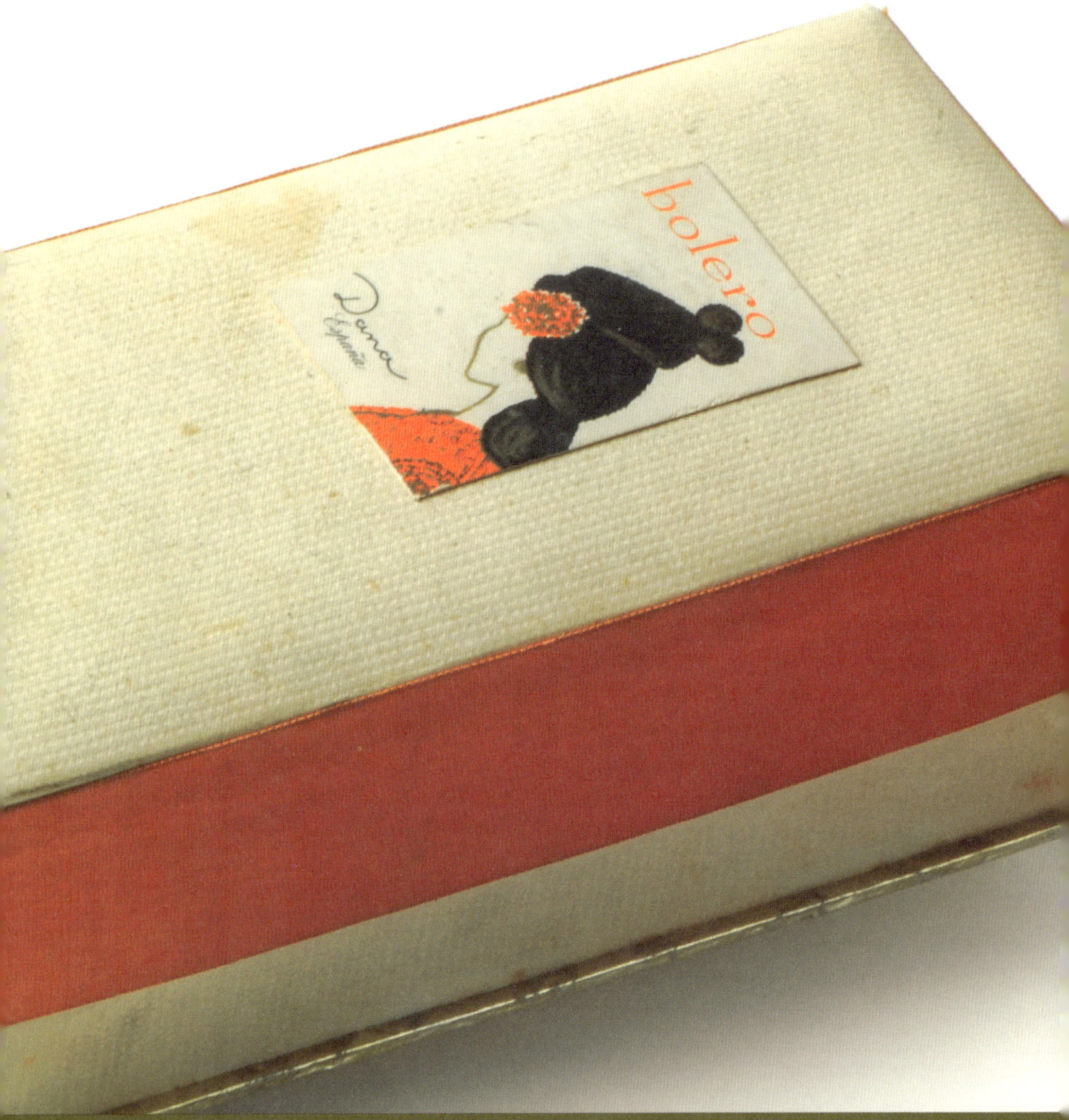
bolero
Dama
España

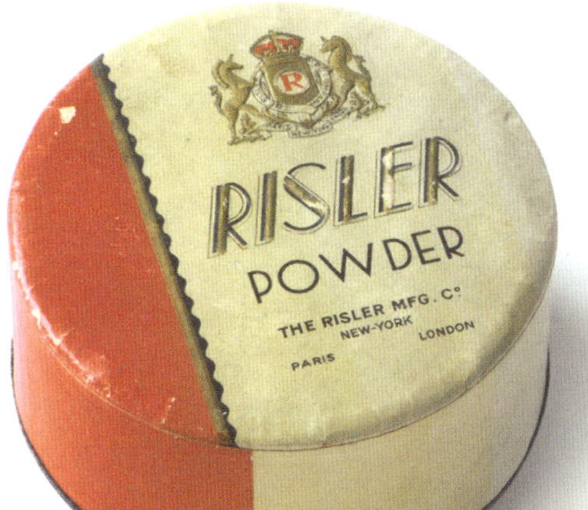
RISLER
POWDER
THE RISLER MFG. C°
NEW-YORK
PARIS
LONDON

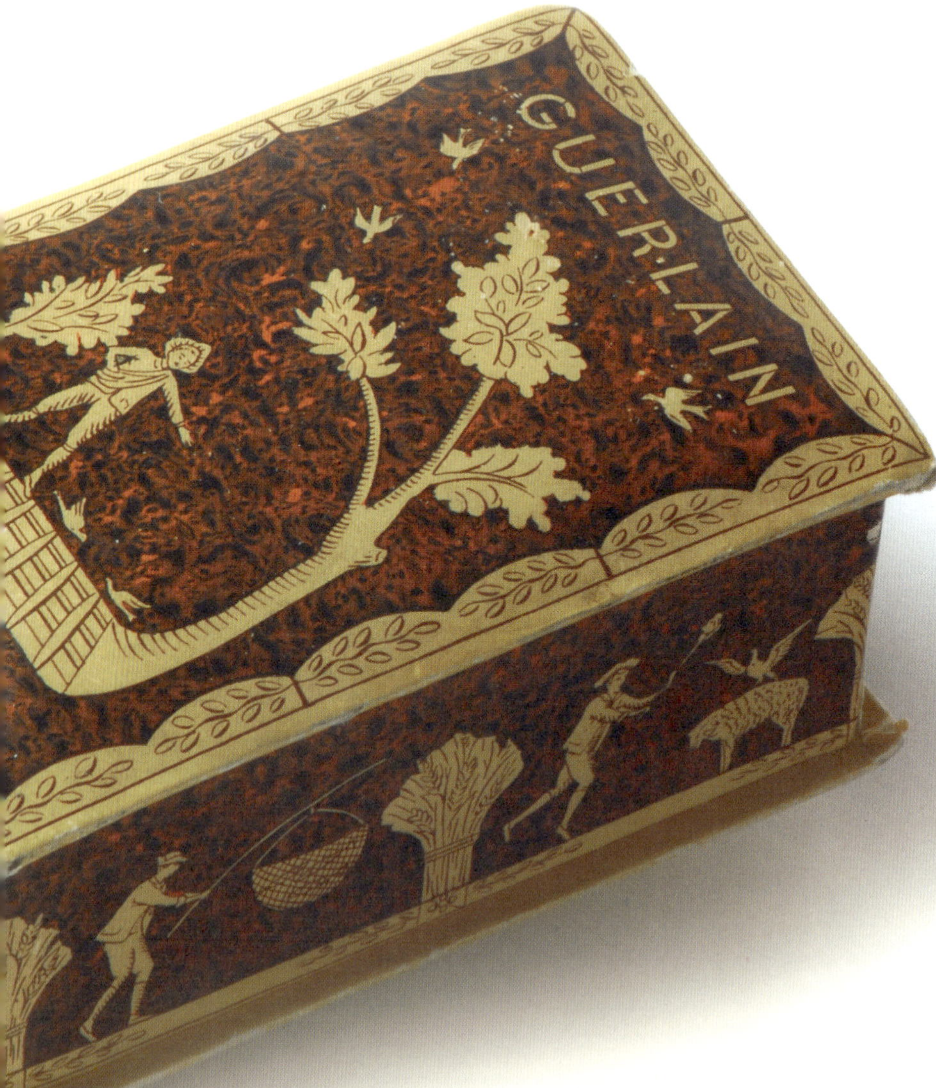
GUERLAIN

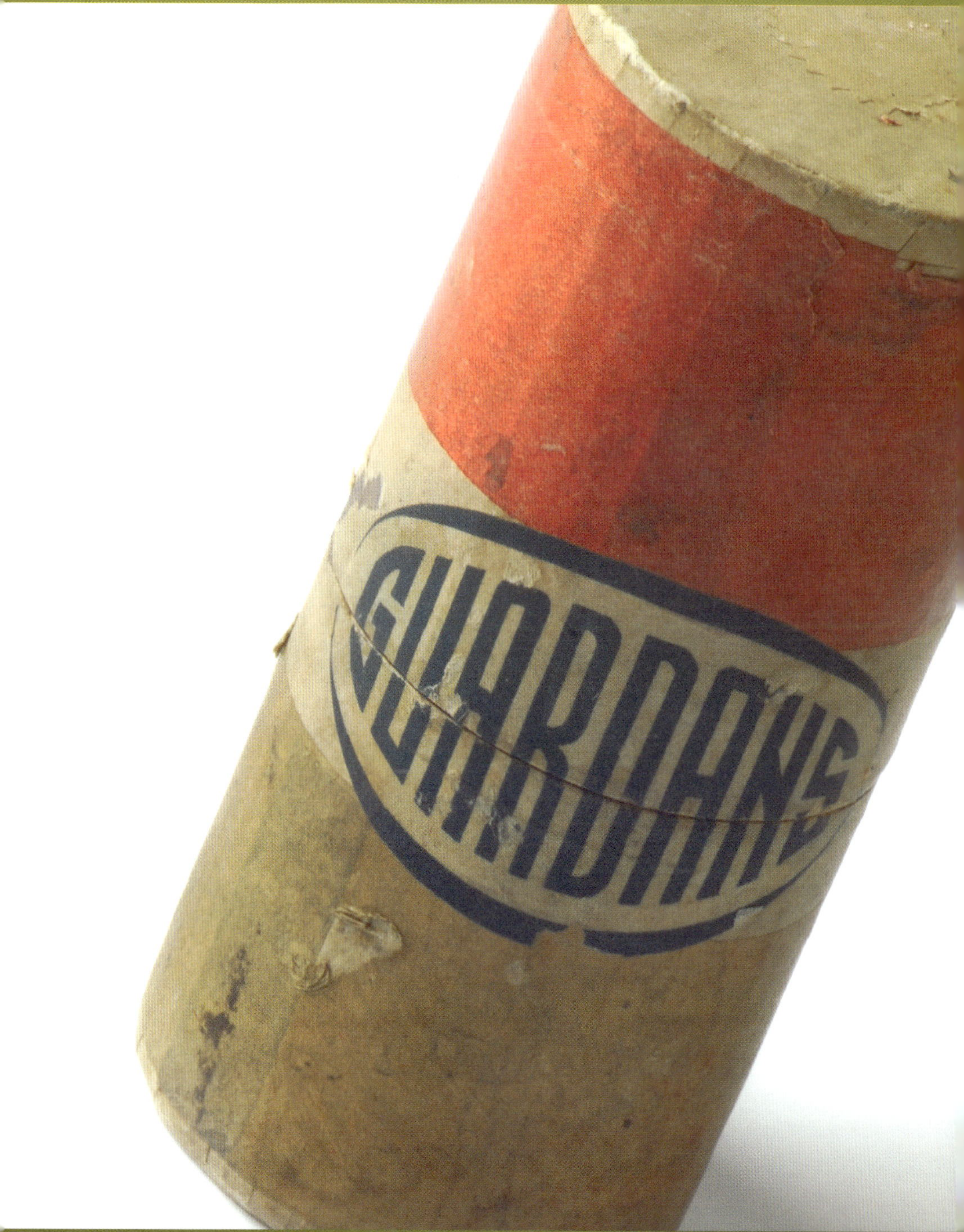
GUARDIANS

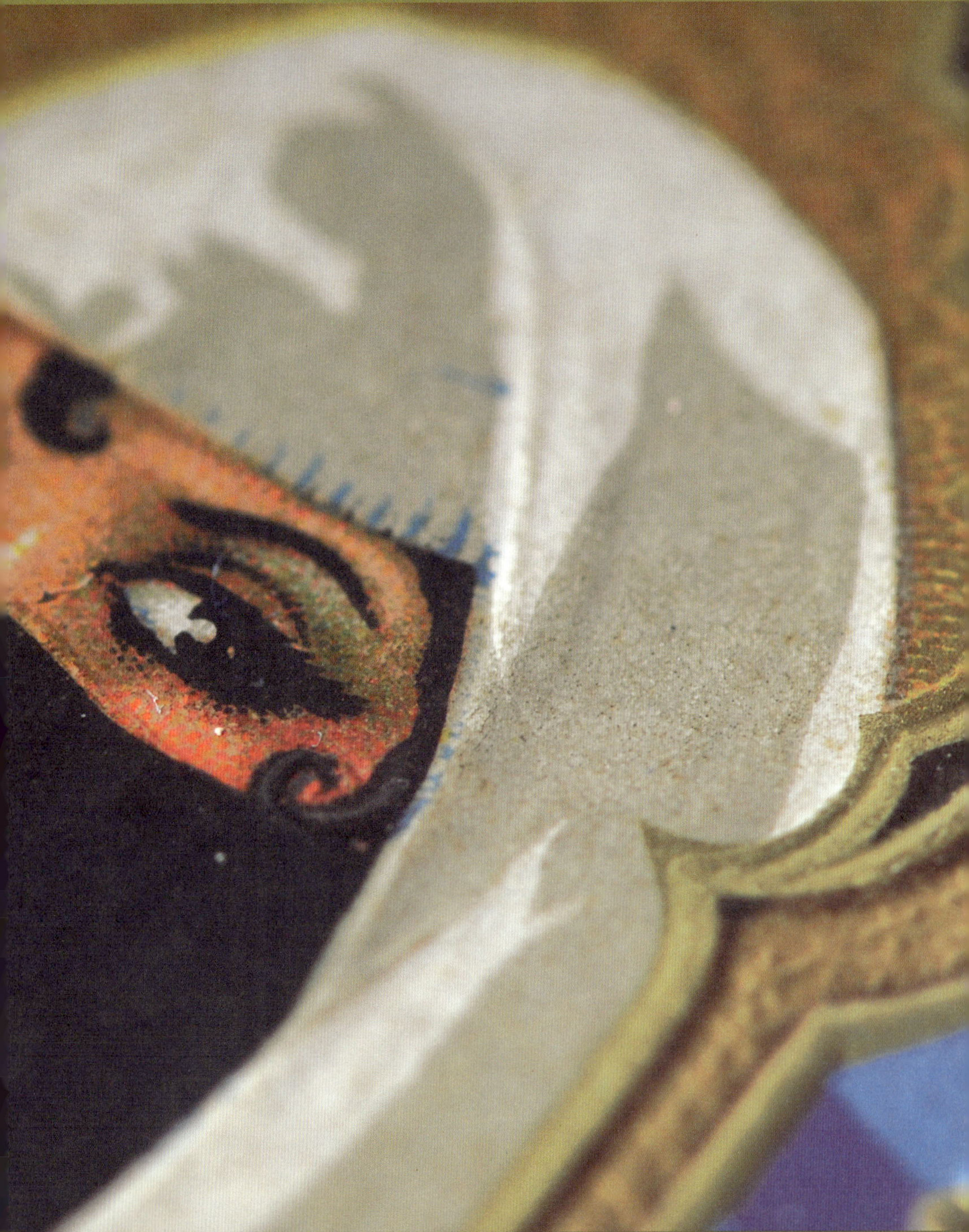

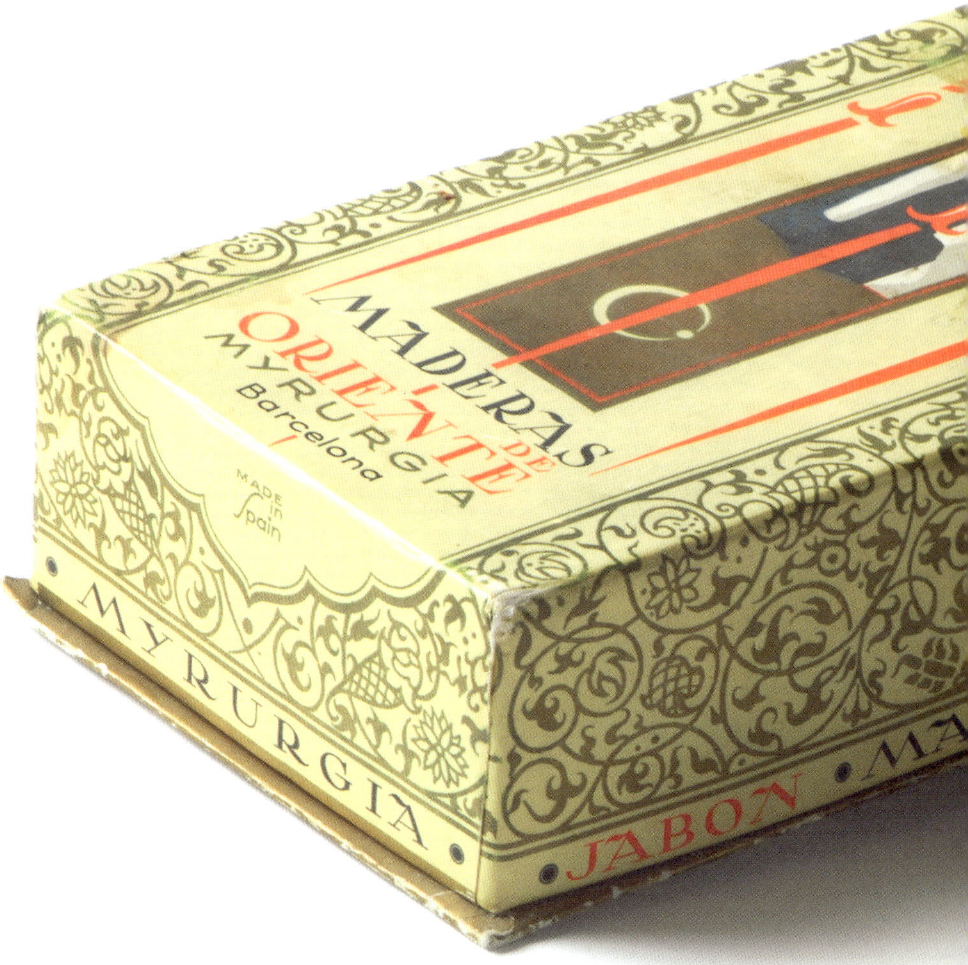
MADERAS
DE
ORIENTE
MYRURGIA
Barcelona
MADE in Spain
MYRURGIA
JABON

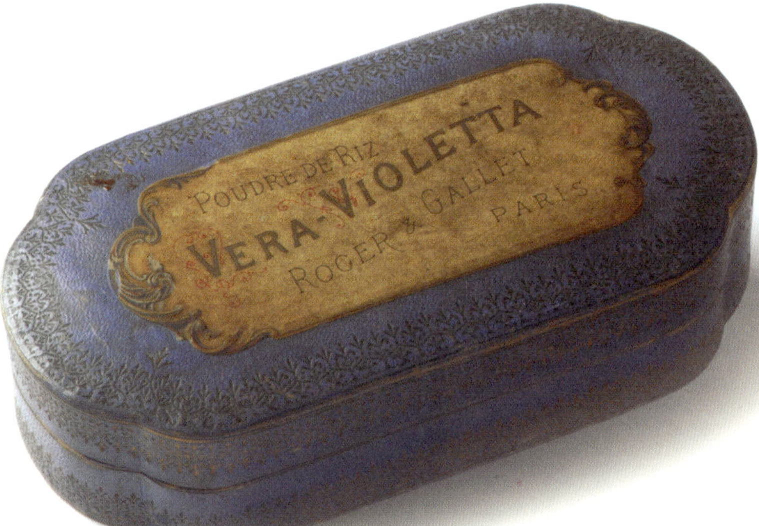
POUDRE DE RIZ
VERA-VIOLETTA
ROGER & GALLET
PARIS

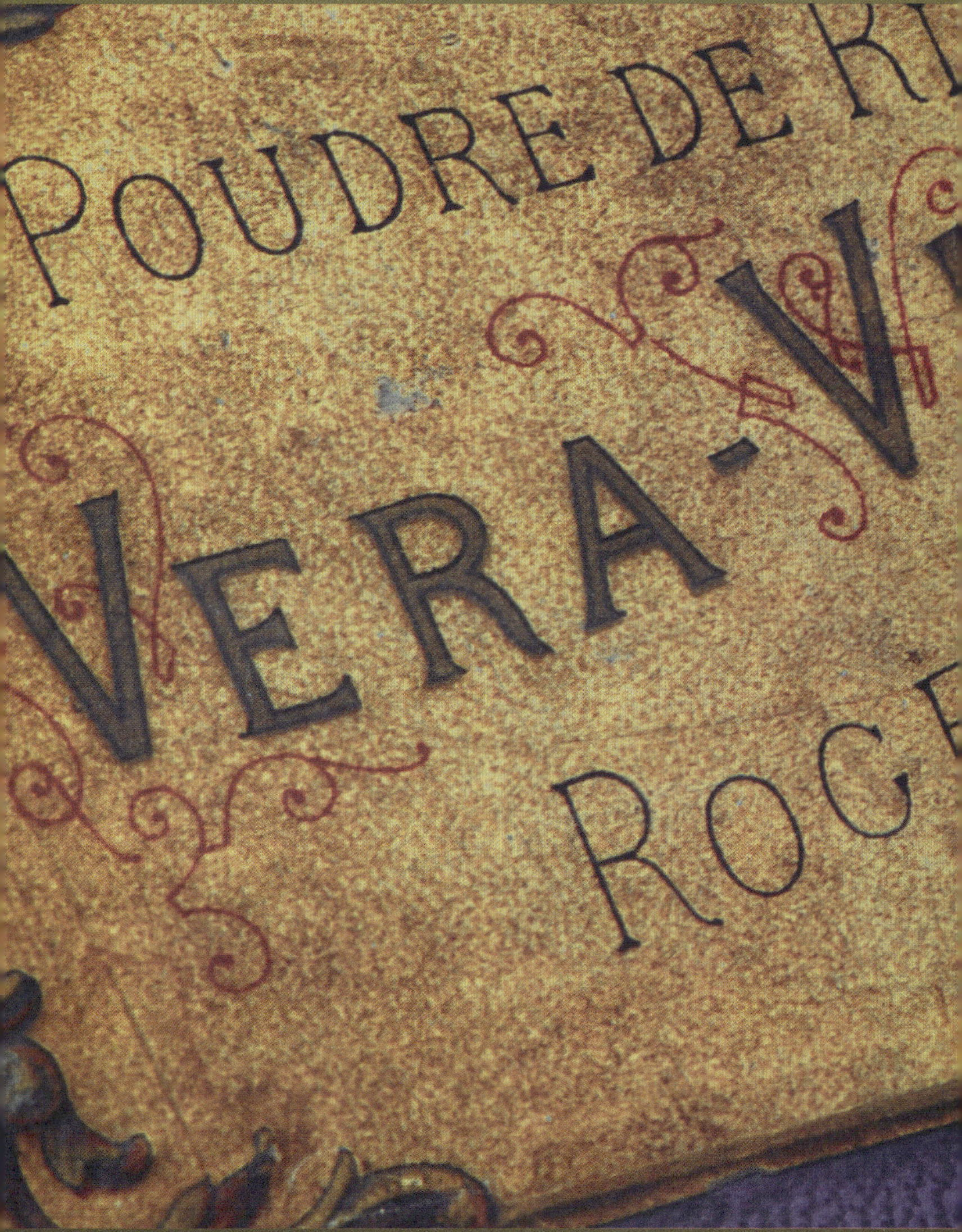
POUDRE DE
VERA-

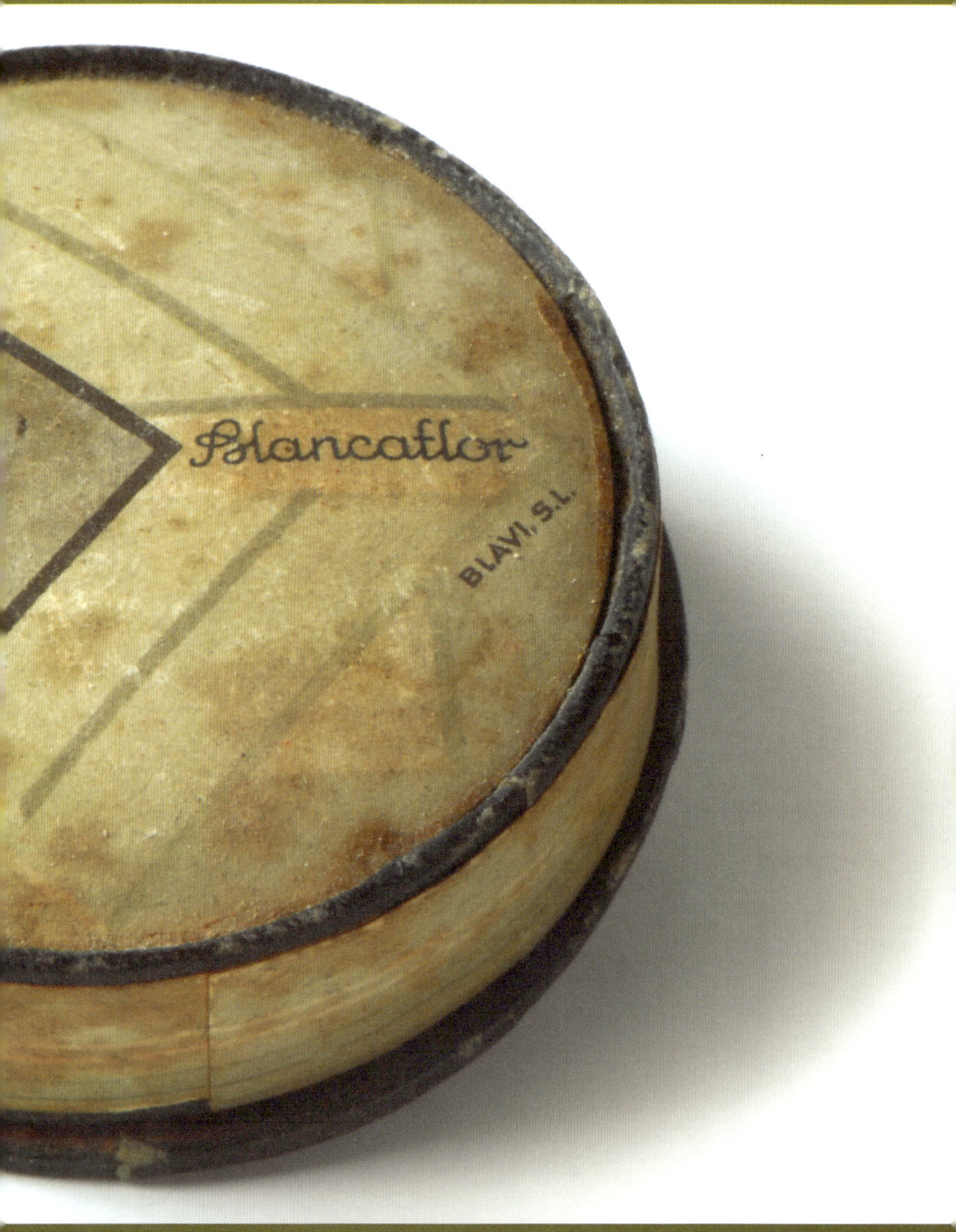
Blancaflor
BLAVI, S.L.

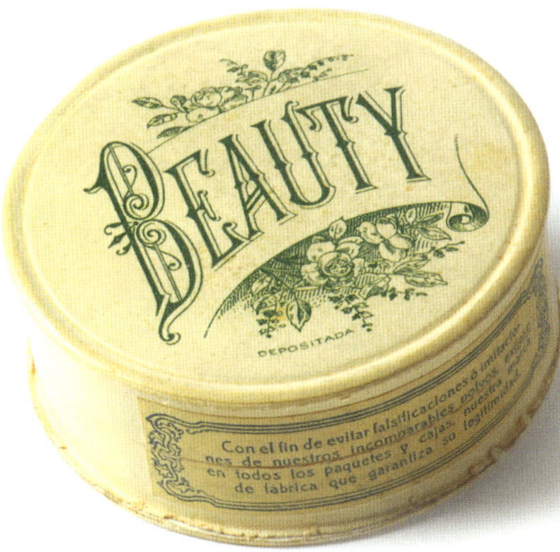
BEAUTY
DEPOSITADA
Con el fin de evitar falsificaciones ó imitacio-
nes de nuestros incomparables polvos
en todos los paquetes y cajas nuestra
de fabrica que garantiza su legitimidad.

MESTICD

OMESTICDOMEST

DOMEST

CDOMESTICDOM

DOMESTIC

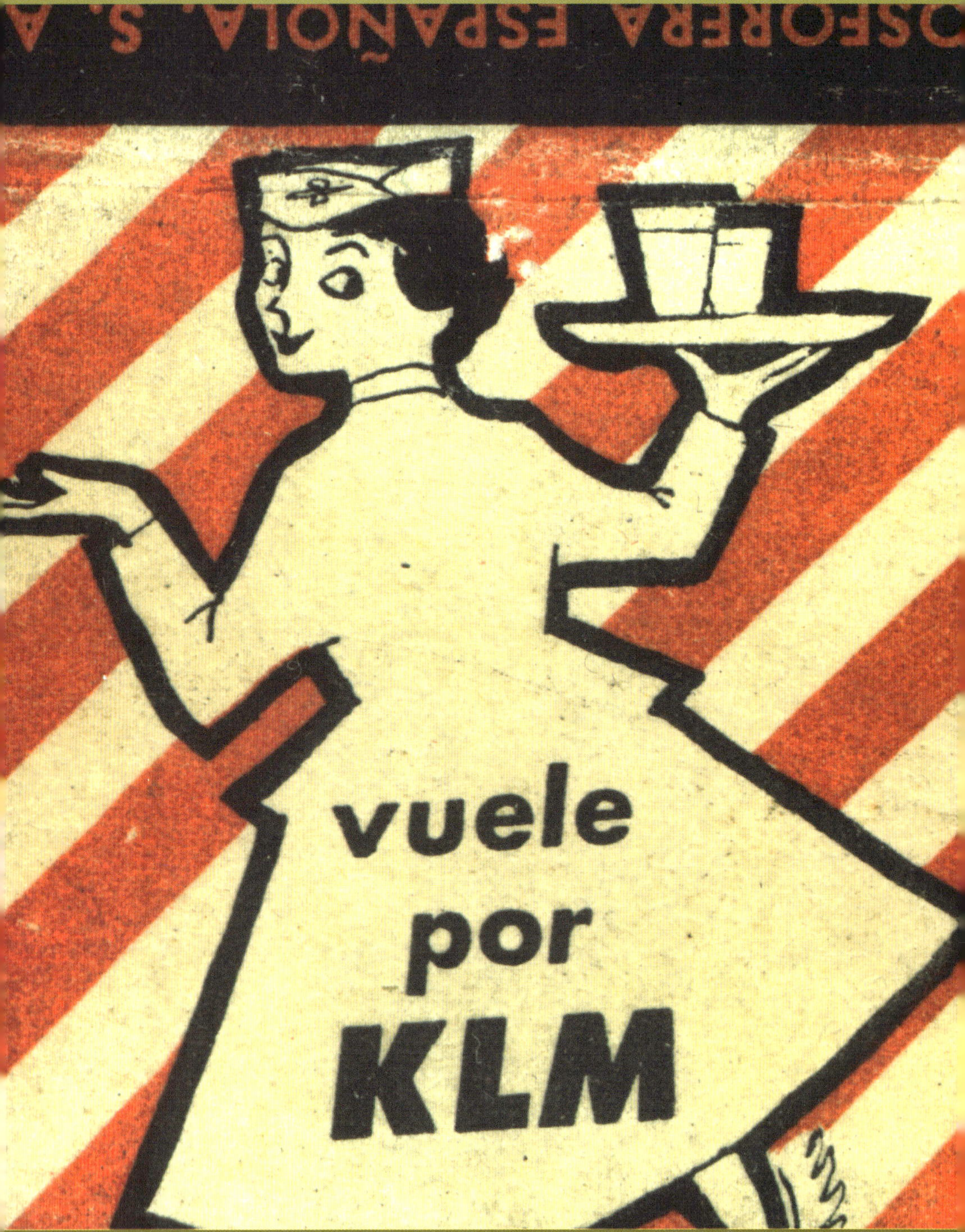
vuele
por
KLM

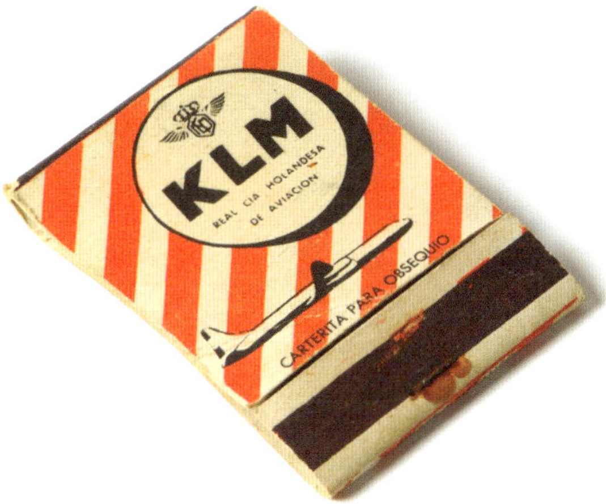
KLM
REAL CIA HOLANDESA DE AVIACION
CARTERITA PARA OBSEQUIO

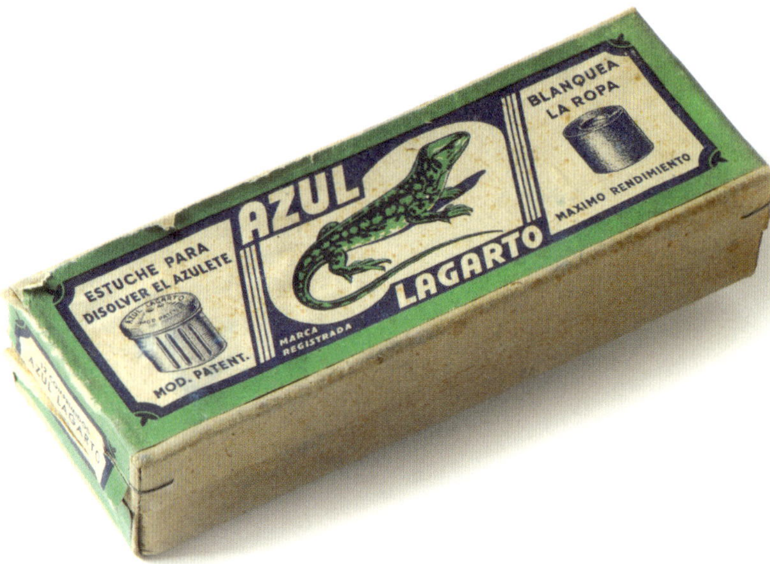
ESTUCHE PARA
DISOLVER EL AZULETE
MOD. PATENT.
AZUL
LAGARTO
MARCA
REGISTRADA
BLANQUEA
LA ROPA
MAXIMO RENDIMIENTO

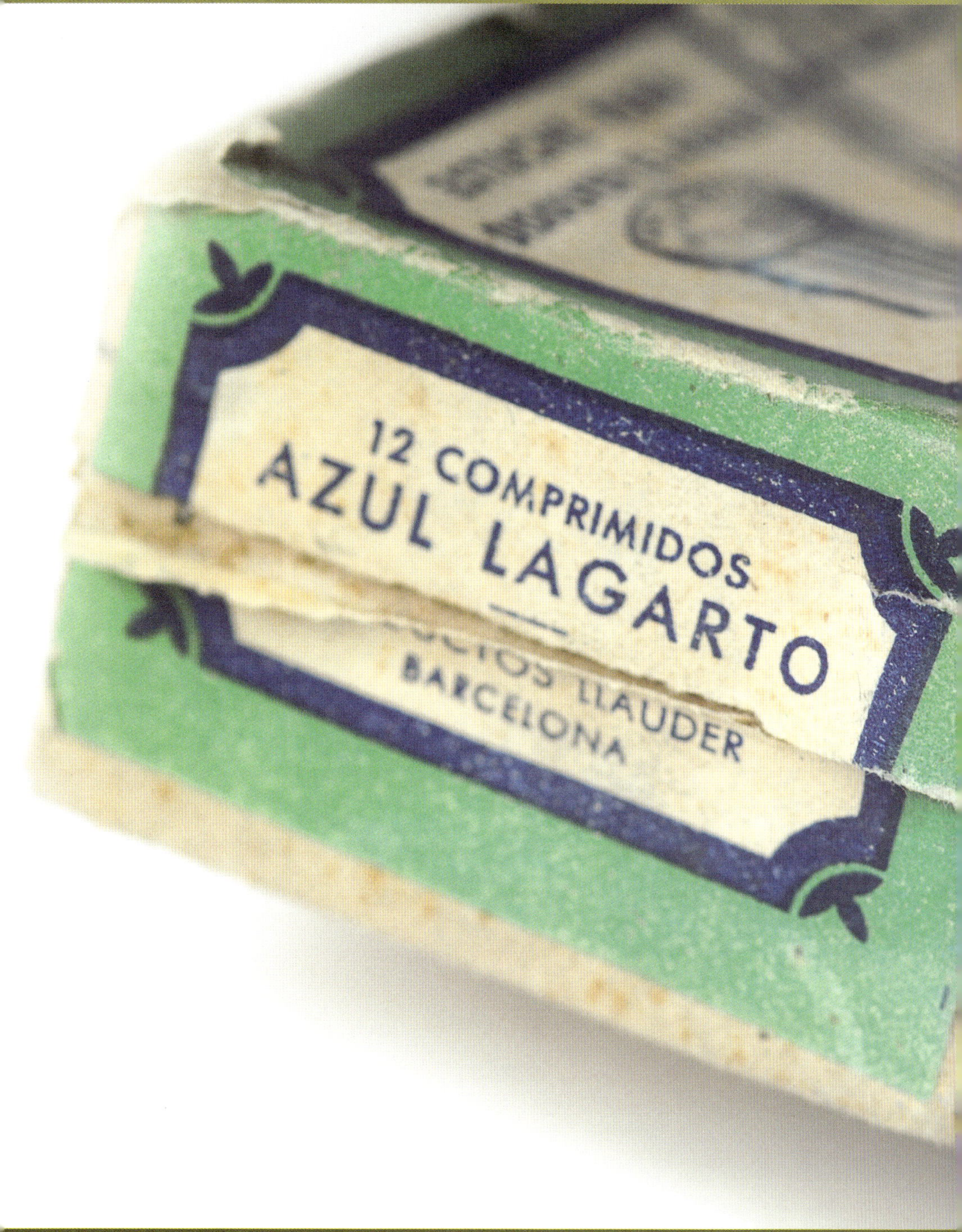
12 COMPRIMIDOS
AZUL LAGARTO
LLAUDER
BARCELONA

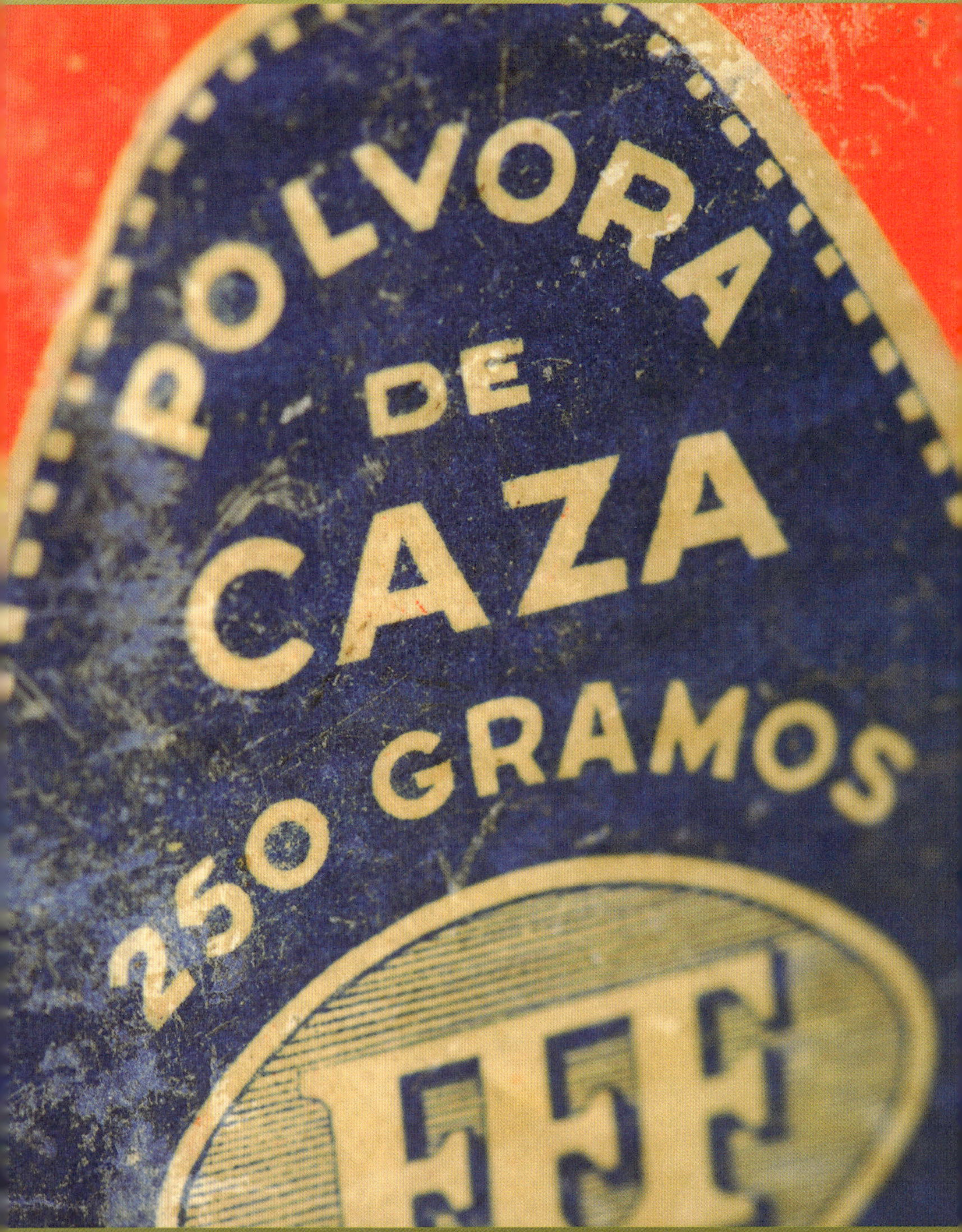
POLVORA
DE
CAZA
250 GRAMOS
EF

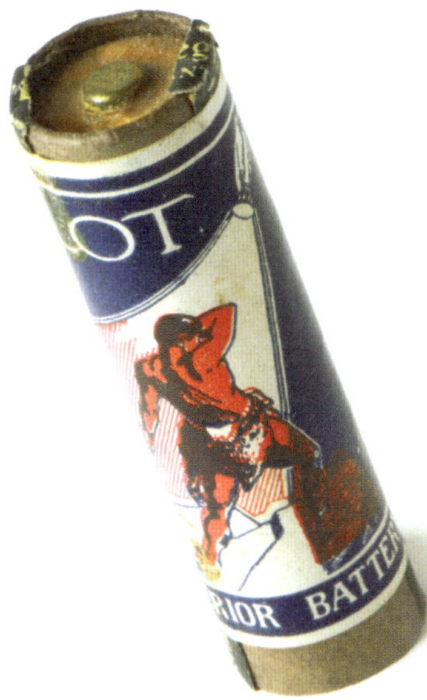
OT
RIOR BATTER

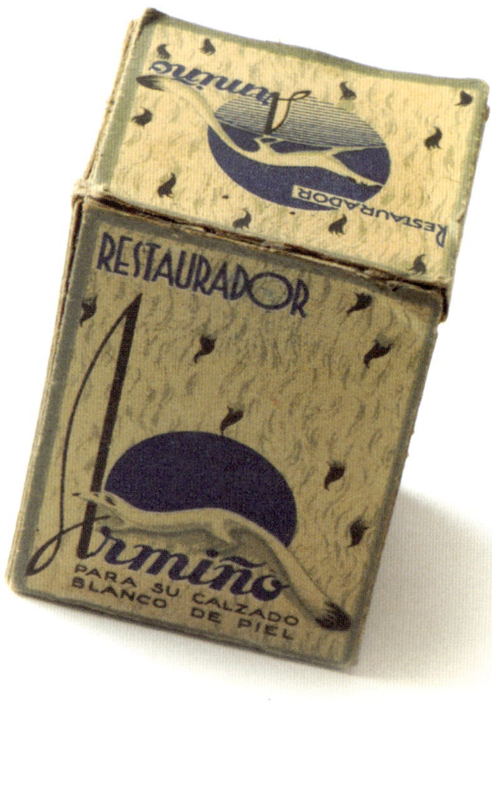
RESTAURADOR
Armiño
PARA SU CALZADO
BLANCO DE PIEL

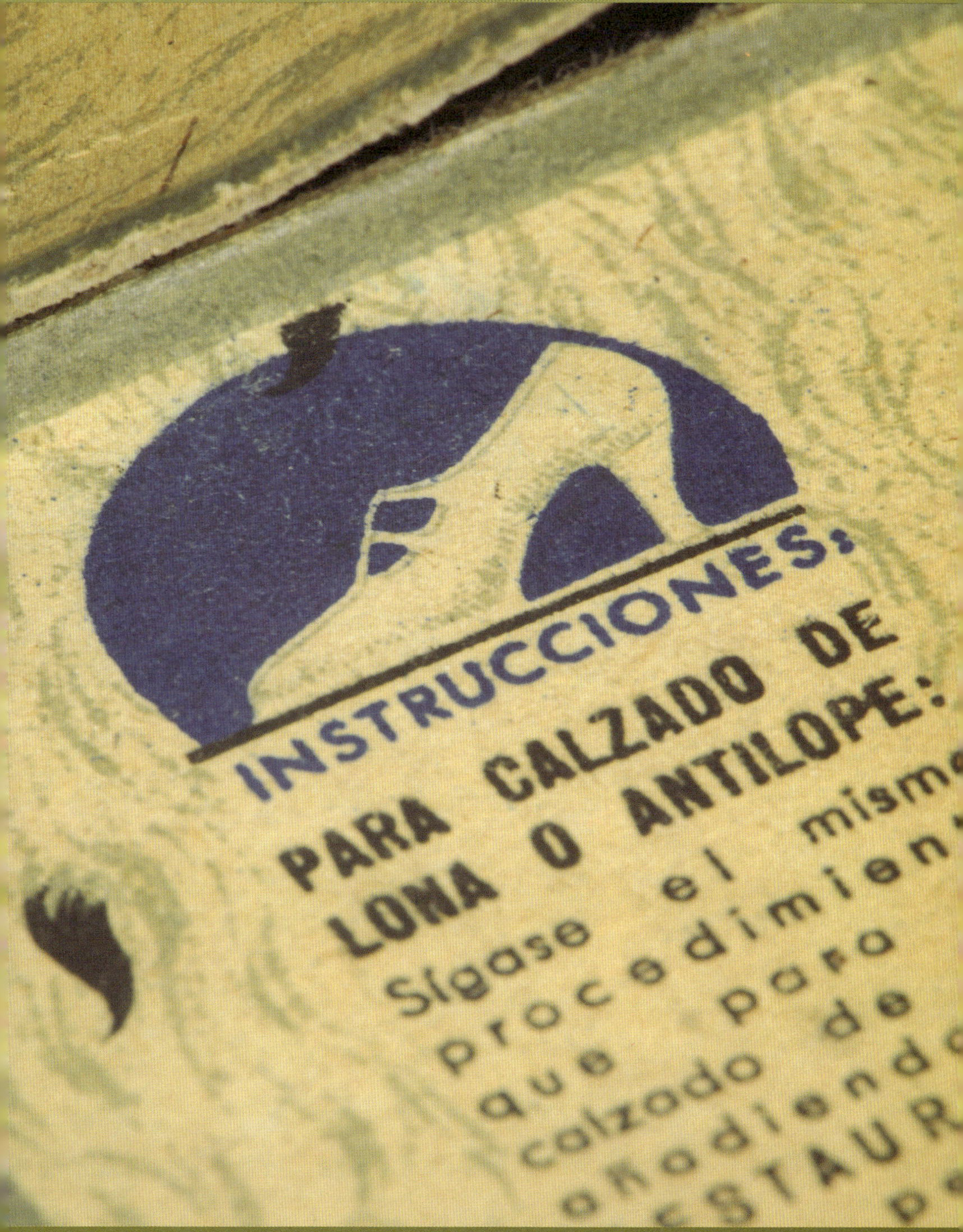
INSTRUCCIONES;
PARA CALZADO DE
LONA O ANTILOPE:
Sígase el
que para
calzado de

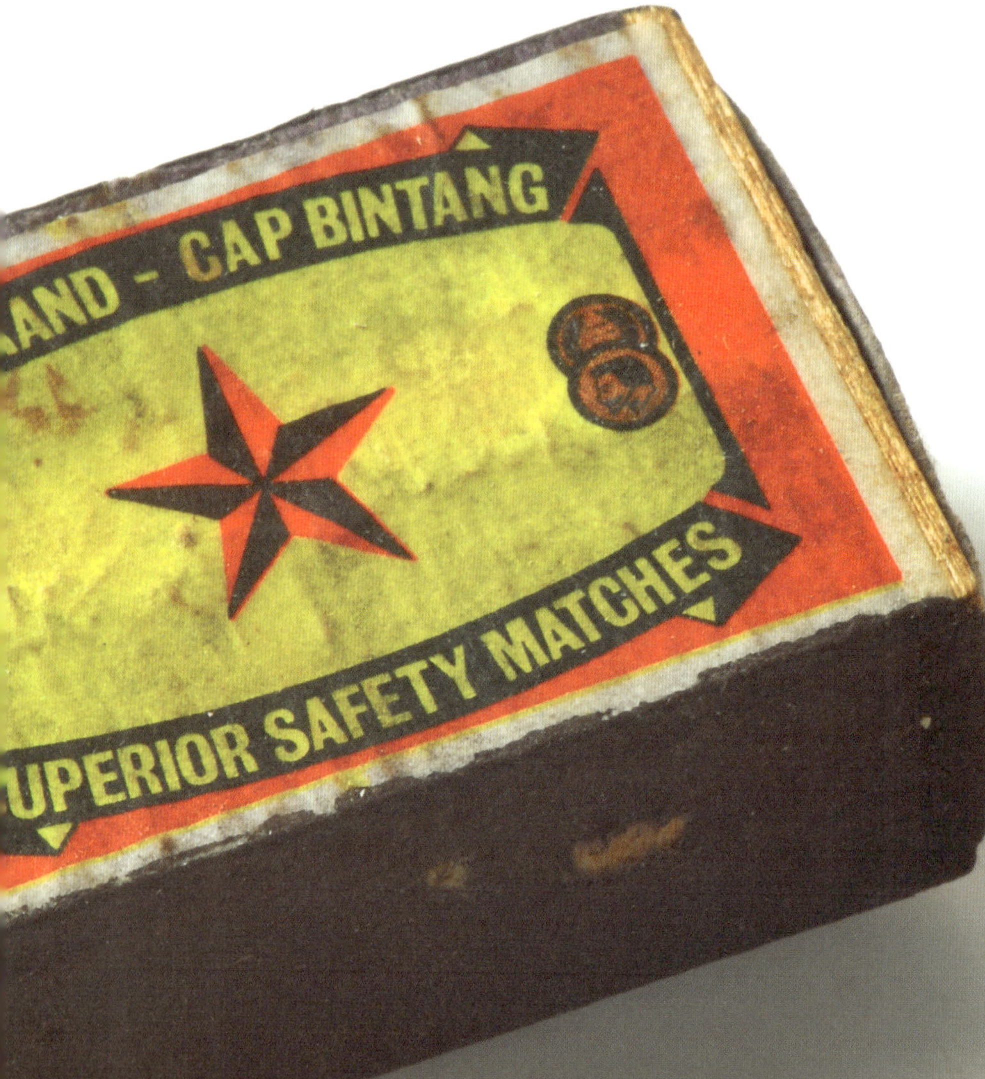
RAND - CAP BINTANG
UPERIOR SAFETY MATCHES

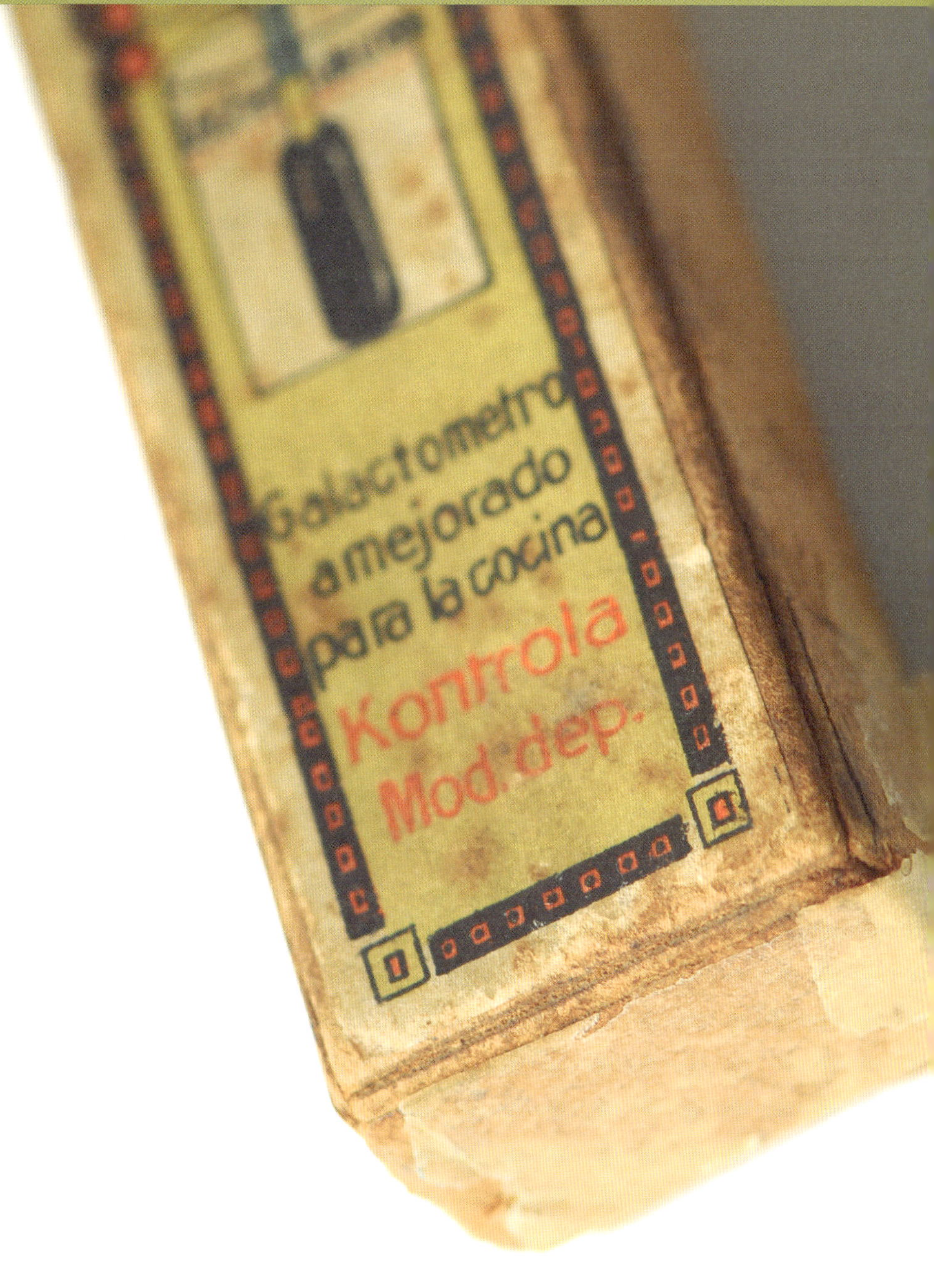
Galactometro
a mejorado
para la cocina
Kontrola
Mod. dep.

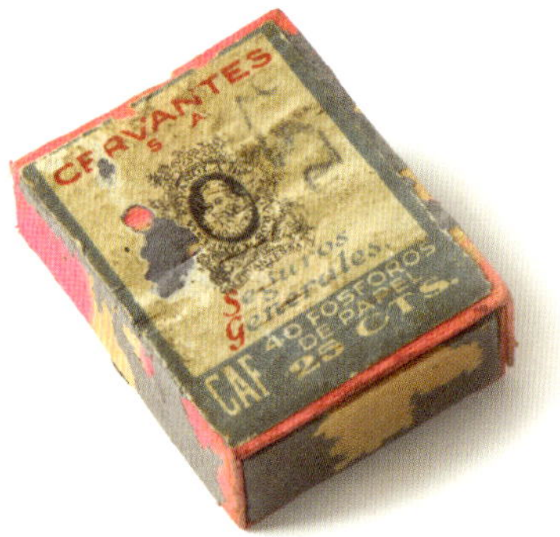
CERVANTES
S. A.
Seguros Generales.
CAF
40 FOSFOROS DE PAPEL
25 CTS.

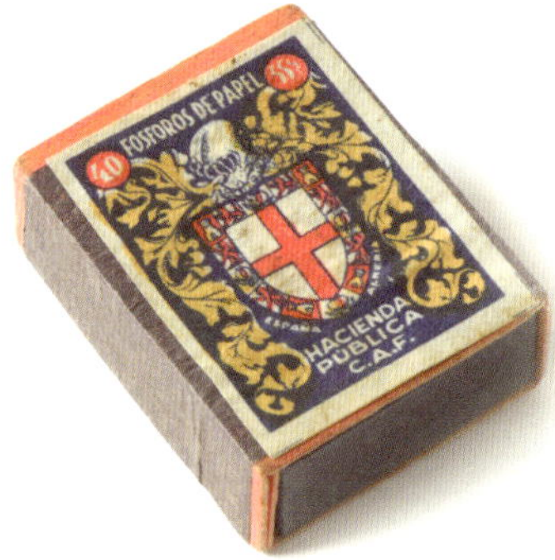
40 FOSFOROS DE PAPEL
HACIENDA PUBLICA
C.A.F.

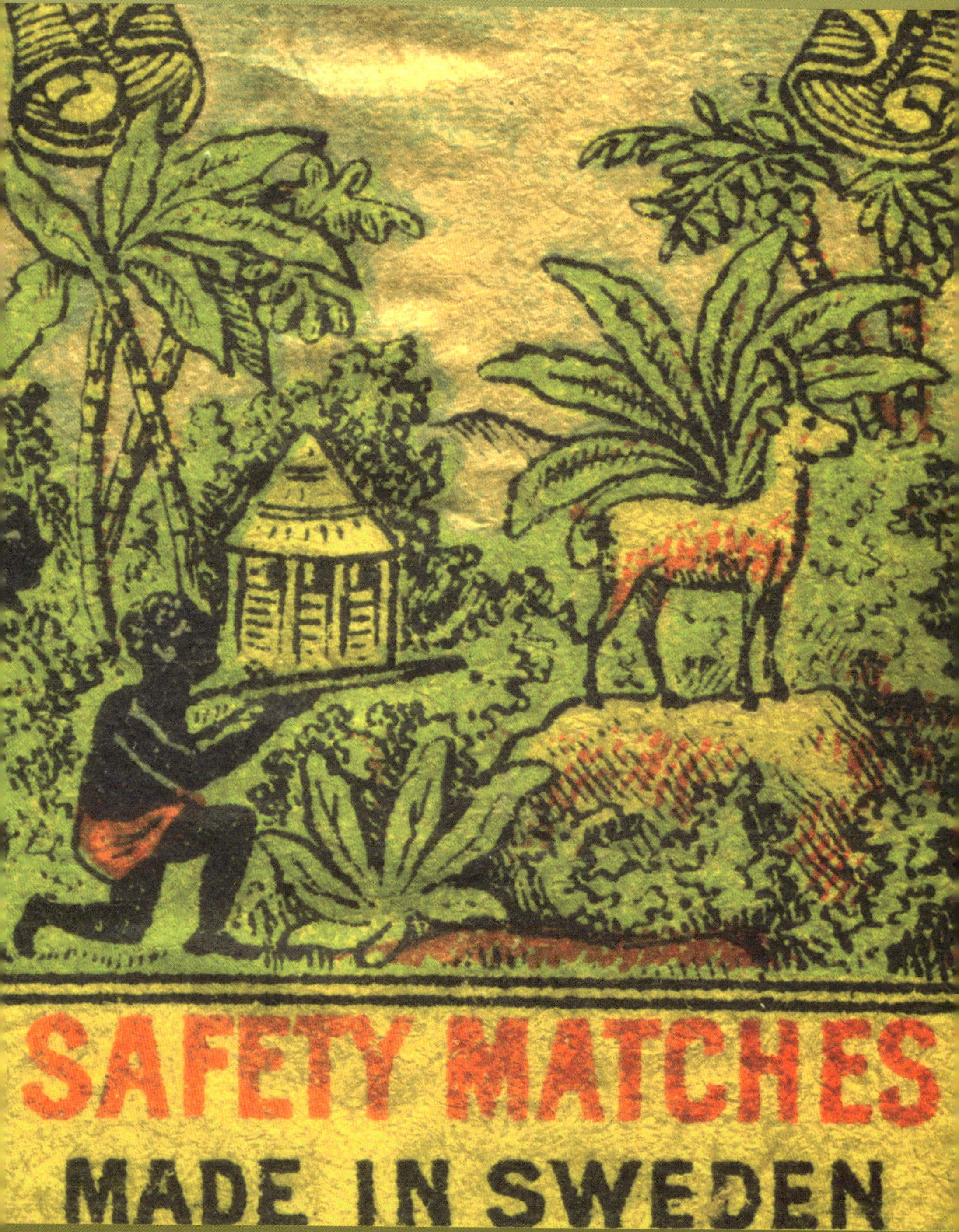
SAFETY MATCHES
MADE IN SWEDEN

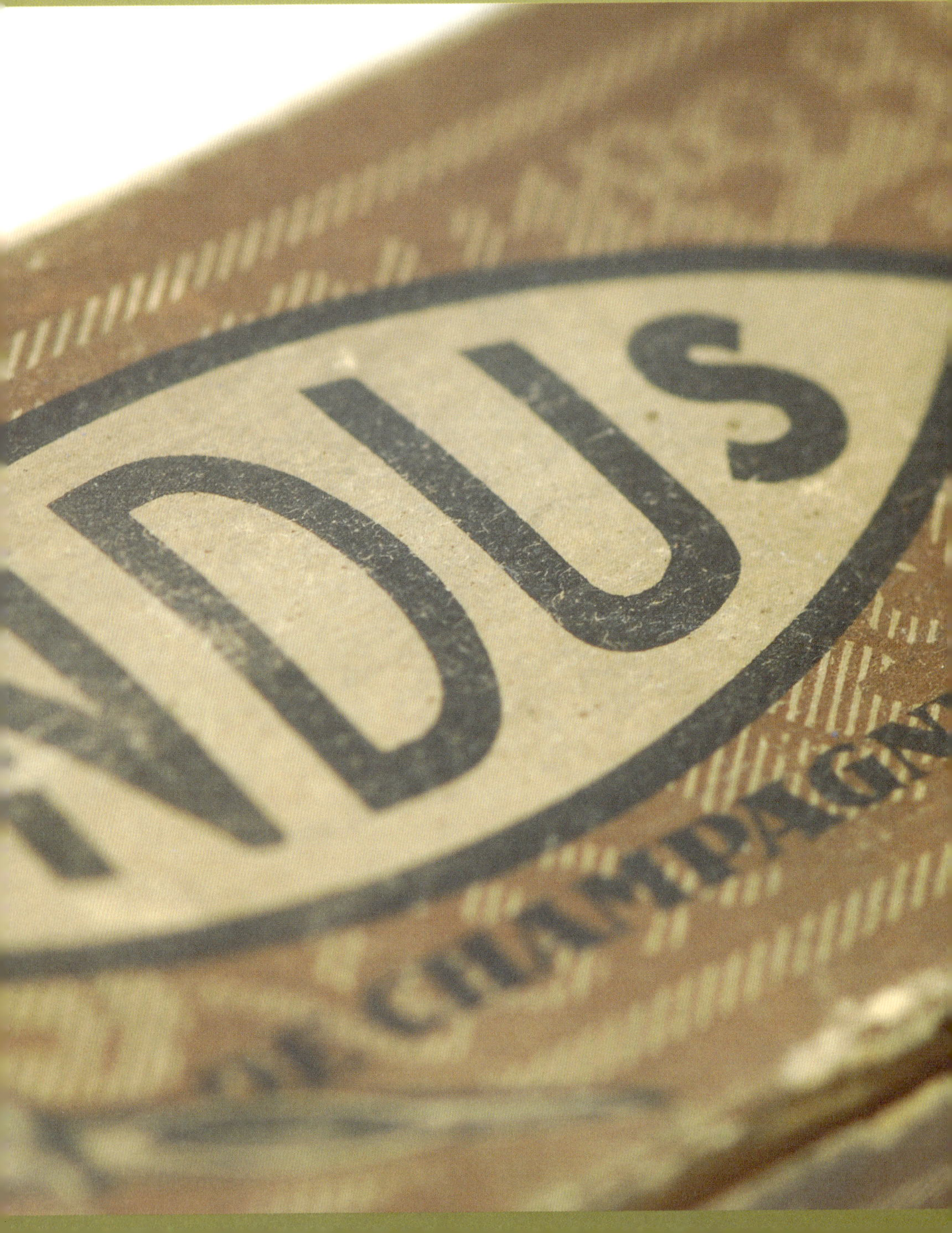
DUS
CHAMPAGN

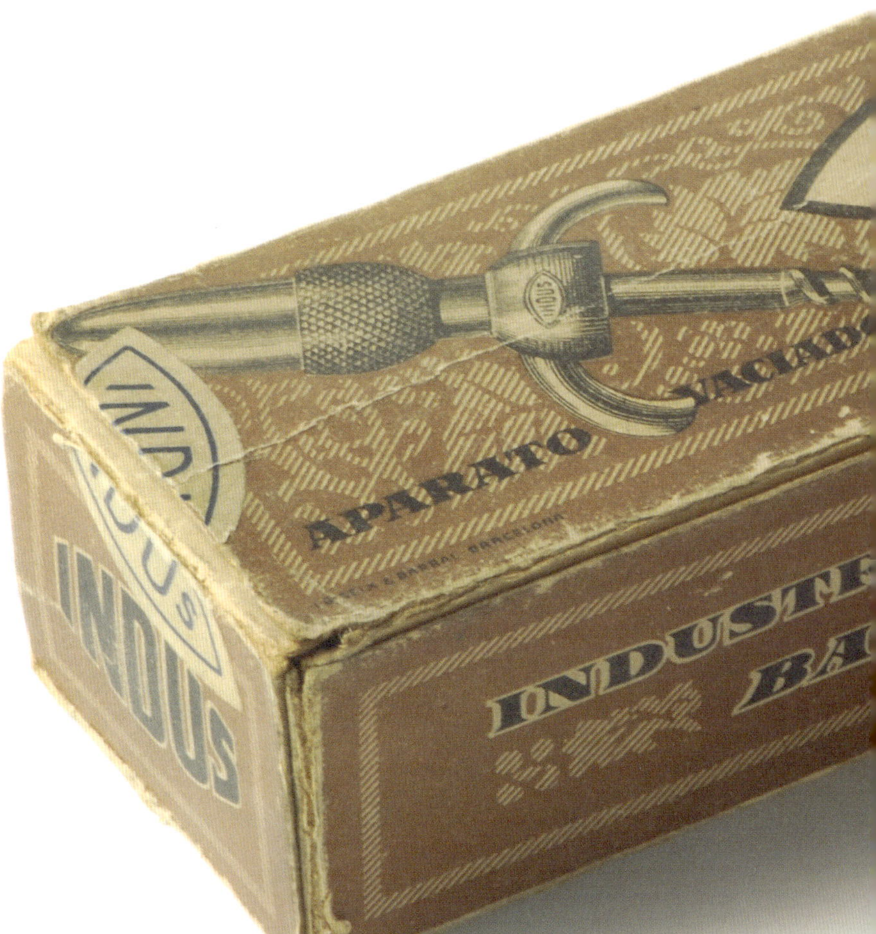
APARATO
INDUS

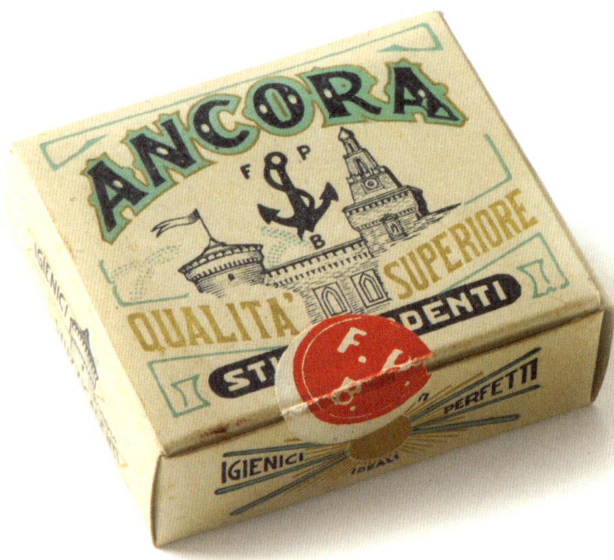
ANCORA
QUALITA' SUPERIORE
IGIENICI
IDEALI
PERFETTI

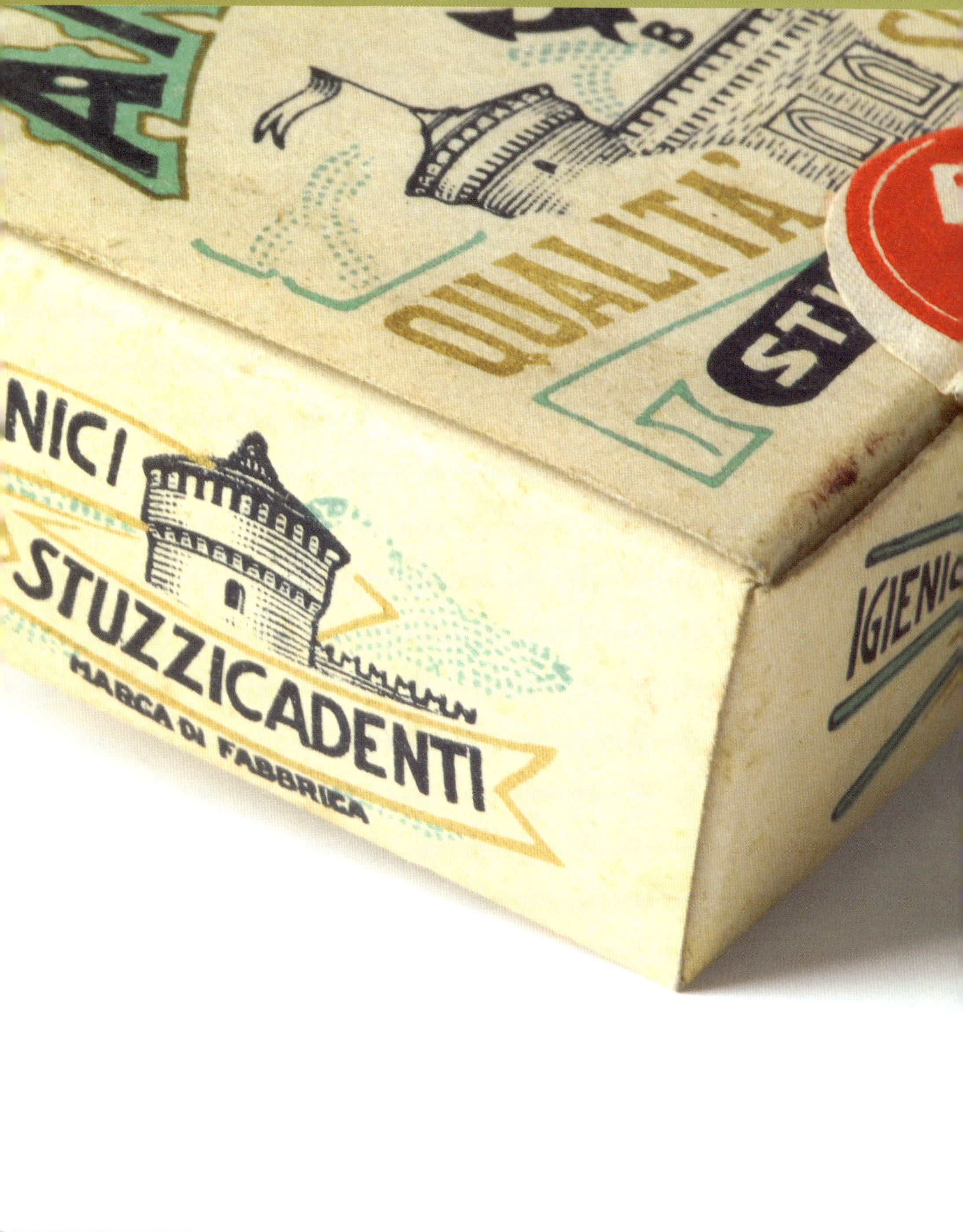
QUALITA'
STUZZICADENTI
MARCA DI FABBRICA

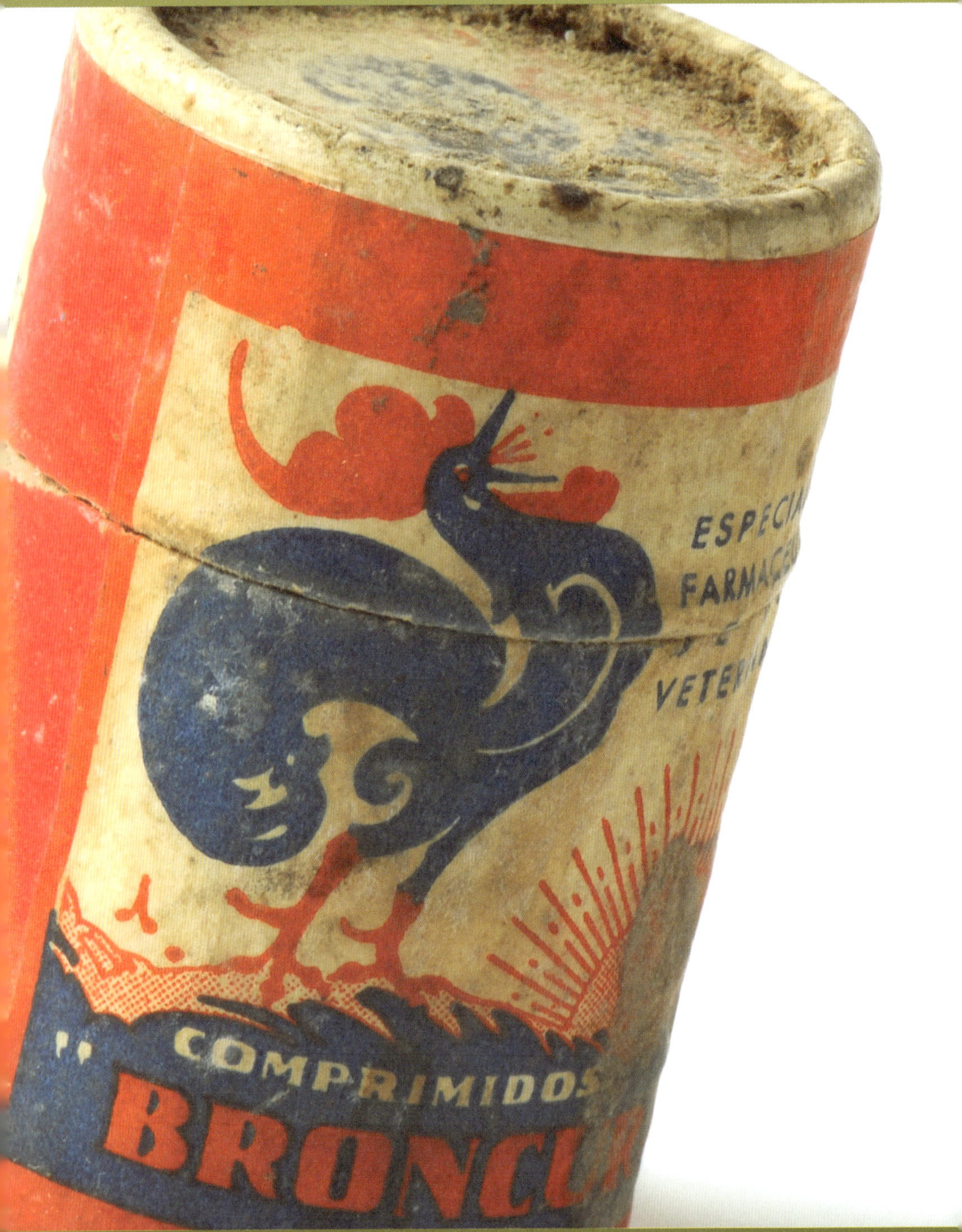
COMPRIMIDOS

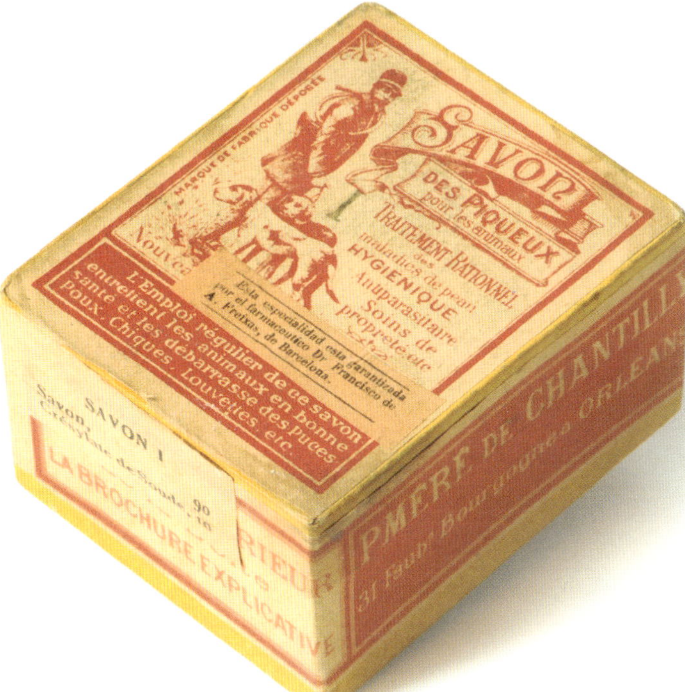
MARQUE DE FABRIQUE DÉPOSÉE
SAVON
DES PIQUEUX
pour les animaux
TRAITEMENT RATIONNEL
des
maladies de peau
HYGIÉNIQUE
Antiparasitaire
Soins de
propreté etc
Esta especialidad esta garantizada
por el farmacéutico Dr Francisco de
A. Freixas, de Barcelona.
L'Emploi régulier de ce savon
entretient les animaux en bonne
santé et les débarrasse des puces
Poux, Chiques, Louvettes, etc.
SAVON I
Savon
Crésylate de Soude
90
10
LA BROCHURE EXPLICATIVE
PMÈRE DE CHANTILLY
31 Faubg Bourgogne ORLEANS

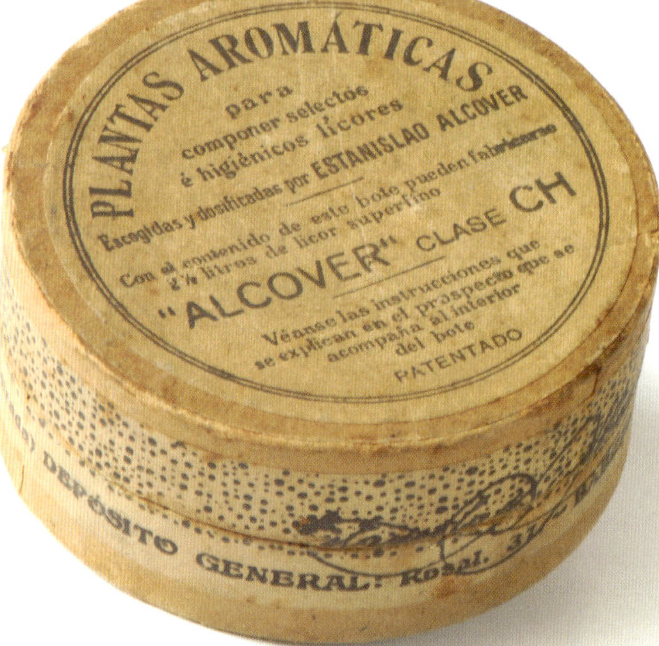
PLANTAS AROMÁTICAS
para
componer selectos
é higiénicos licores
Escogidas y dosificadas por ESTANISLAO ALCOVER
Con el contenido de este bote pueden fabricarse
2½ litros de licor superfino
"ALCOVER" CLASE CH
Véanse las instrucciones que
se explican en el prospecto que se
acompaña al interior
del bote
PATENTADO
DEPÓSITO GENERAL: Rosal. 31

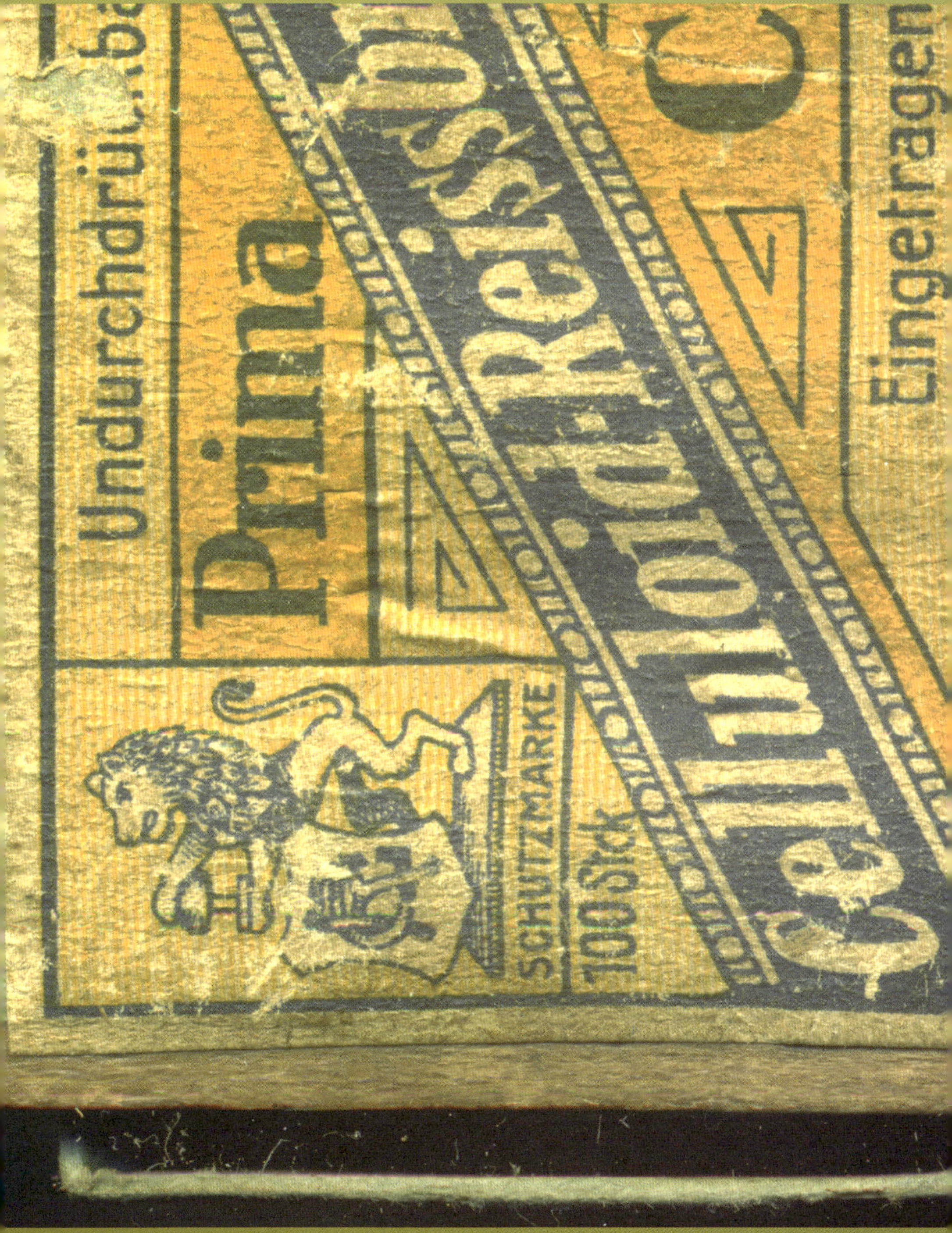
Prima
SCHUTZMARKE
100 Stck
Eingetragen

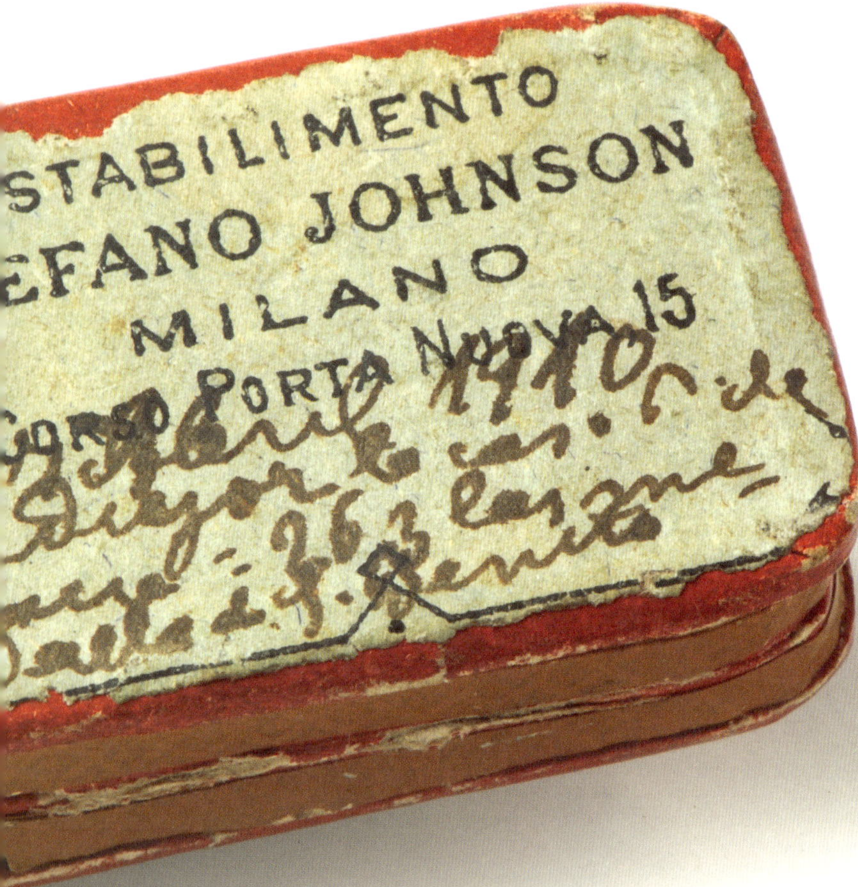
STABILIMENTO
EFANO JOHNSON
MILANO
CORSO PORTA NUOVA 15

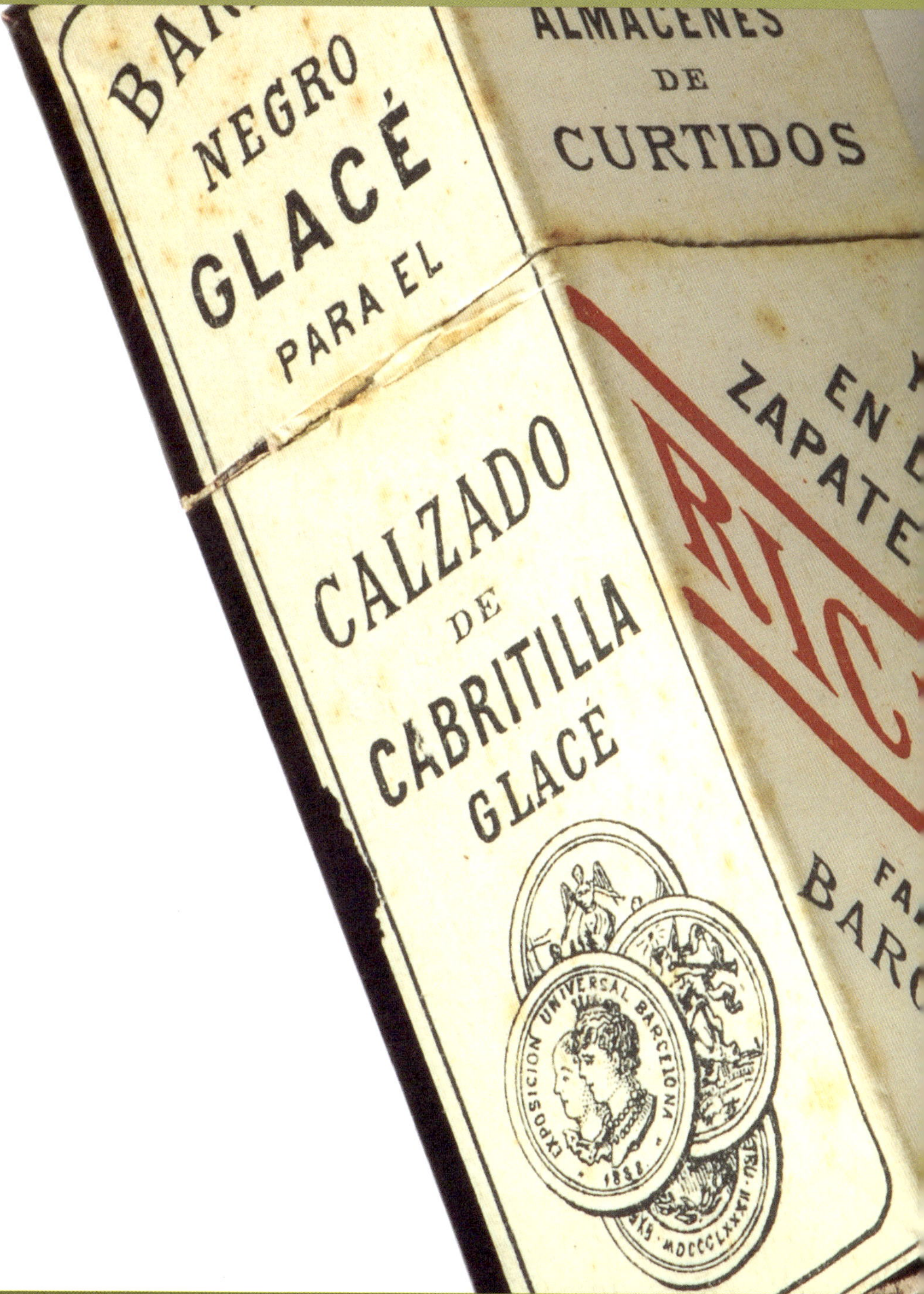
NEGRO
GLACÉ
PARA EL
ALMACENES
DE
CURTIDOS
CALZADO
DE
CABRITILLA
GLACÉ
EXPOSICION UNIVERSAL BARCELONA
1888

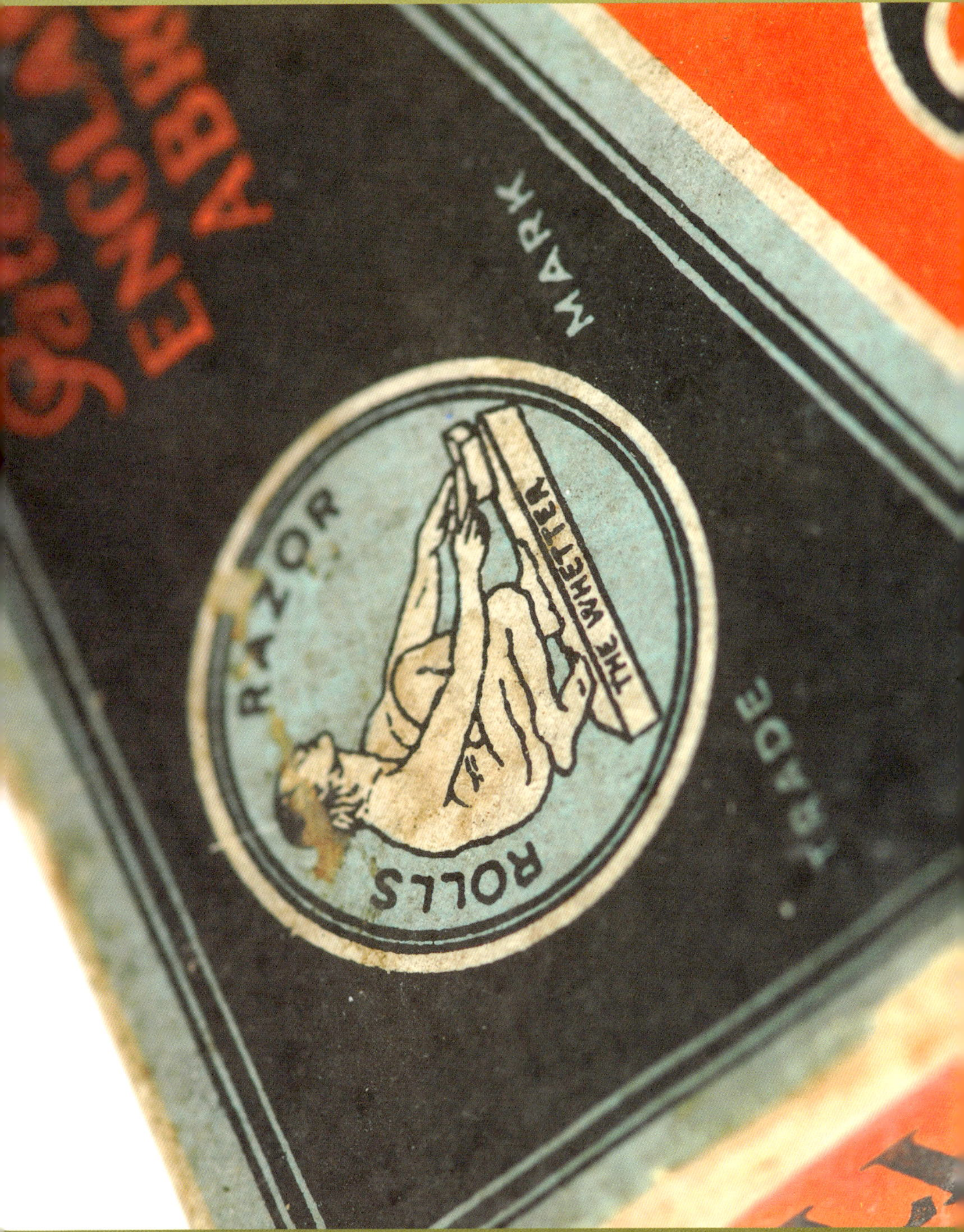
MARK
RAZOR
THE WHETTER
ROLLS
TRADE

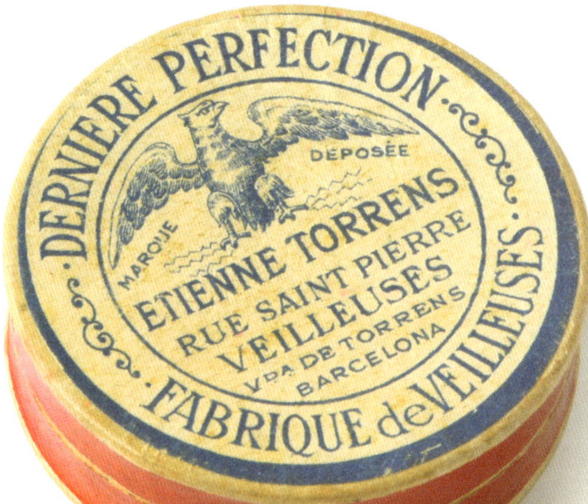
·DERNIERE PERFECTION·
MARQUE
DÉPOSÉE
ETIENNE TORRENS
RUE SAINT PIERRE
VEILLEUSES
Vda DE TORRENS
BARCELONA
·FABRIQUE de VEILLEUSES·

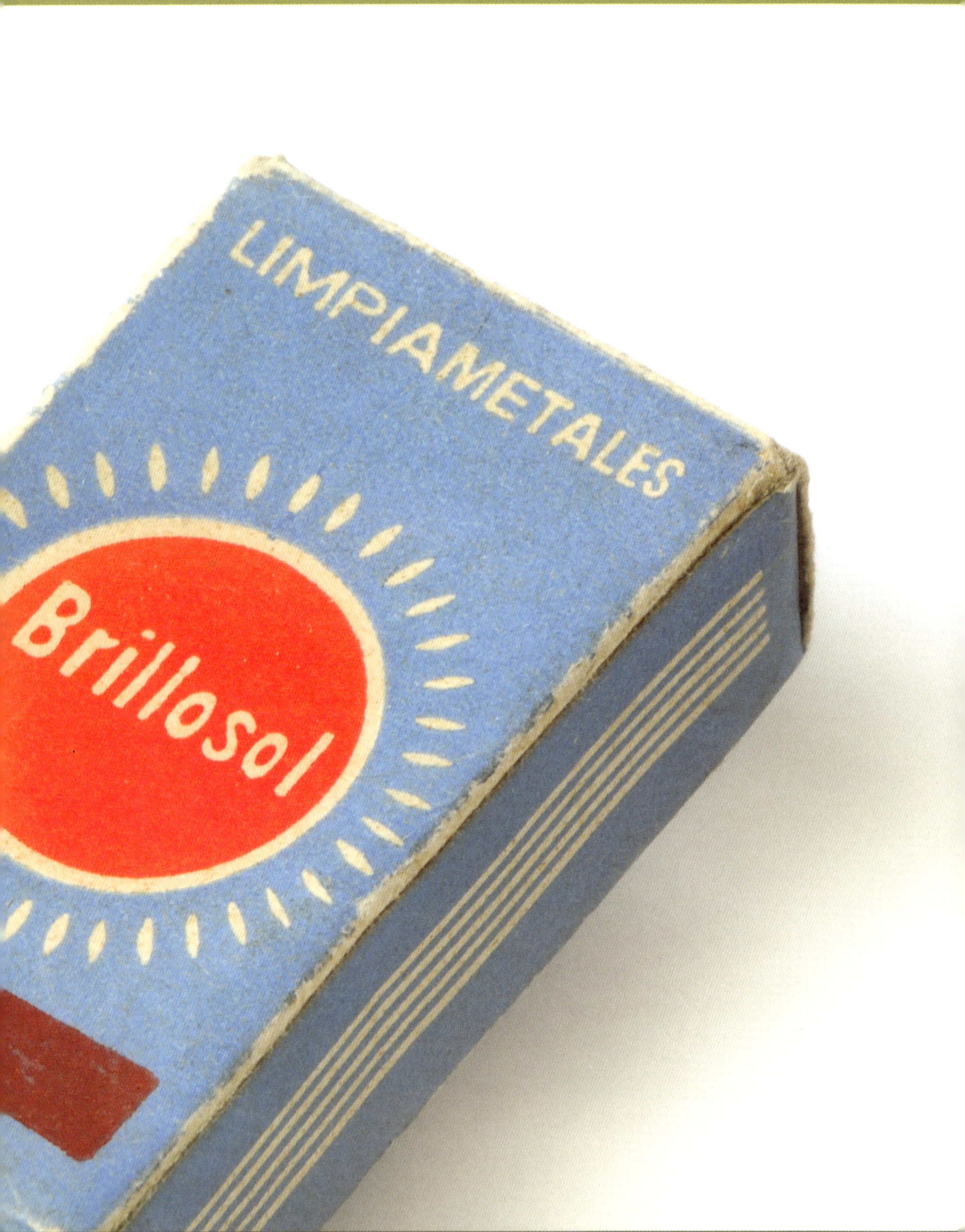
LIMPIAMETALES
Brillosol

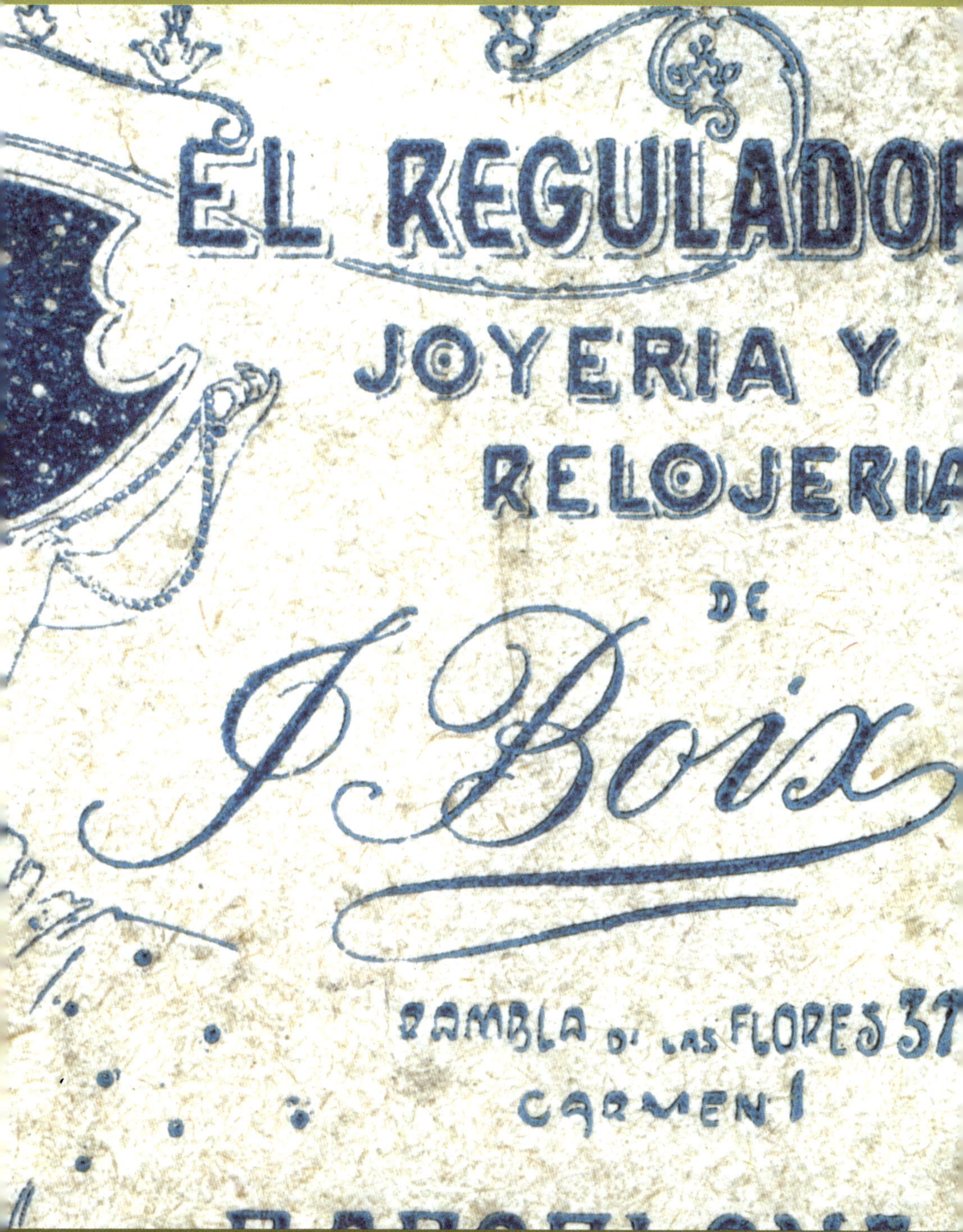
EL REGULADOR
JOYERIA Y
RELOJERIA
DE
J. Boix
RAMBLA D' LAS FLORES 37
CARMEN 1

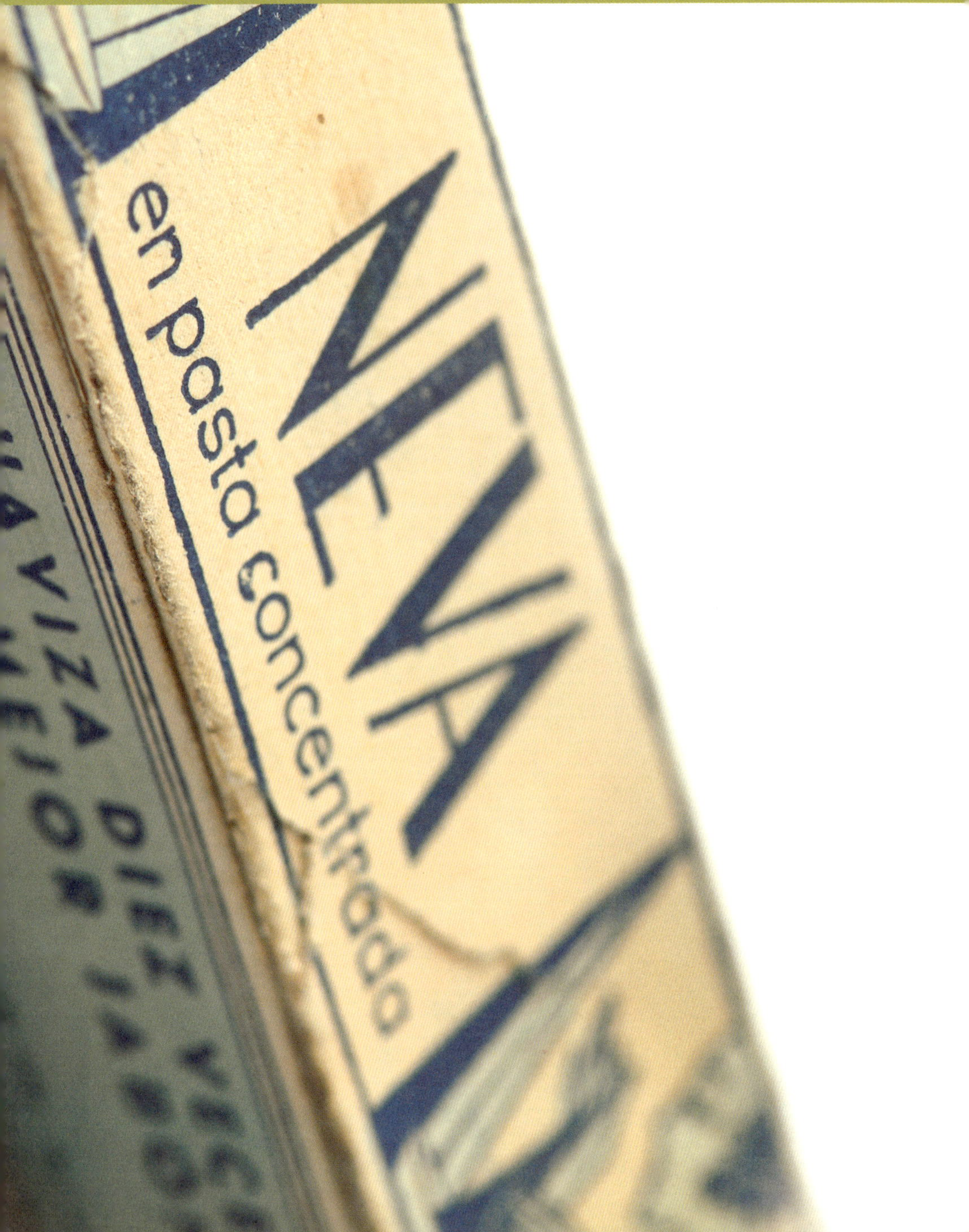
NEVA
en pasta concentrada

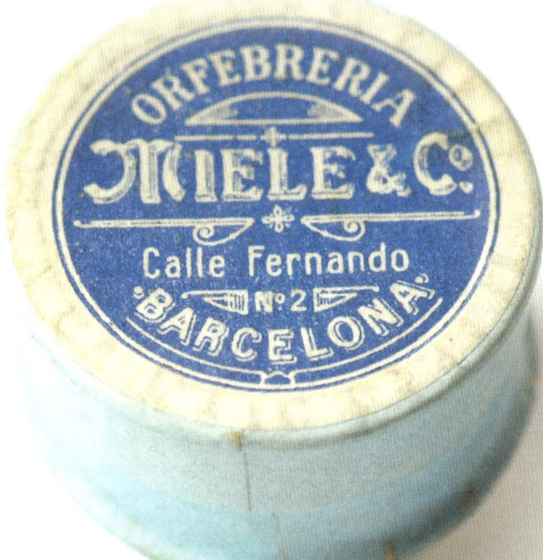
ORFEBRERIA
MIELE & Co.
Calle Fernando
Nº 2
BARCELONA

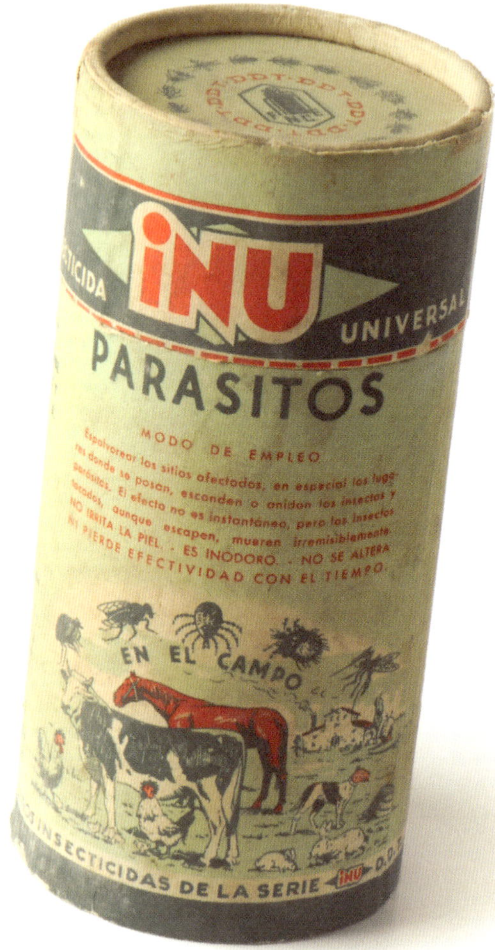
INU
UNIVERSAL
PARASITOS
MODO DE EMPLEO
EN EL CAMPO
INSECTICIDAS DE LA SERIE INU

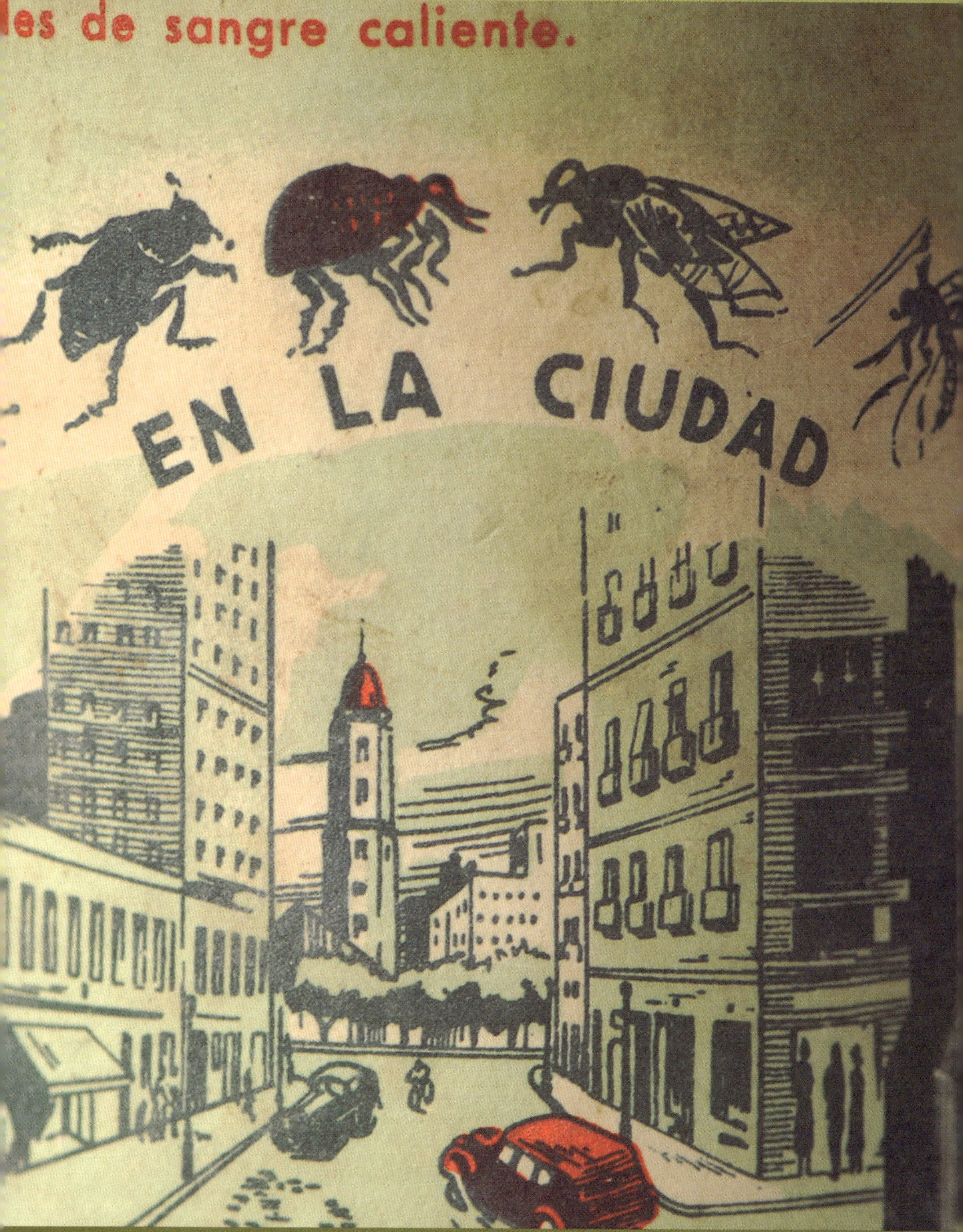
des de sangre caliente.
EN LA CIUDAD

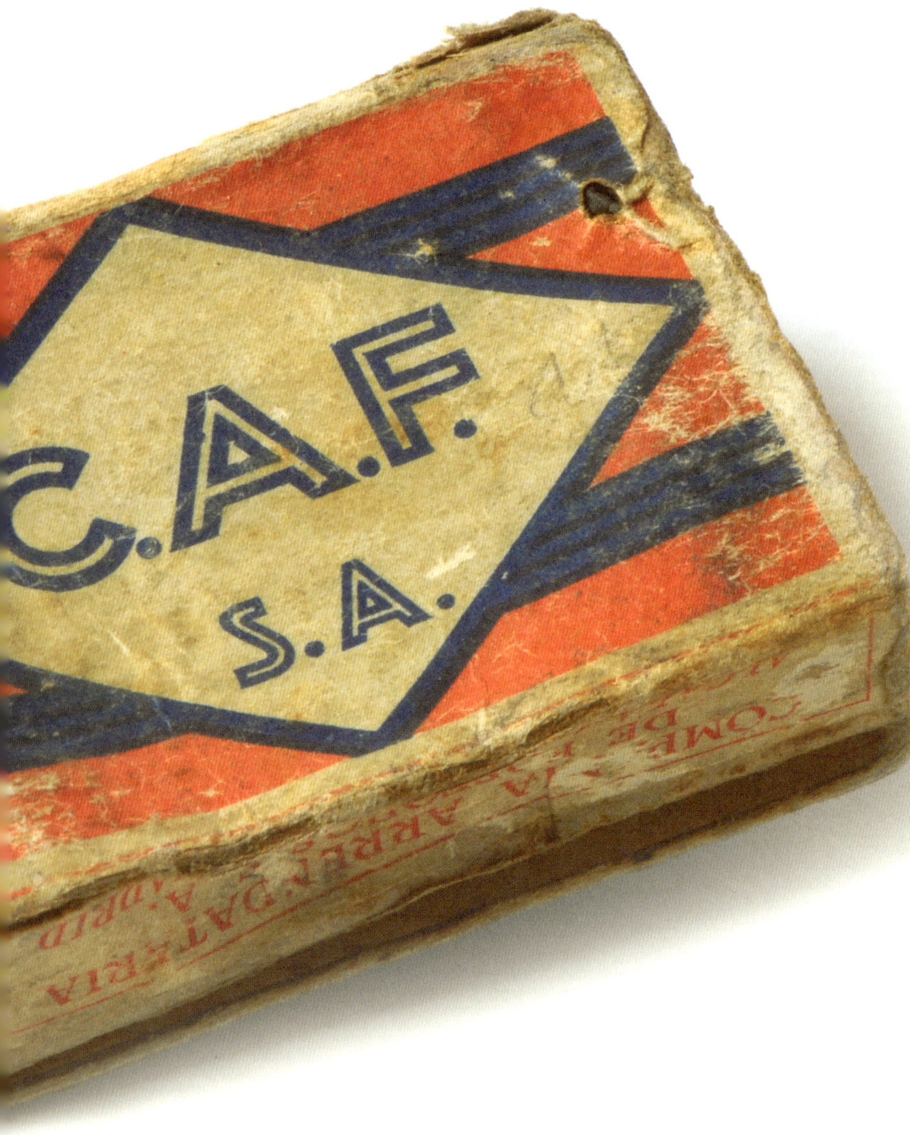
C.A.F.
S.A.

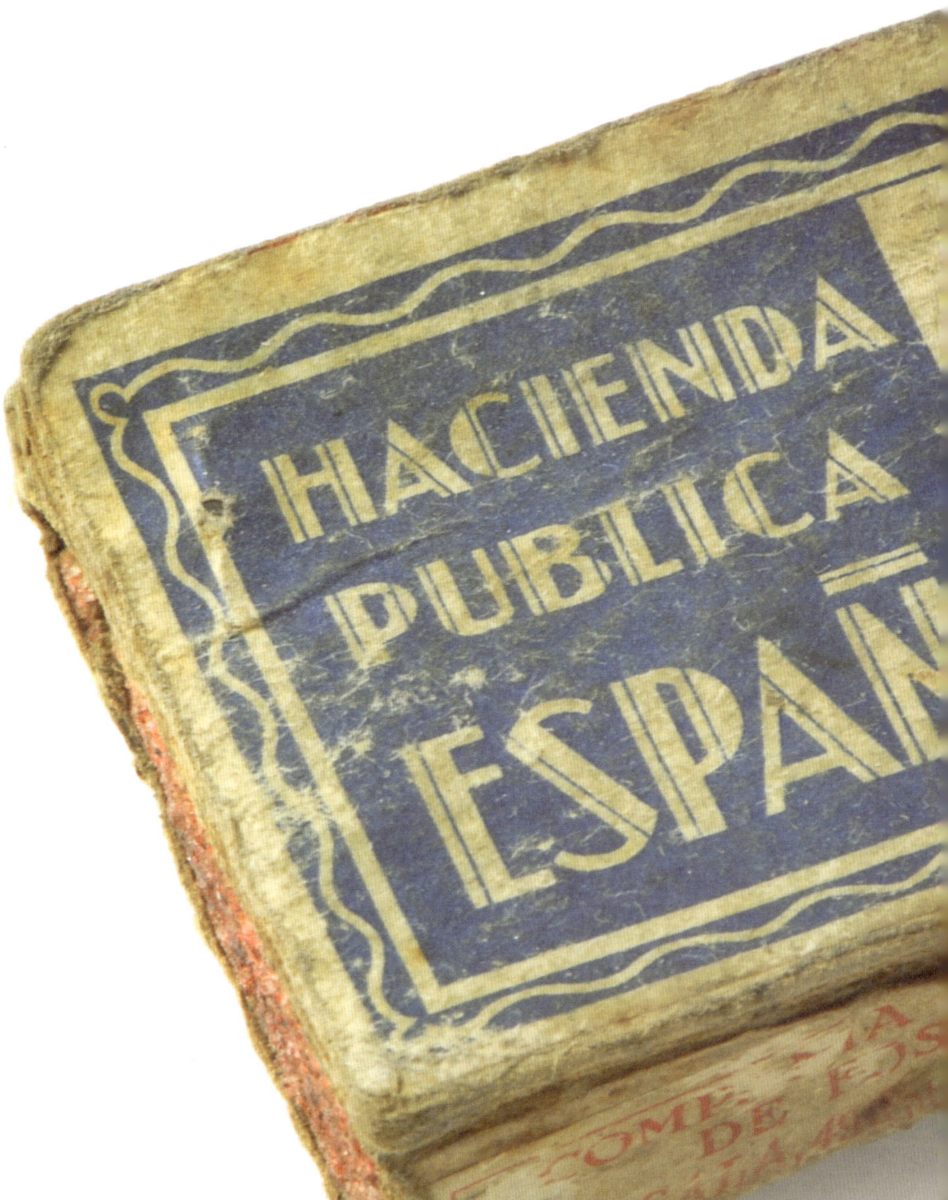
HACIENDA
PUBLICA
ESPAÑ

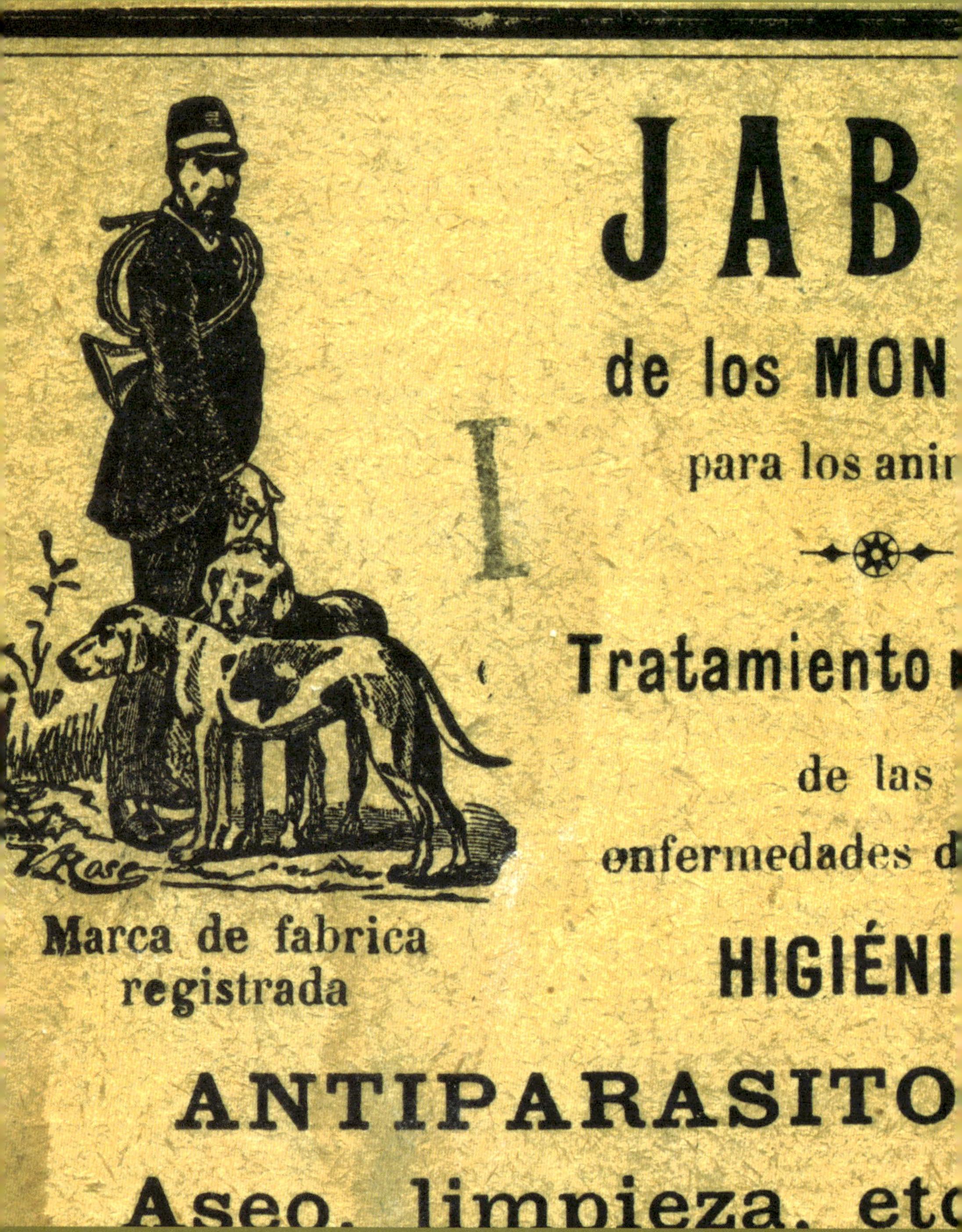
JAB
de los MON
para los ani
I
Tratamiento
de las
enfermedades d
HIGIÉNI
Marca de fabrica
registrada
ANTIPARASITO
Aseo, limpieza, et

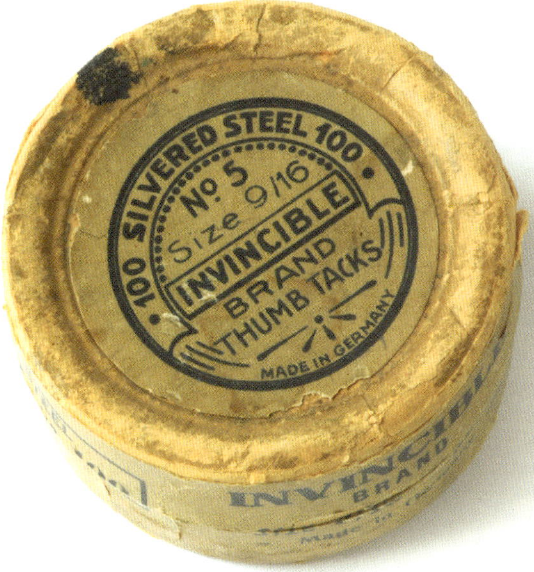
100 SILVERED STEEL 100
No. 5
Size 9/16
INVINCIBLE
BRAND
THUMB TACKS
MADE IN GERMANY
INVINCIBLE
BRAND

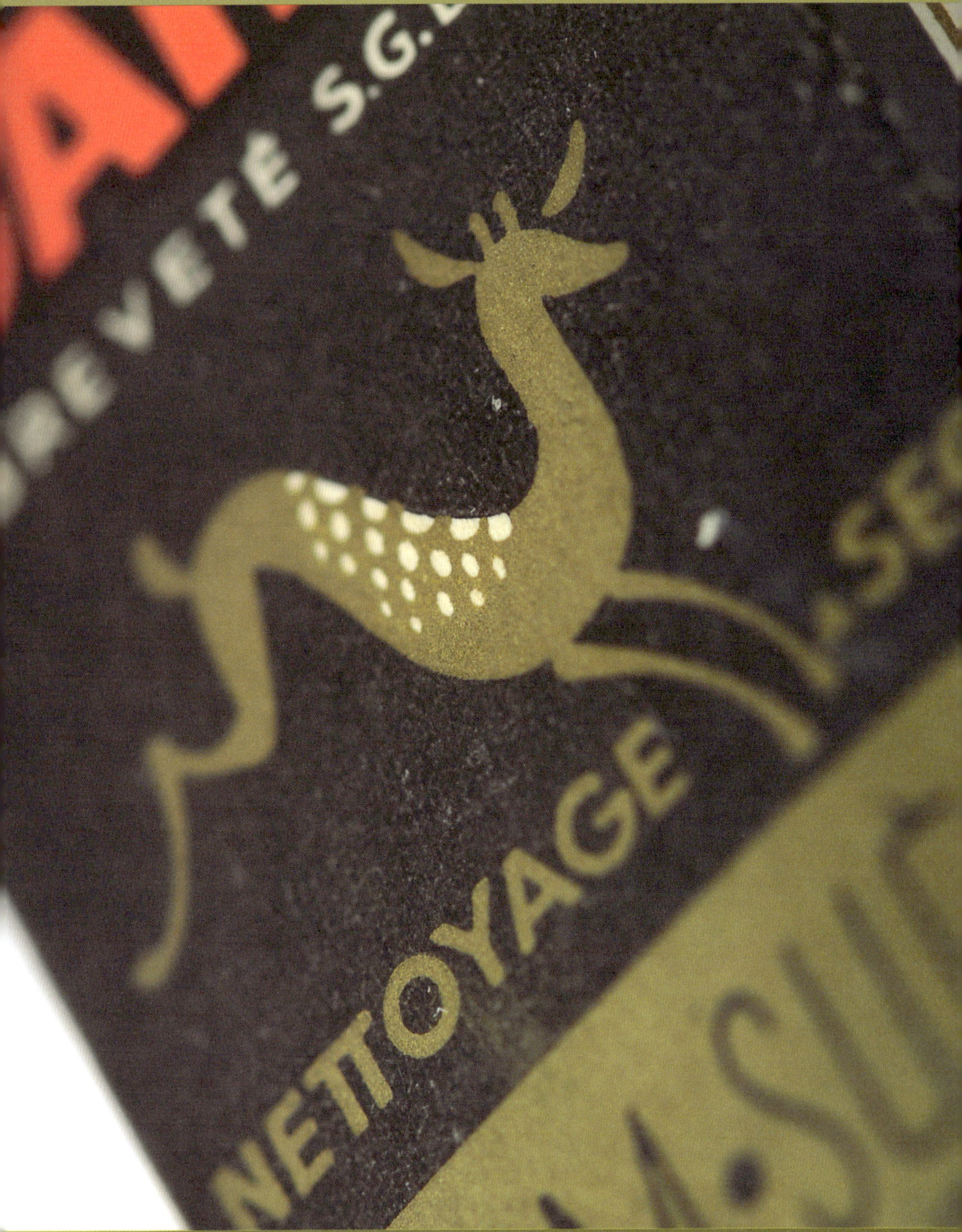
S.G.
BREVETÉ
NETTOYAGE

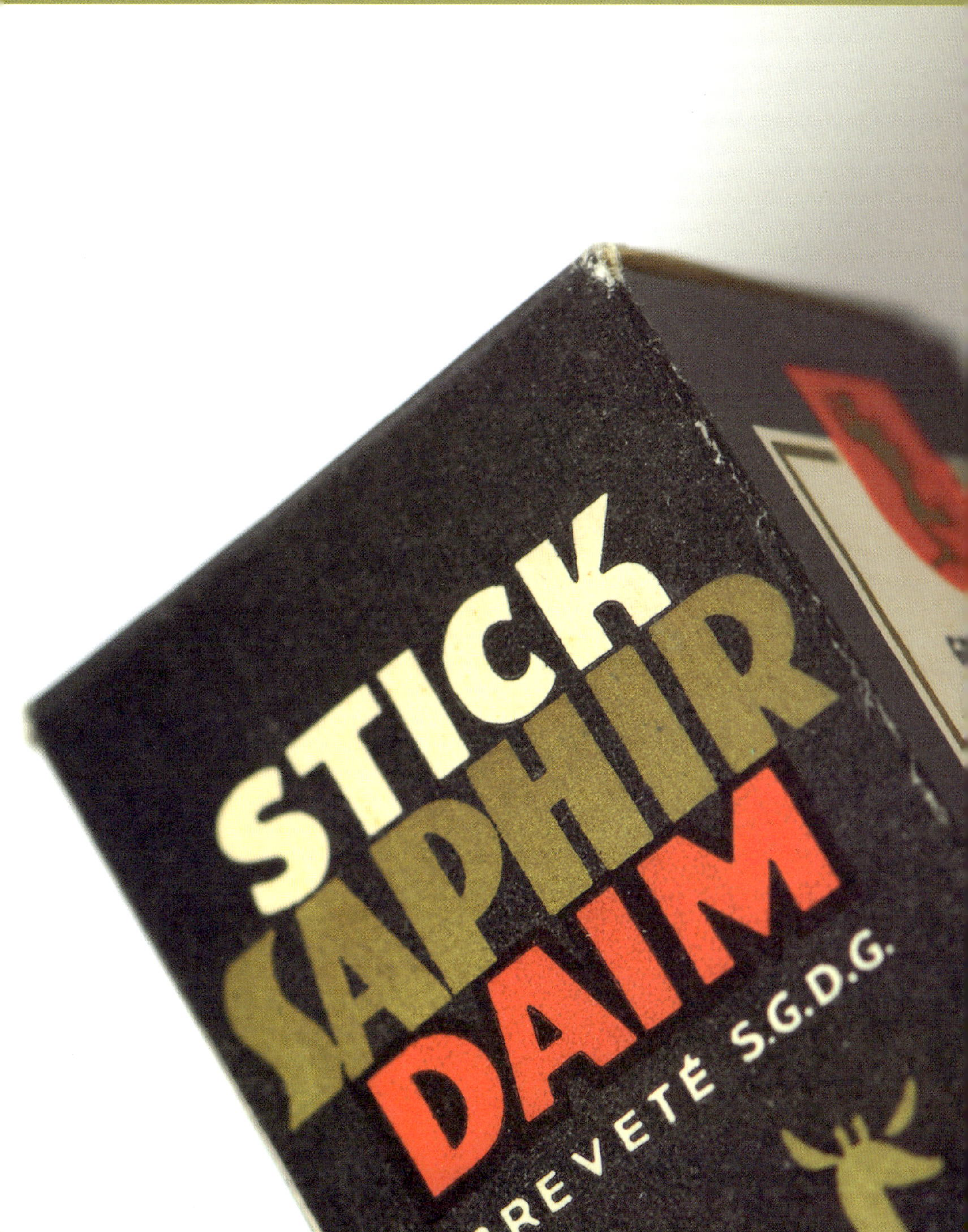
STICK
SAPHIR
DAIM
REVETÉ S.G.D.G.

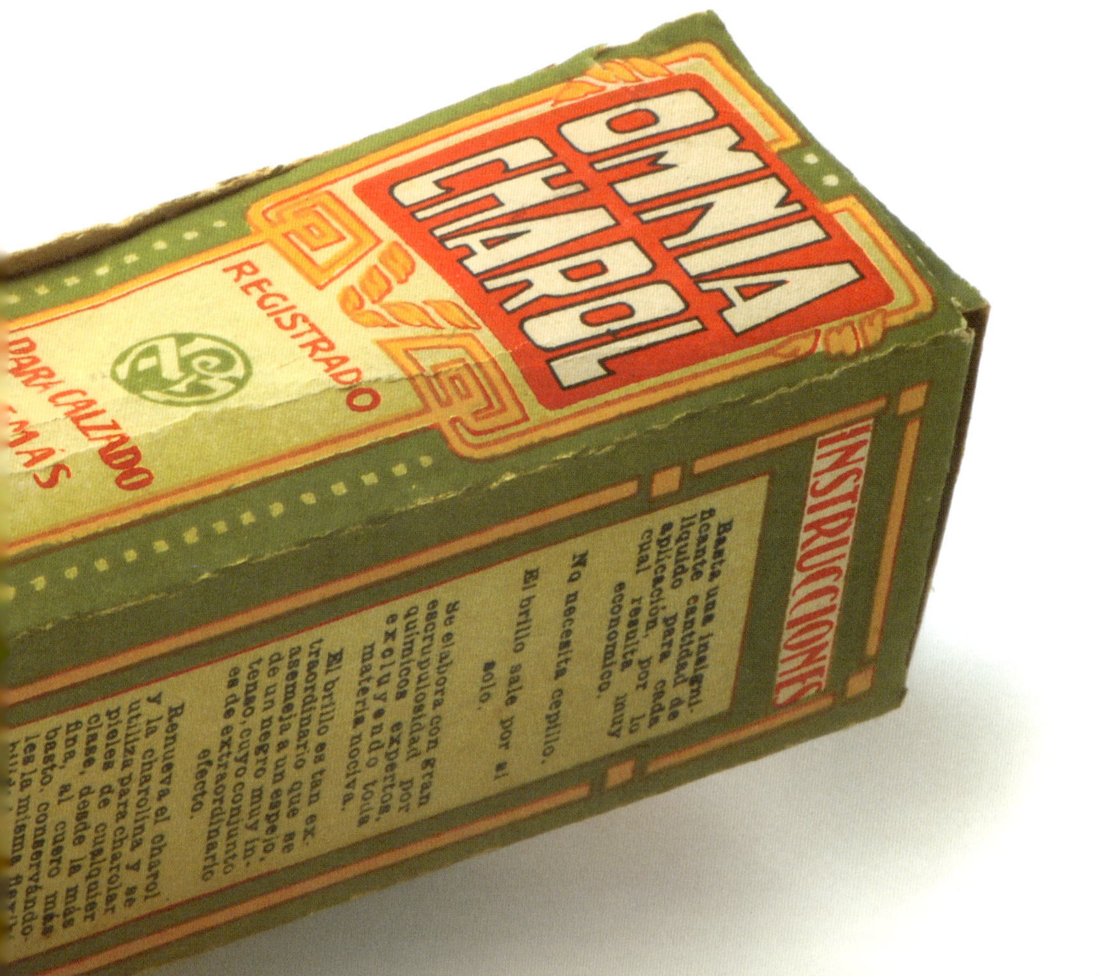
OMNIA
CHAROL
REGISTRADO
PARA CALZADO
INSTRUCCIONES

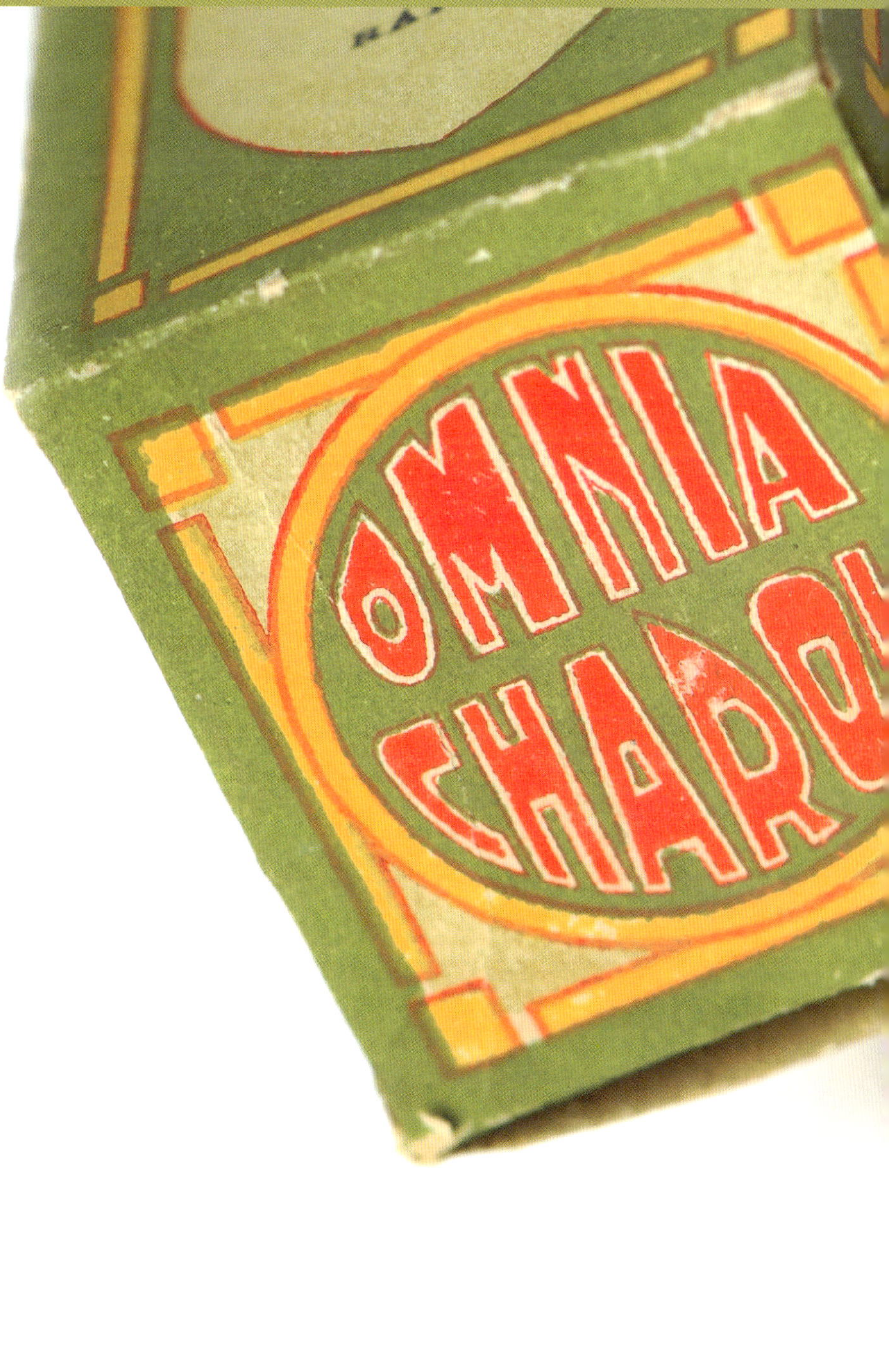
OMNIA

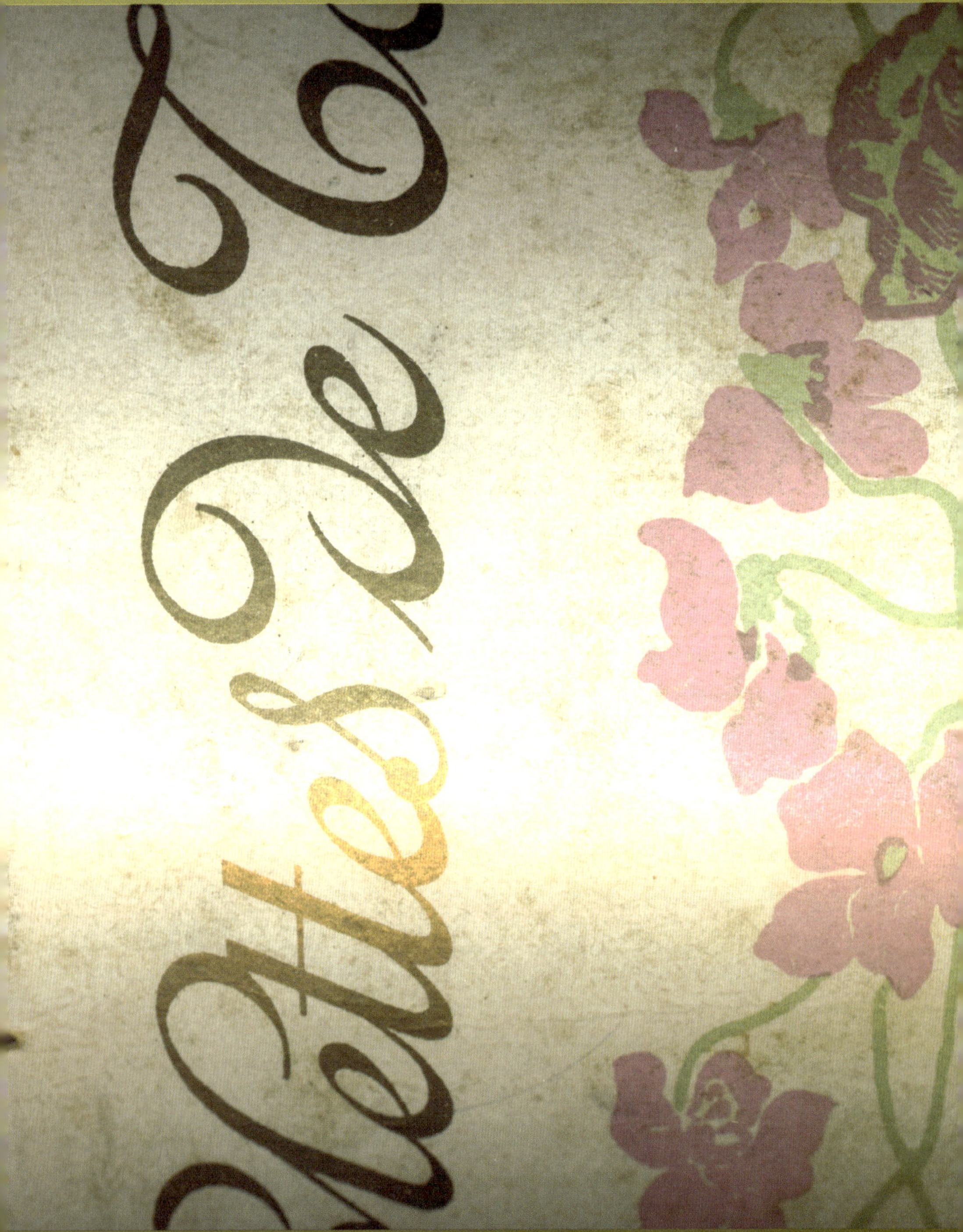

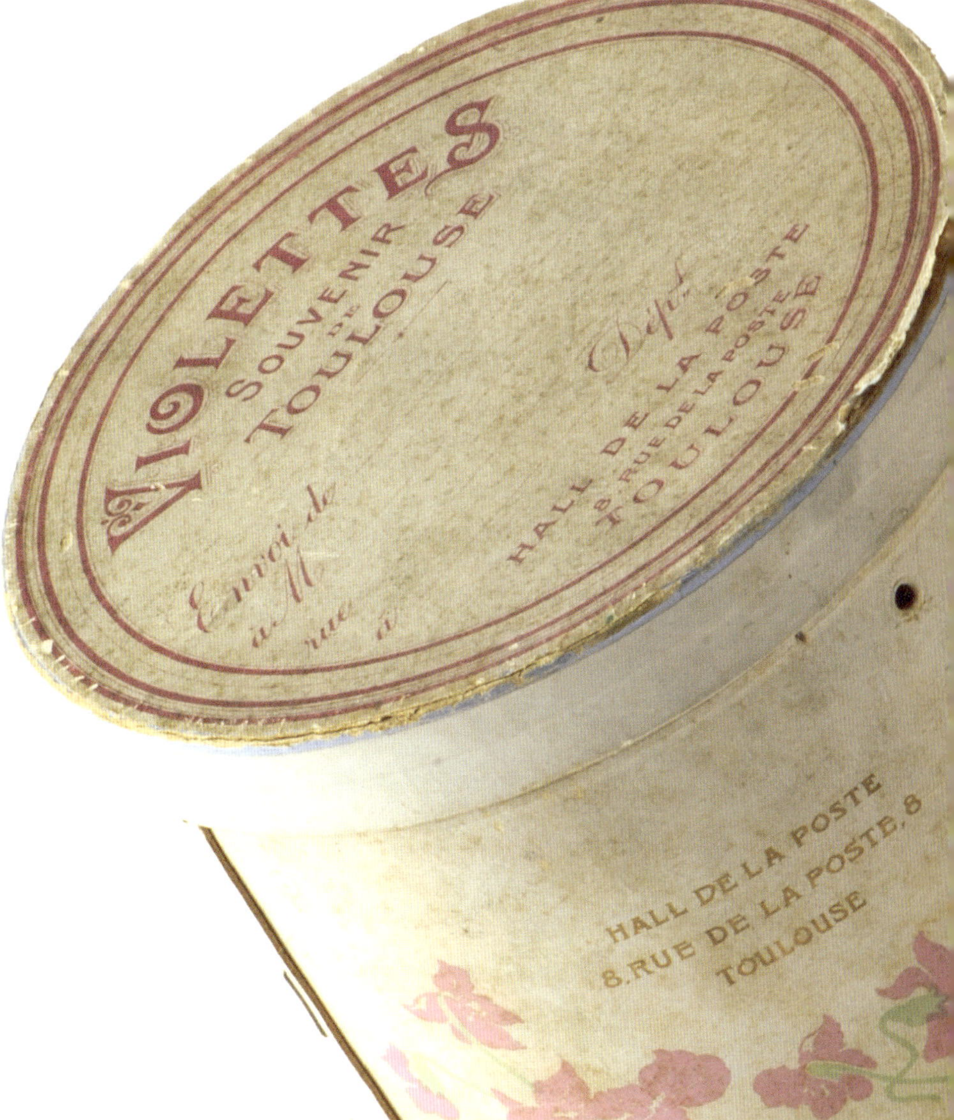
VIOLETTES
SOUVENIR DE TOULOUSE
Envoi de
à M
rue
à
Dépôt
HALL DE LA POSTE
8 RUE DE LA POSTE
TOULOUSE
HALL DE LA POSTE
8, RUE DE LA POSTE, 8
TOULOUSE

OODFO

OODFOODFOO

FOOD

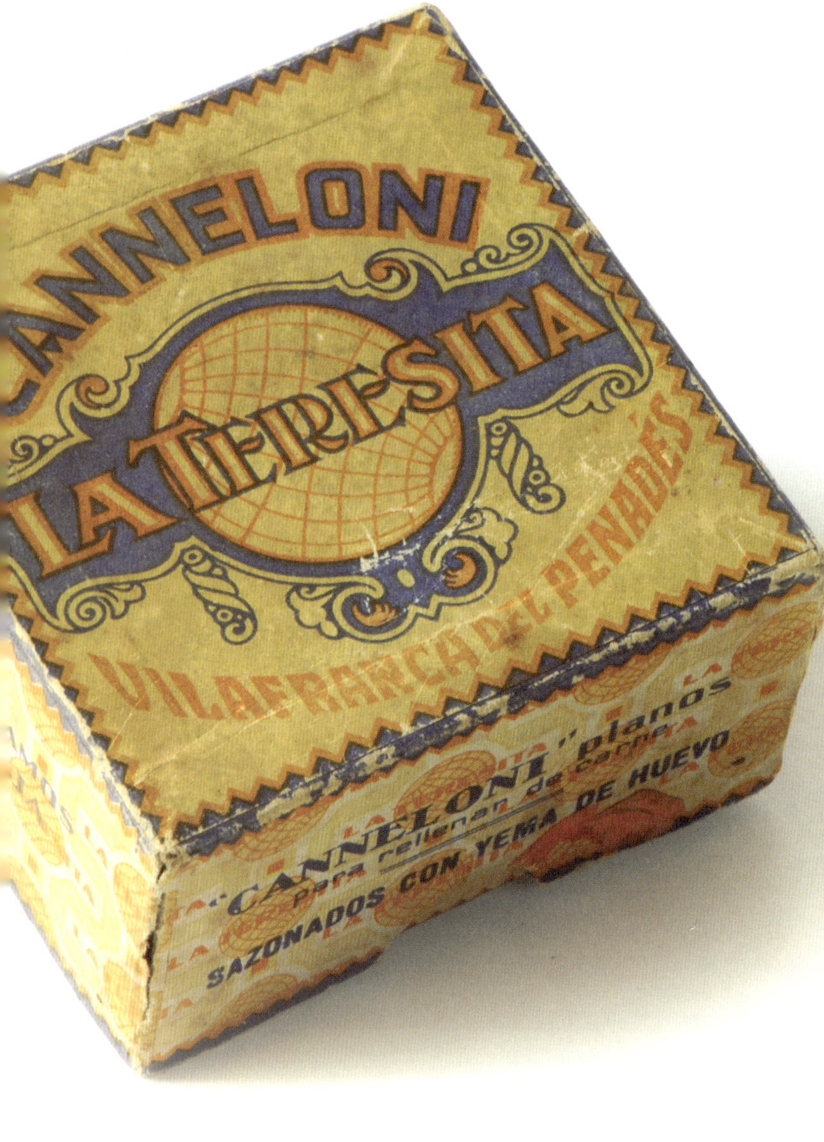
CANNELONI
LA TERESITA
VILAFRANCA DEL PENADÉS
"CANNELONI" planos
para rellenar de carne
SAZONADOS CON YEMA DE HUEVO

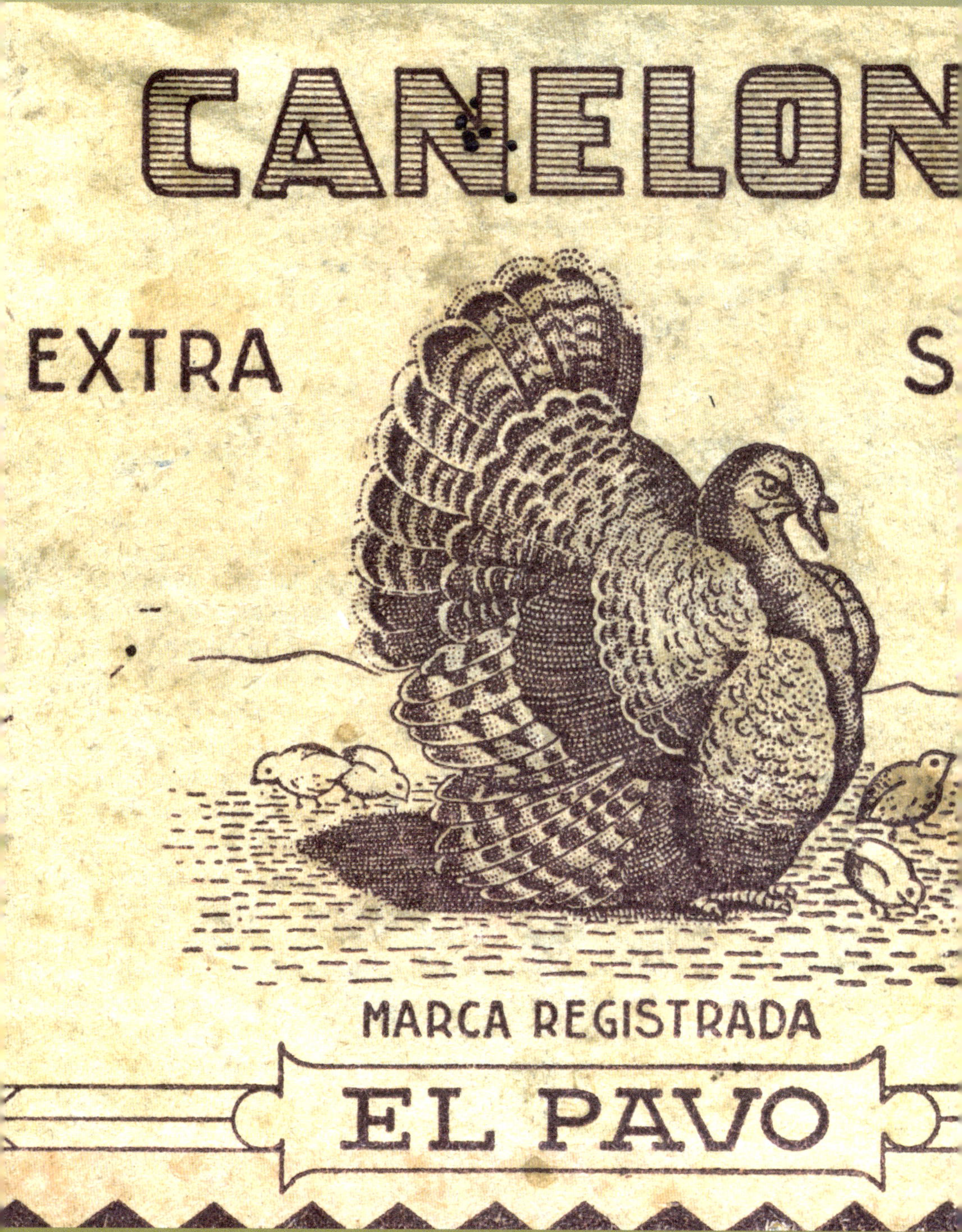
CANELON
EXTRA
S
MARCA REGISTRADA
EL PAVO

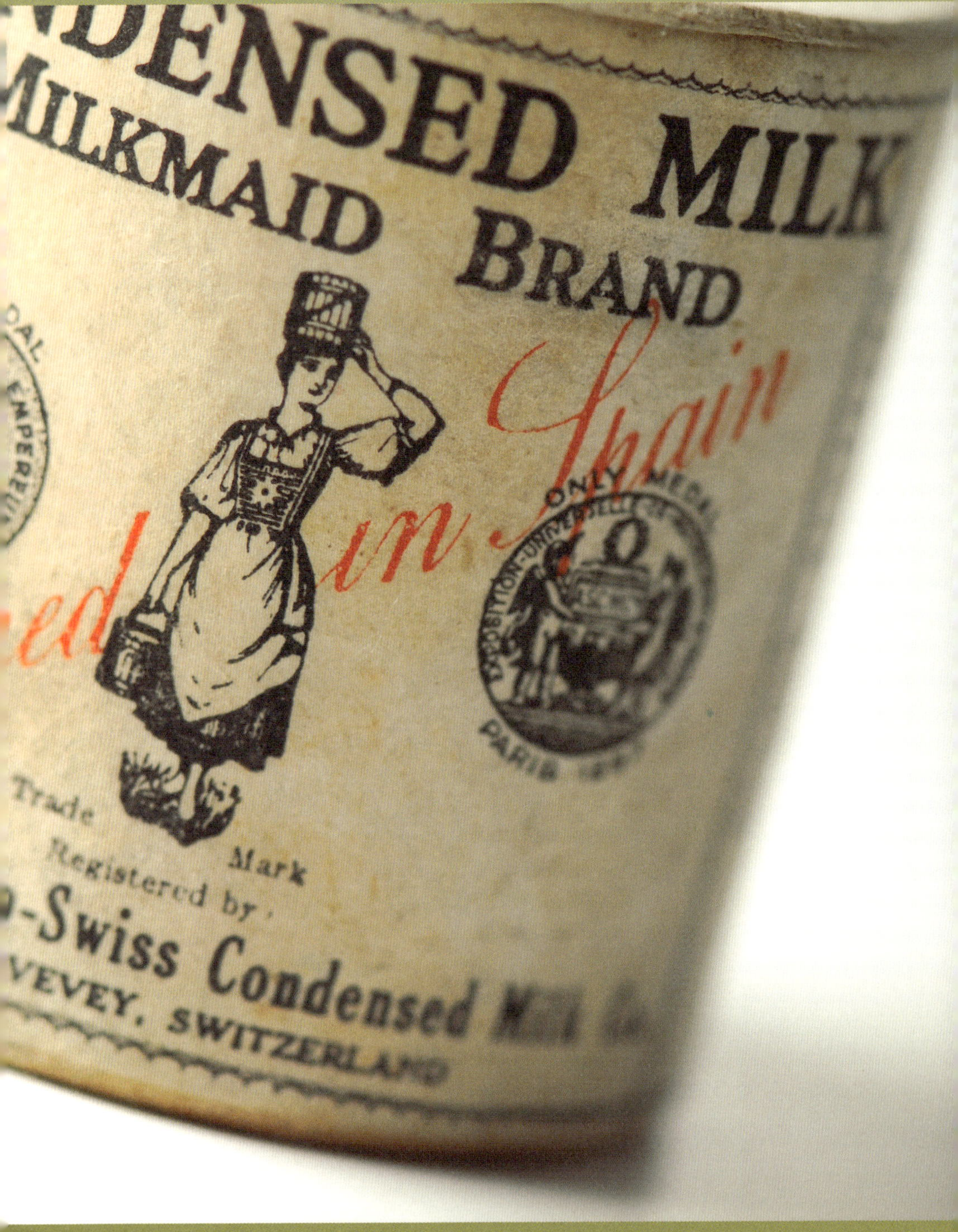
NDENSED MILK
MILKMAID BRAND
in Spain
Trade Mark
Registered by
-Swiss Condensed
VEVEY, SWITZERLAND
ONLY MEDAL
PARIS

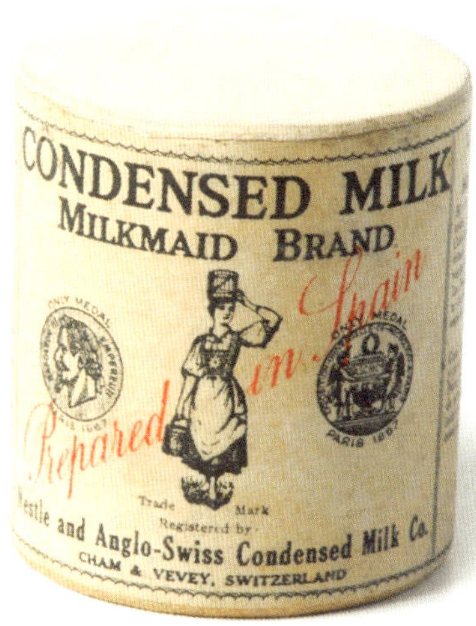
CONDENSED MILK
MILKMAID BRAND
Prepared in Spain
ONLY MEDAL
PARIS 1867
Trade Mark
Registered by
Nestle and Anglo-Swiss Condensed Milk Co.
CHAM & VEVEY, SWITZERLAND

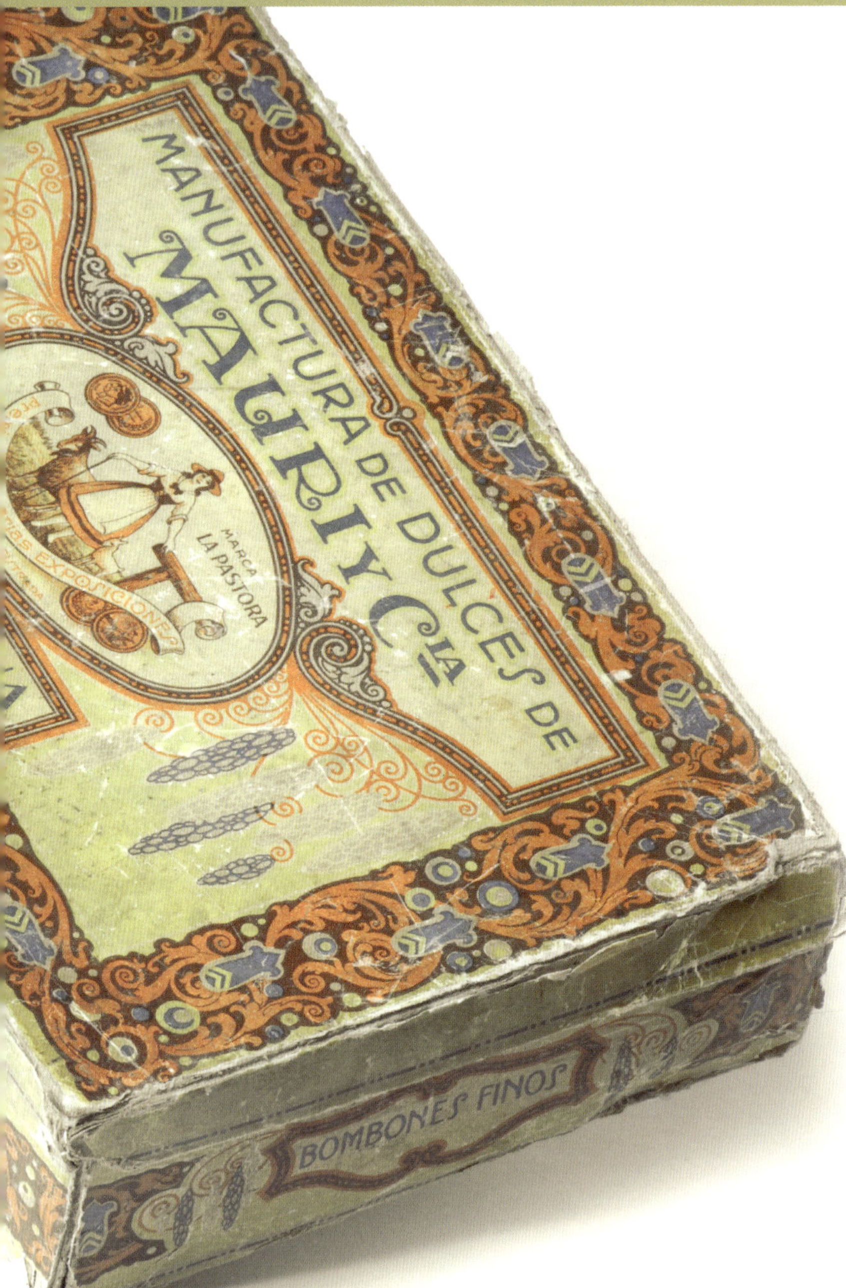
MANUFACTURA DE DULCES DE
MAURI Y Cía
MARCA
LA PASTORA
BOMBONES FINOS

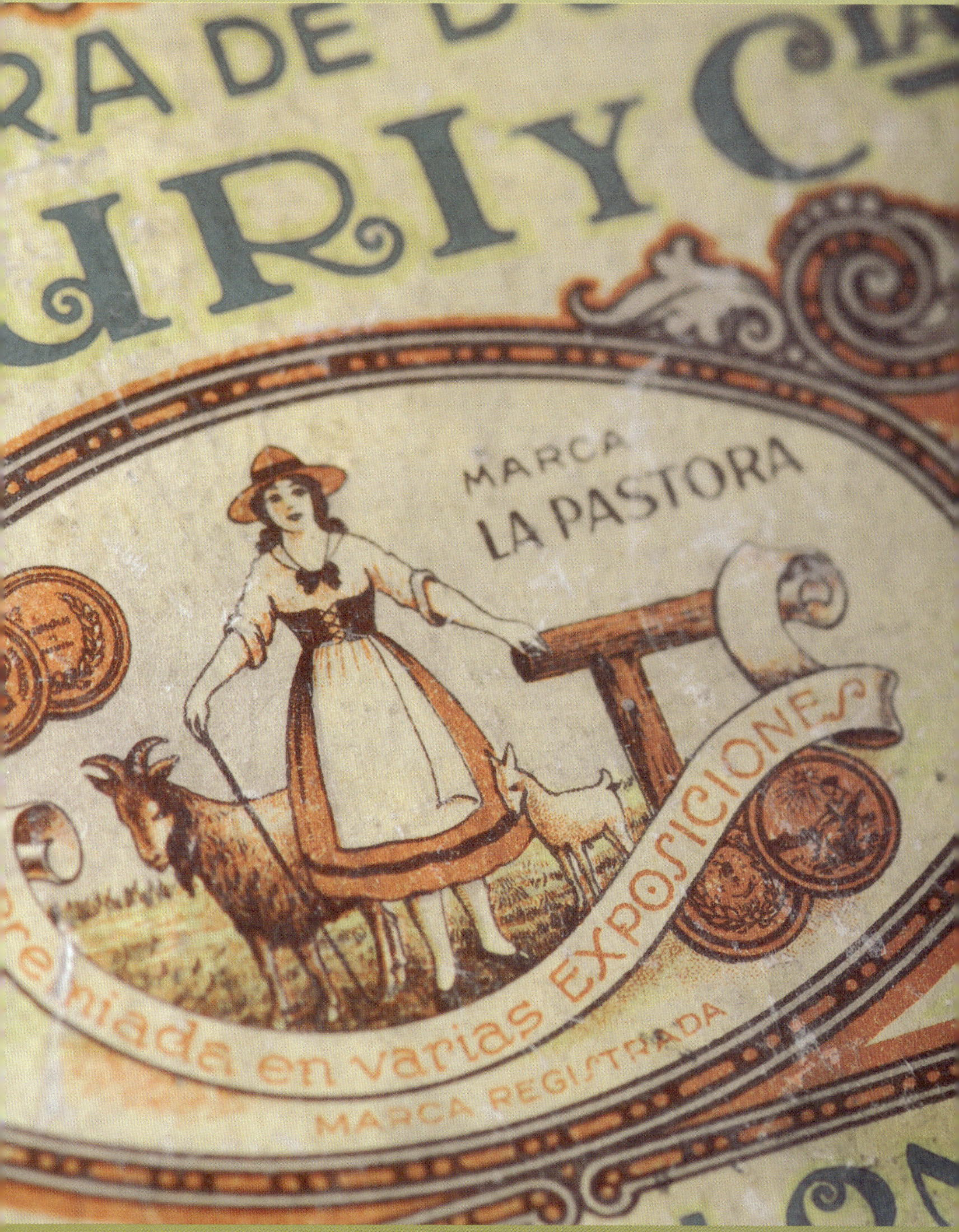
MARCA
LA PASTORA
premiada en varias EXPOSICIONES
MARCA REGISTRADA

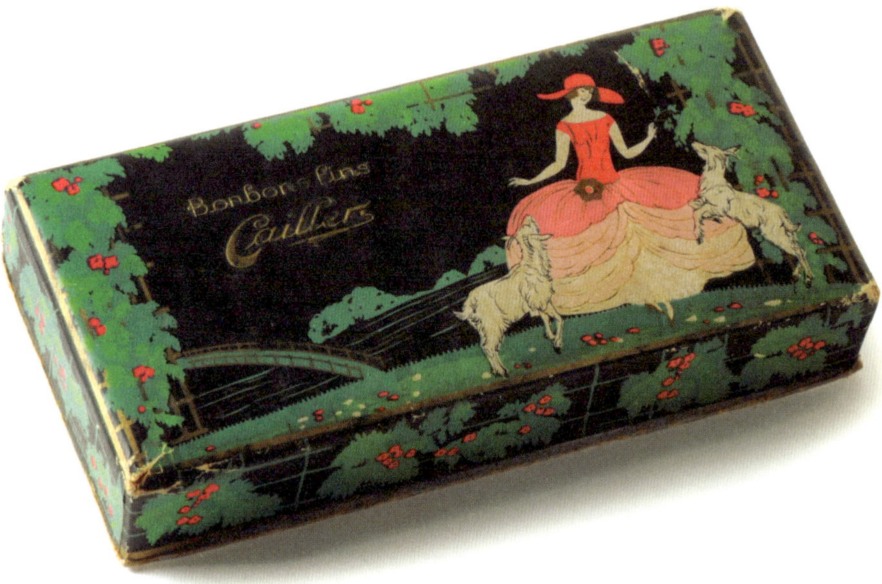
Bonbons fins
Cailler

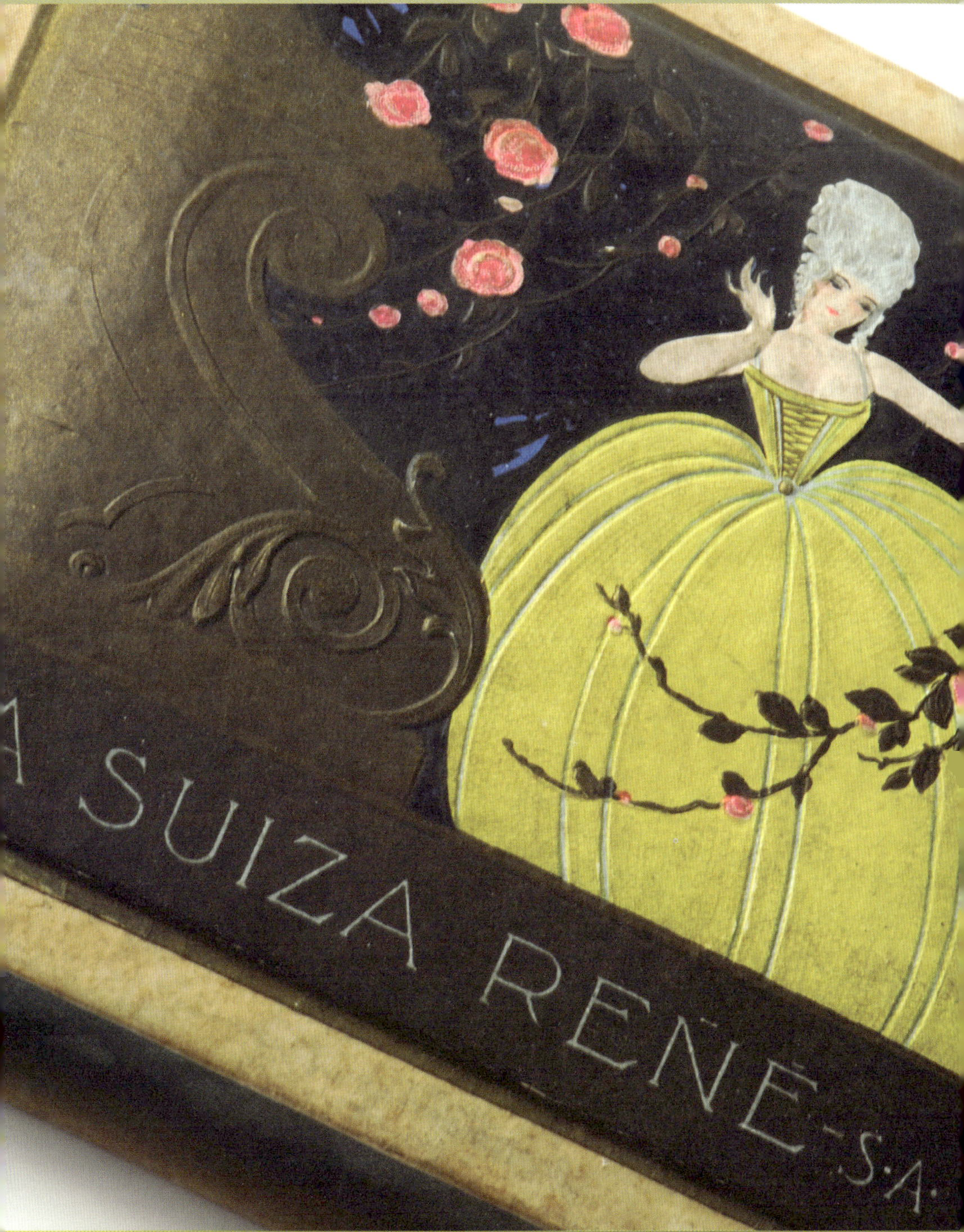
A SUIZA RENÉ S.A.

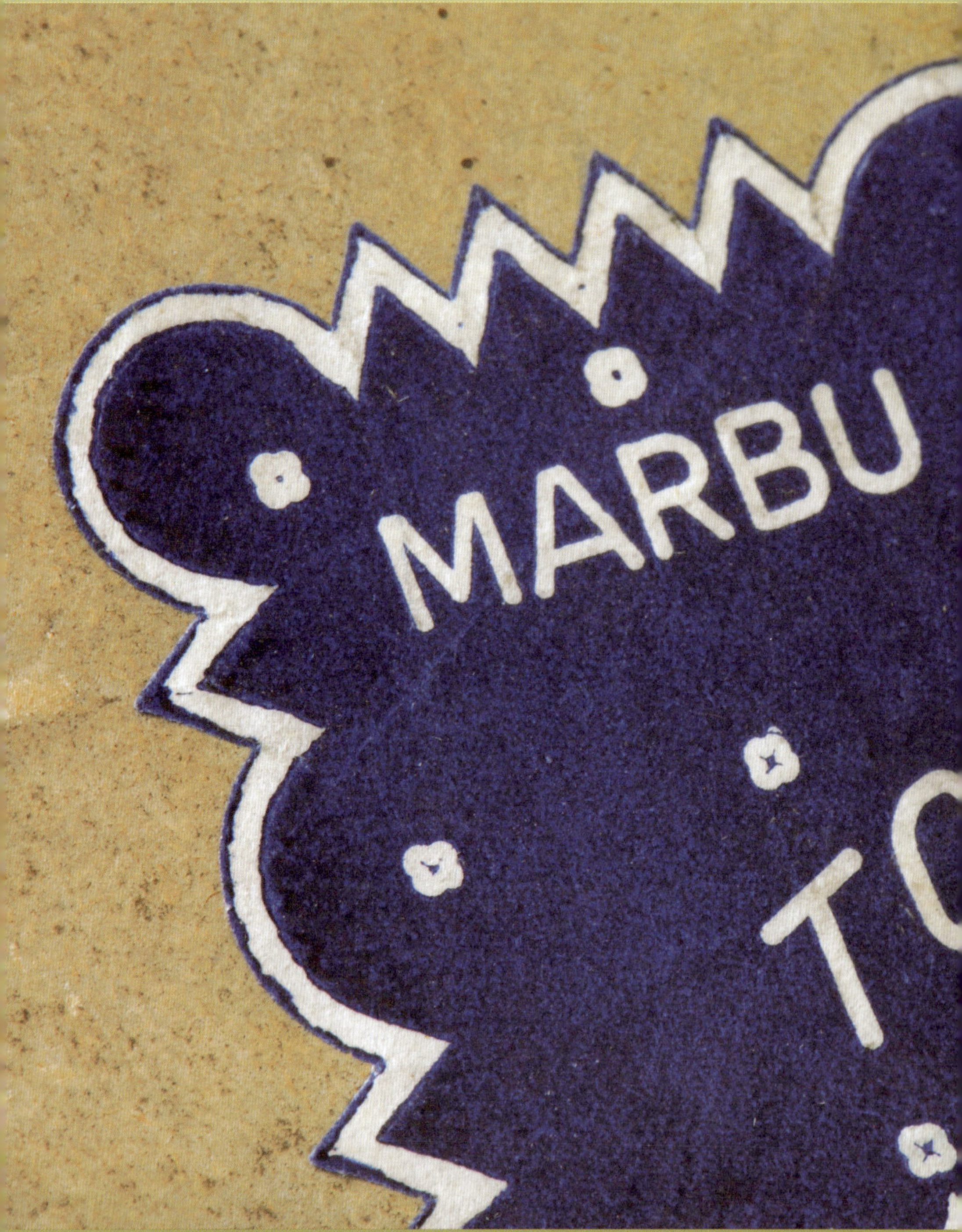
MARBU
TO

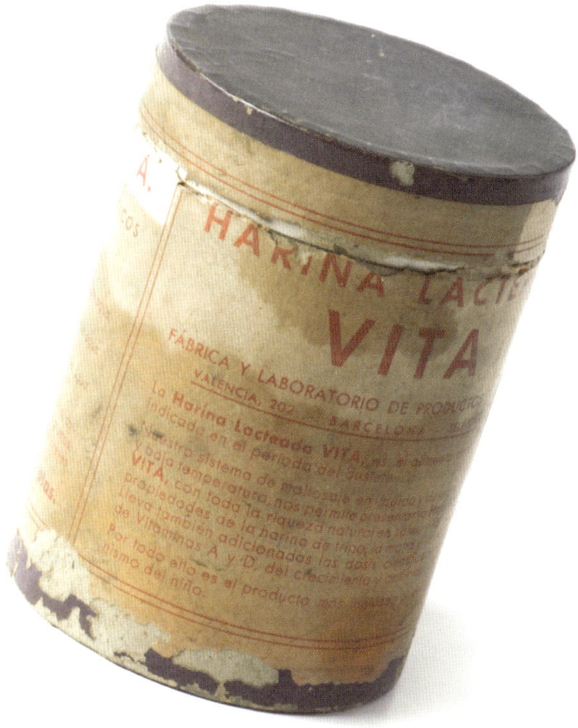
VITA
FÁBRICA Y LABORATORIO DE
La Harina Lacteada VITA

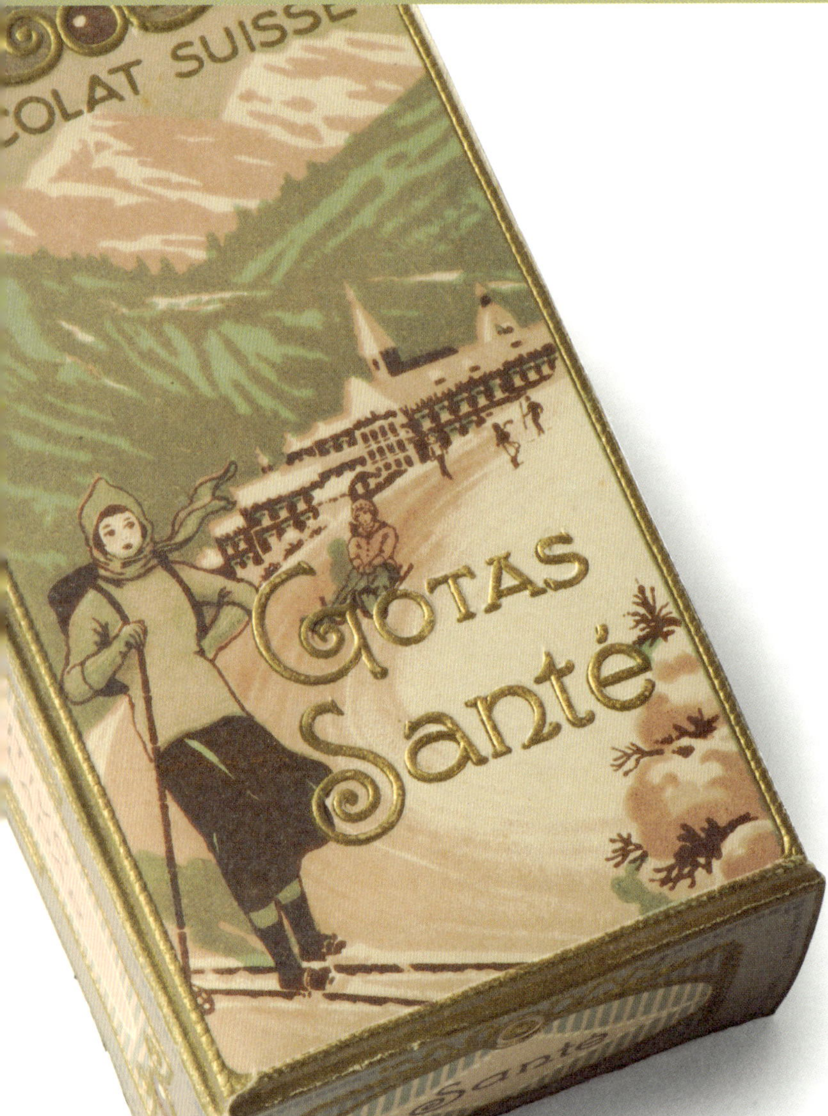
COLAT SUISSE
Gotas
Santé
Santé

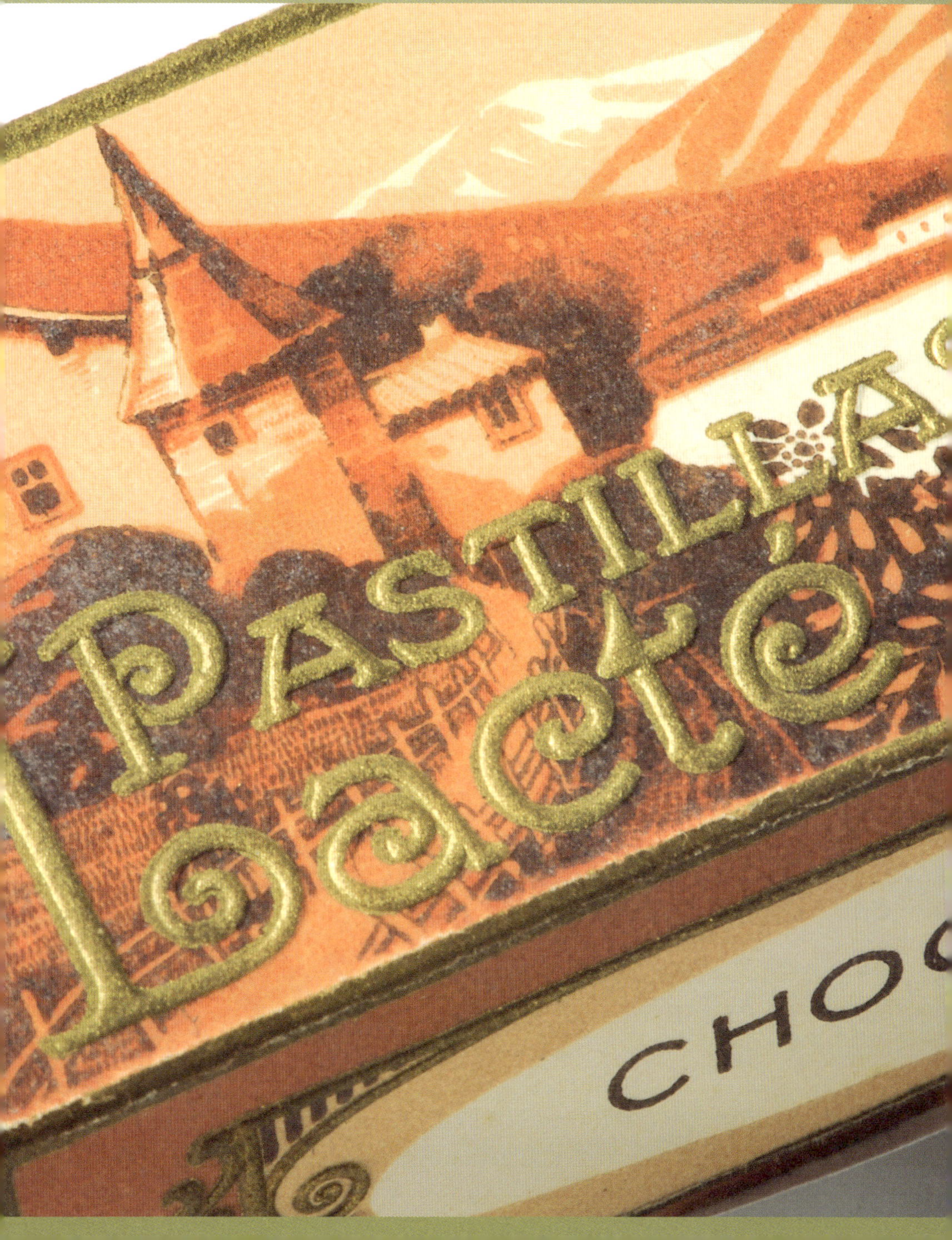
PASTILLAS
lacté
CHOC

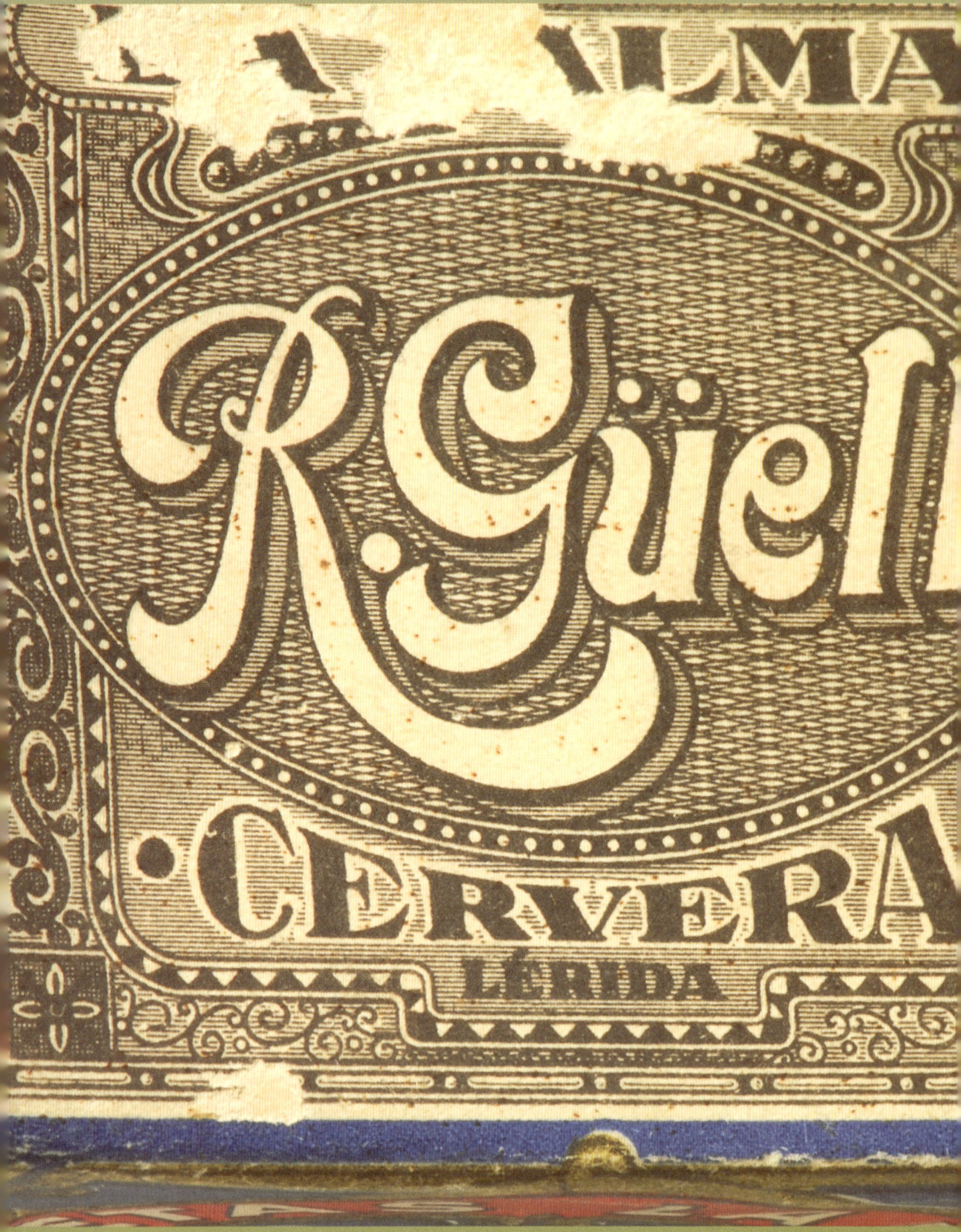
CERVERA
LÉRIDA

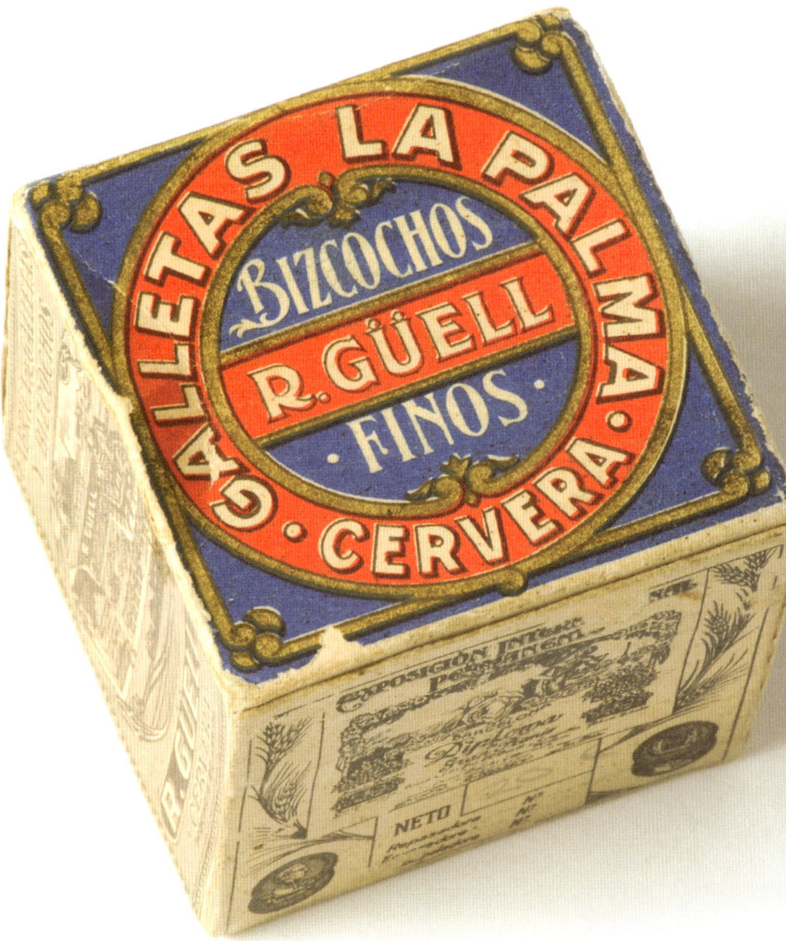
GALLETAS LA PALMA
BIZCOCHOS
R. GÜELL
FINOS
CERVERA
EXPOSICIÓN INTER.
NETO

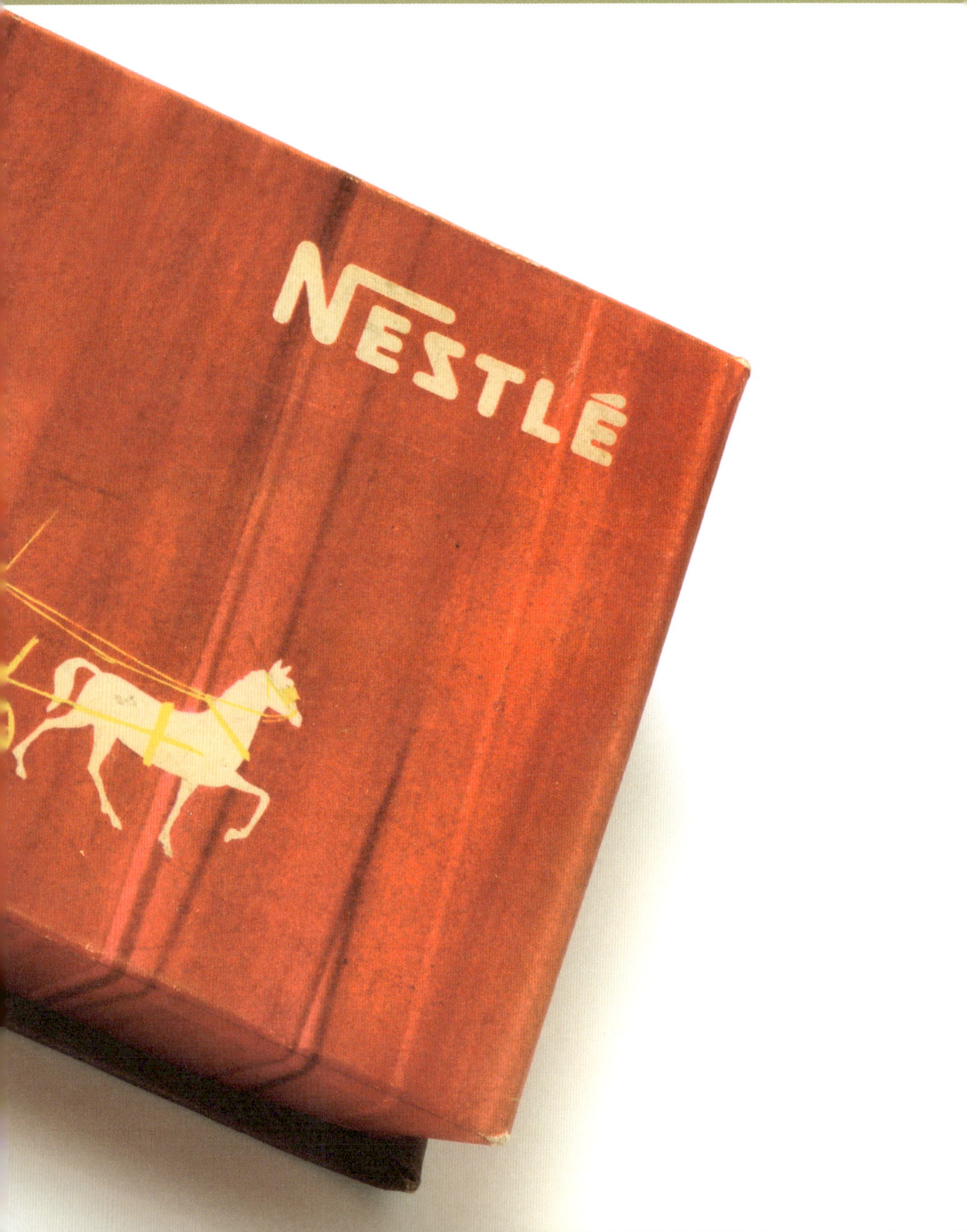
NESTLÉ

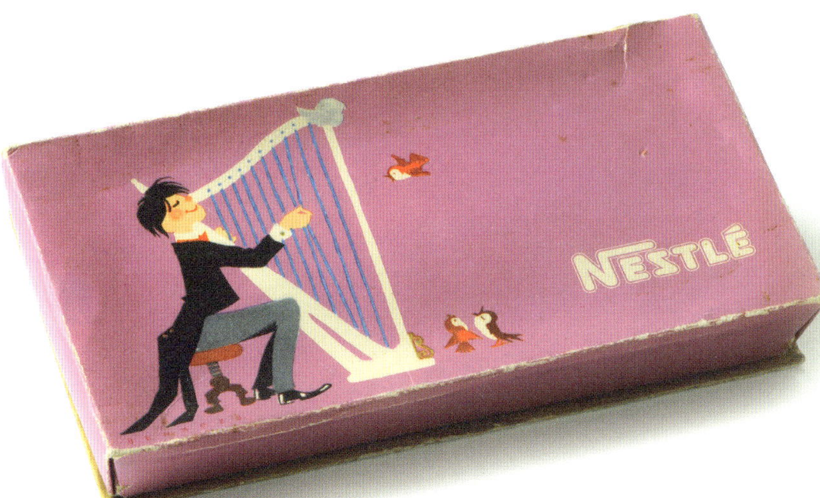
NESTLÉ

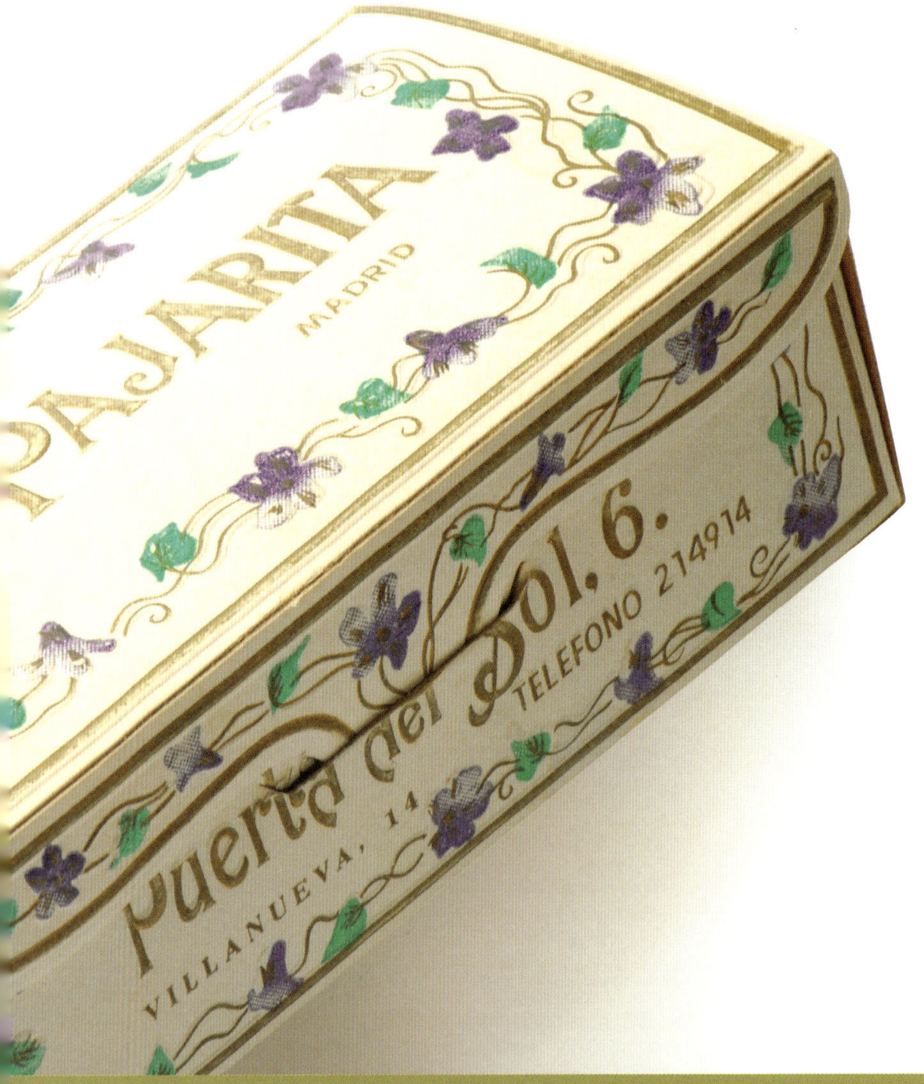
PAJARITA
MADRID
Puerta del Sol, 6.
TELEFONO 214914
VILLANUEVA, 14

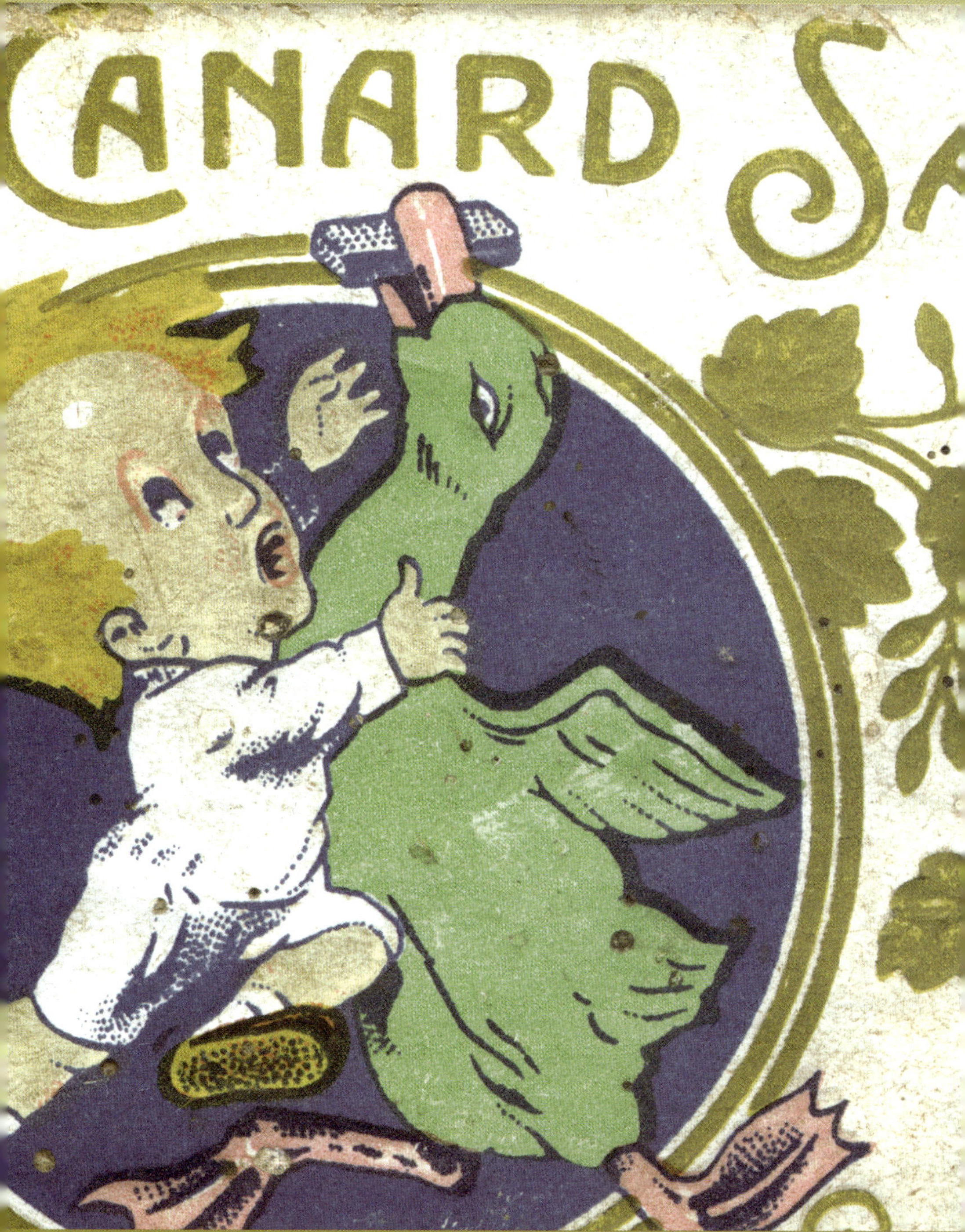
CANARD

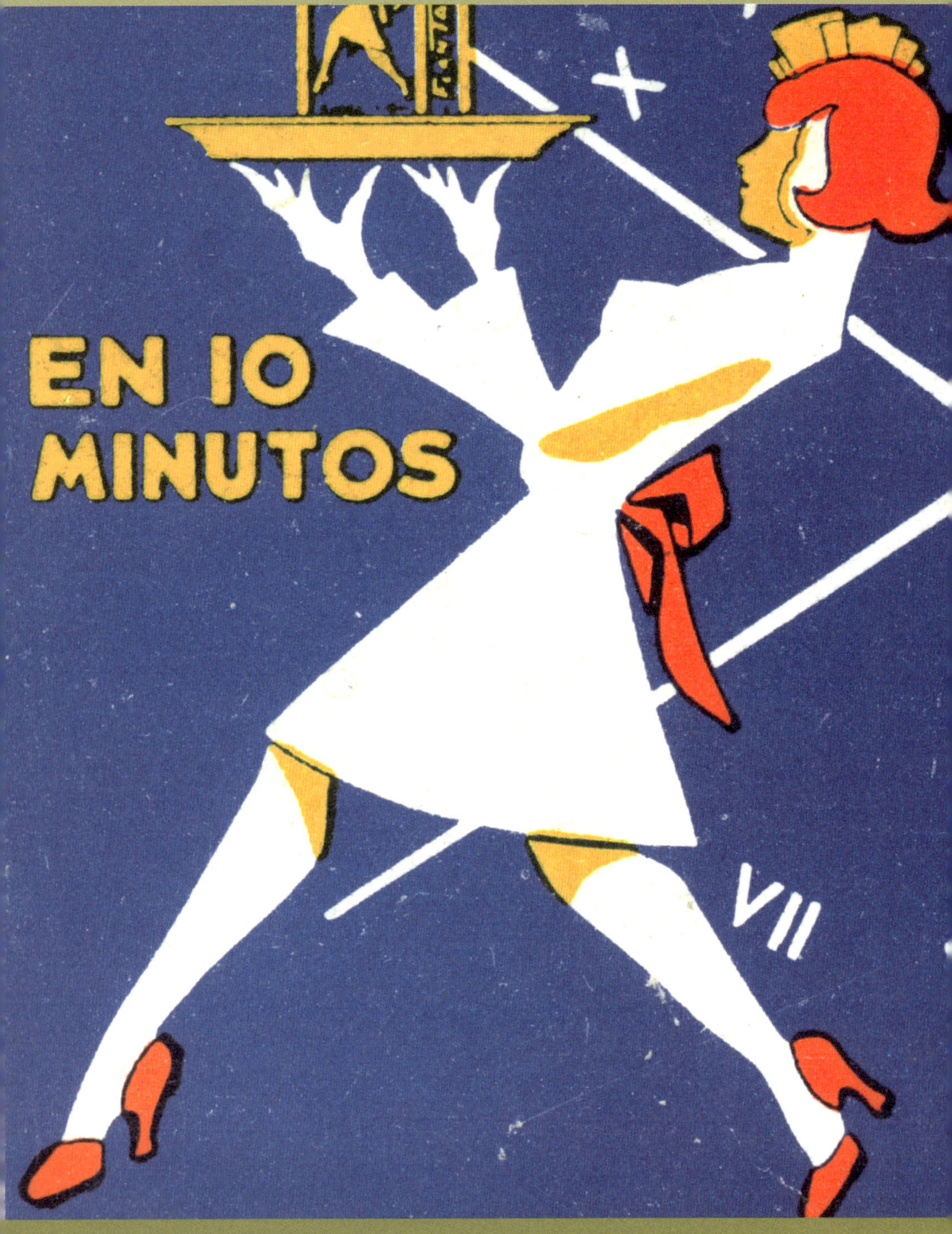
EN 10
MINUTOS
VII

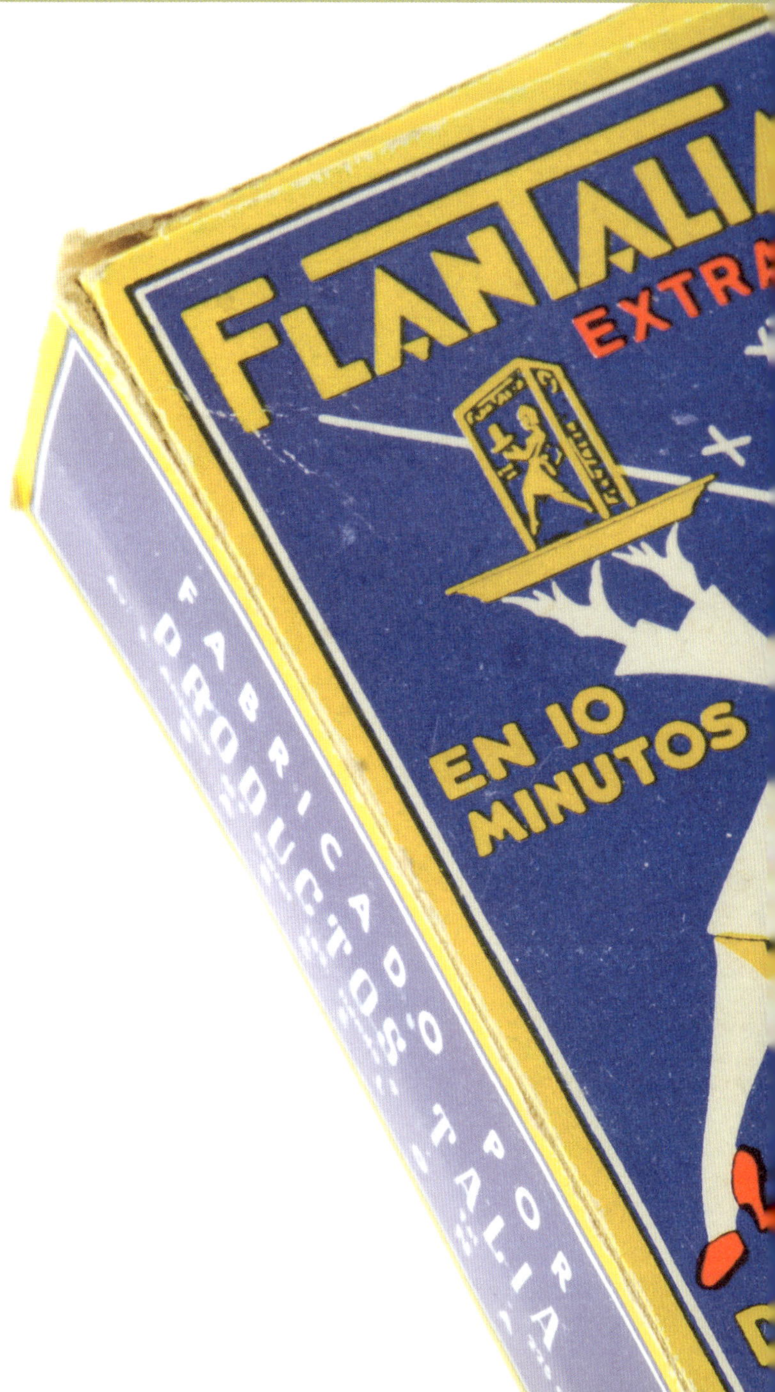
FLANTALIA
EXTRA
EN 10 MINUTOS
FABRICADO POR
PRODUCTOS TALIA

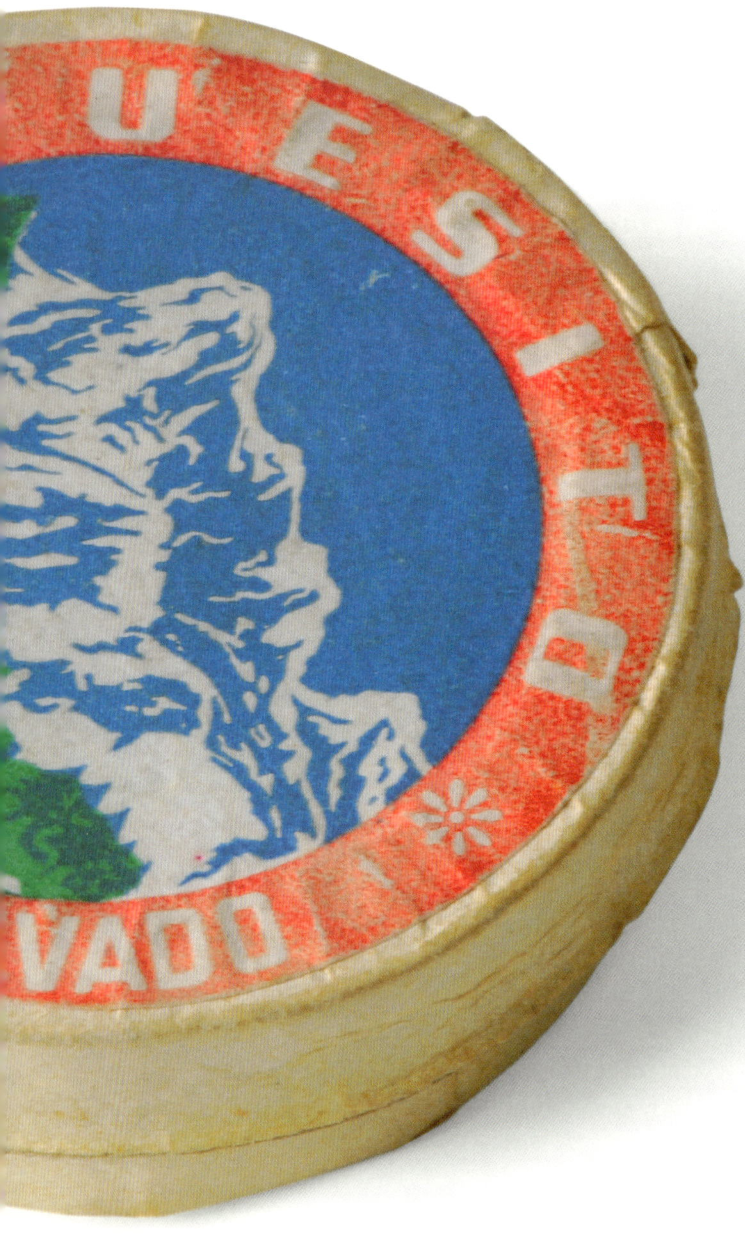
U'ESITO
VADO

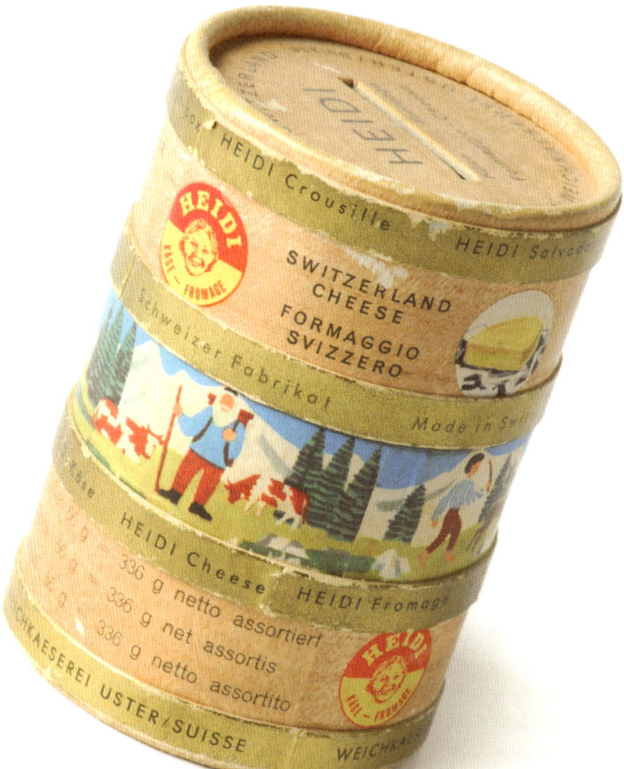
HEIDI Crousille
HEIDI
SWITZERLAND CHEESE
FORMAGGIO SVIZZERO
Schweizer Fabrikat
HEIDI Cheese
HEIDI Fromage
336 g netto assortiert
336 g net assortis
336 g netto assortito
EICHKAESEREI USTER/SUISSE

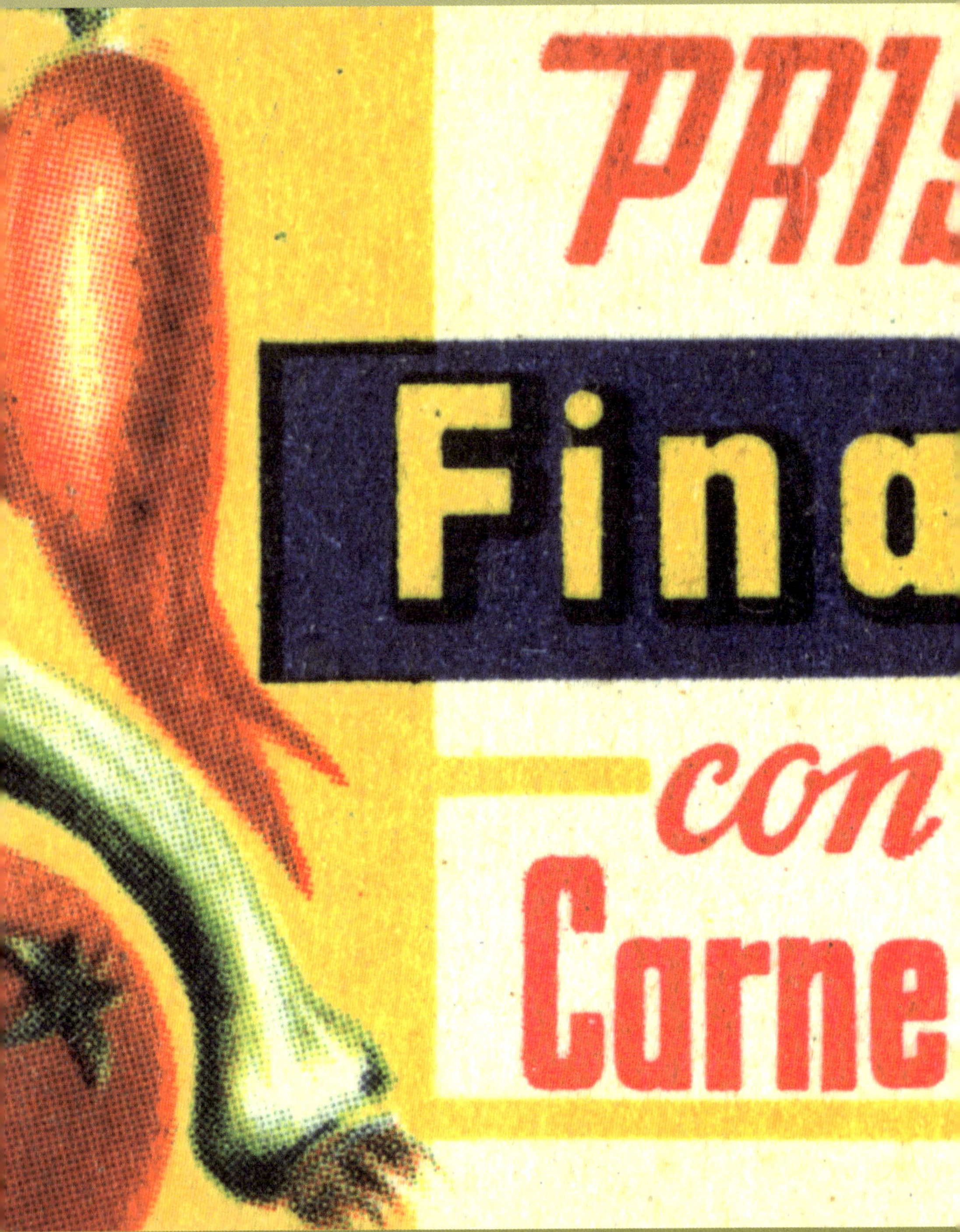
PRIS
Fina
con
Carne

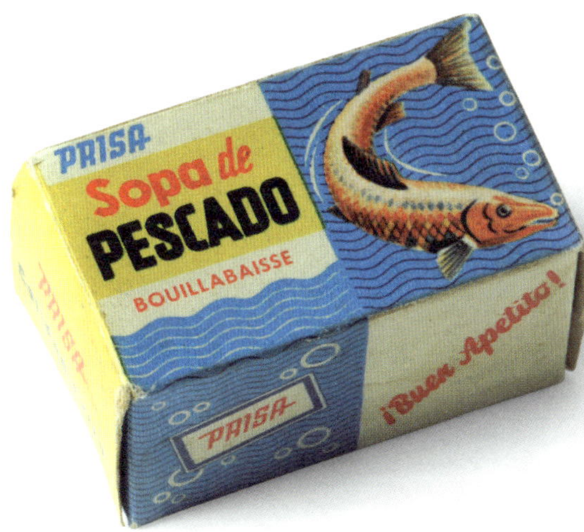
PRISA
Sopa de
PESCADO
BOUILLABAISSE
¡Buen Apetito!
PRISA

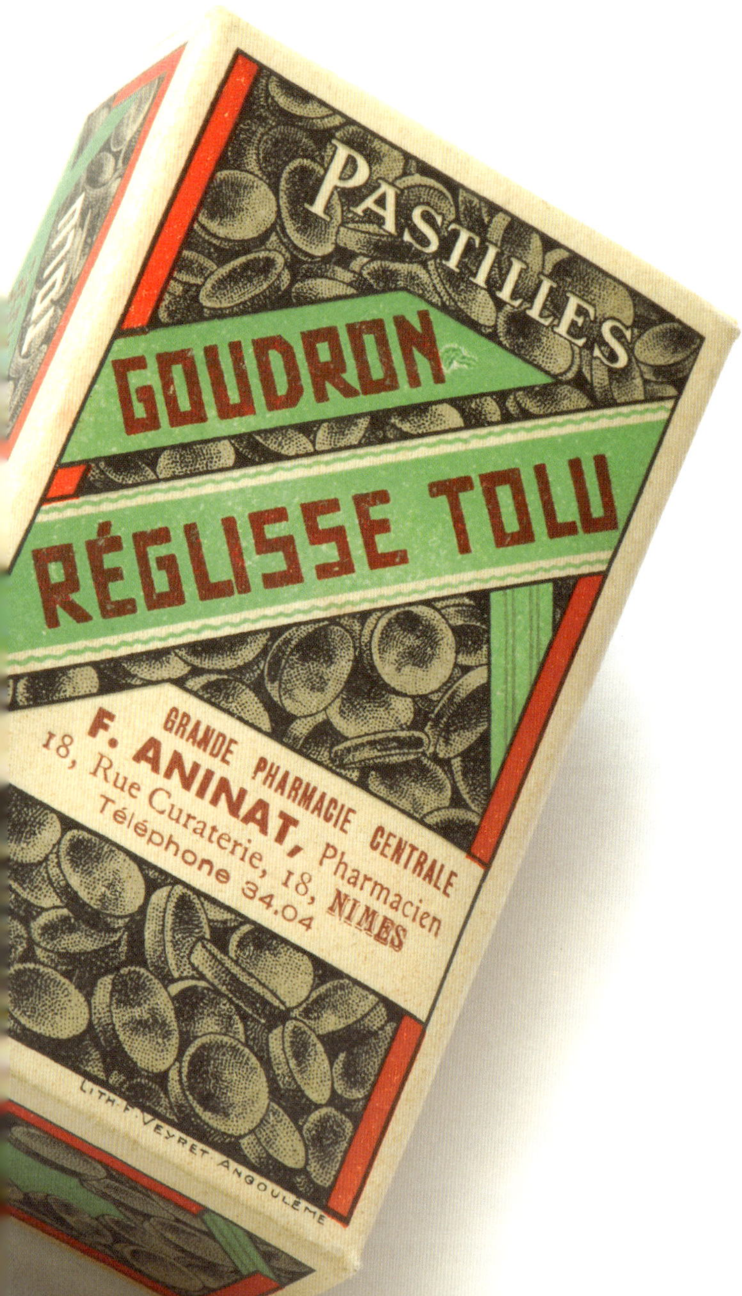
PASTILLES
GOUDRON
RÉGLISSE TOLU
GRANDE PHARMACIE CENTRALE
F. ANINAT, Pharmacien
18, Rue Curaterie, 18, NIMES
Téléphone 34.04
LITH. F. VEYRET ANGOULÊME

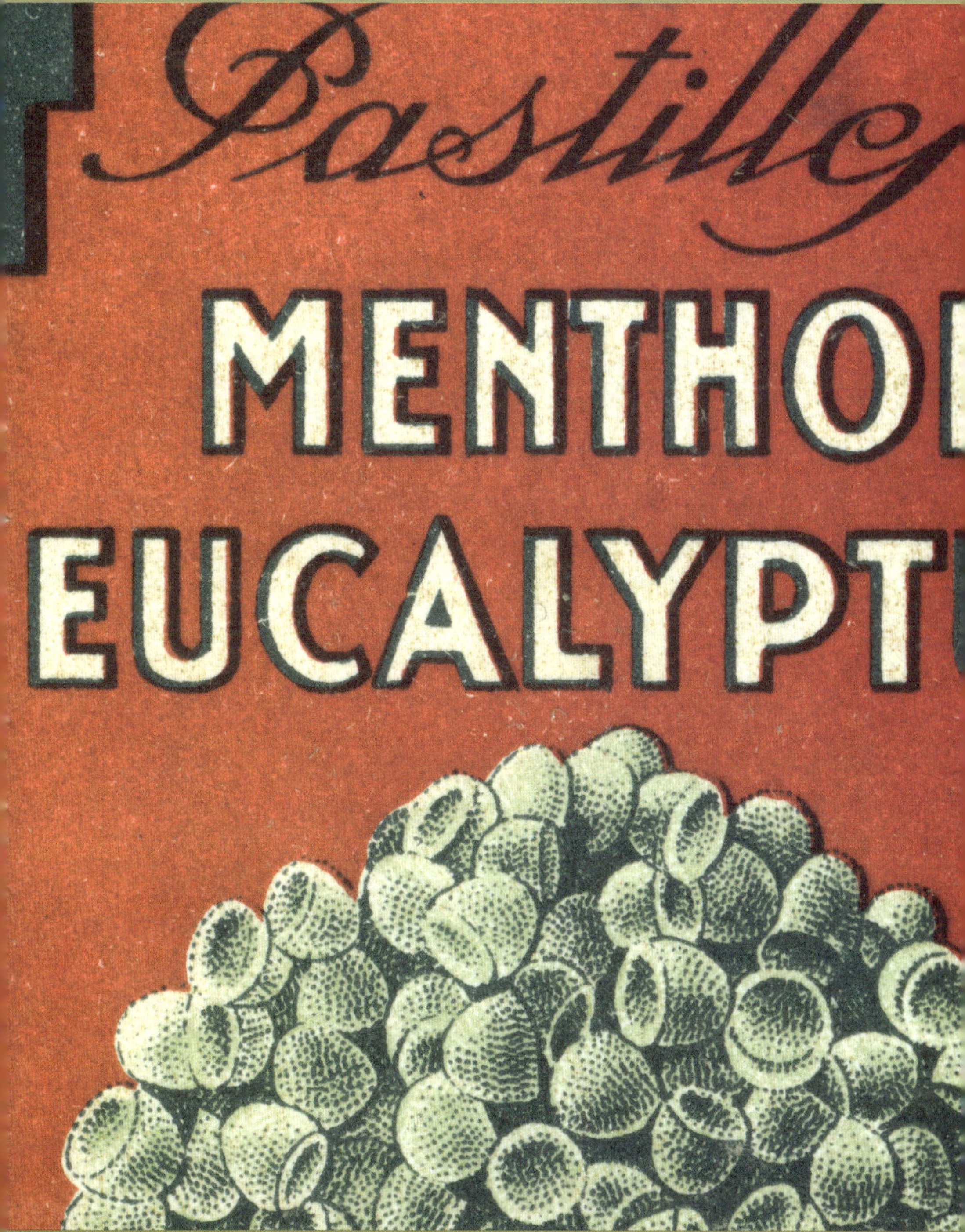
MENTHO
EUCALYPT

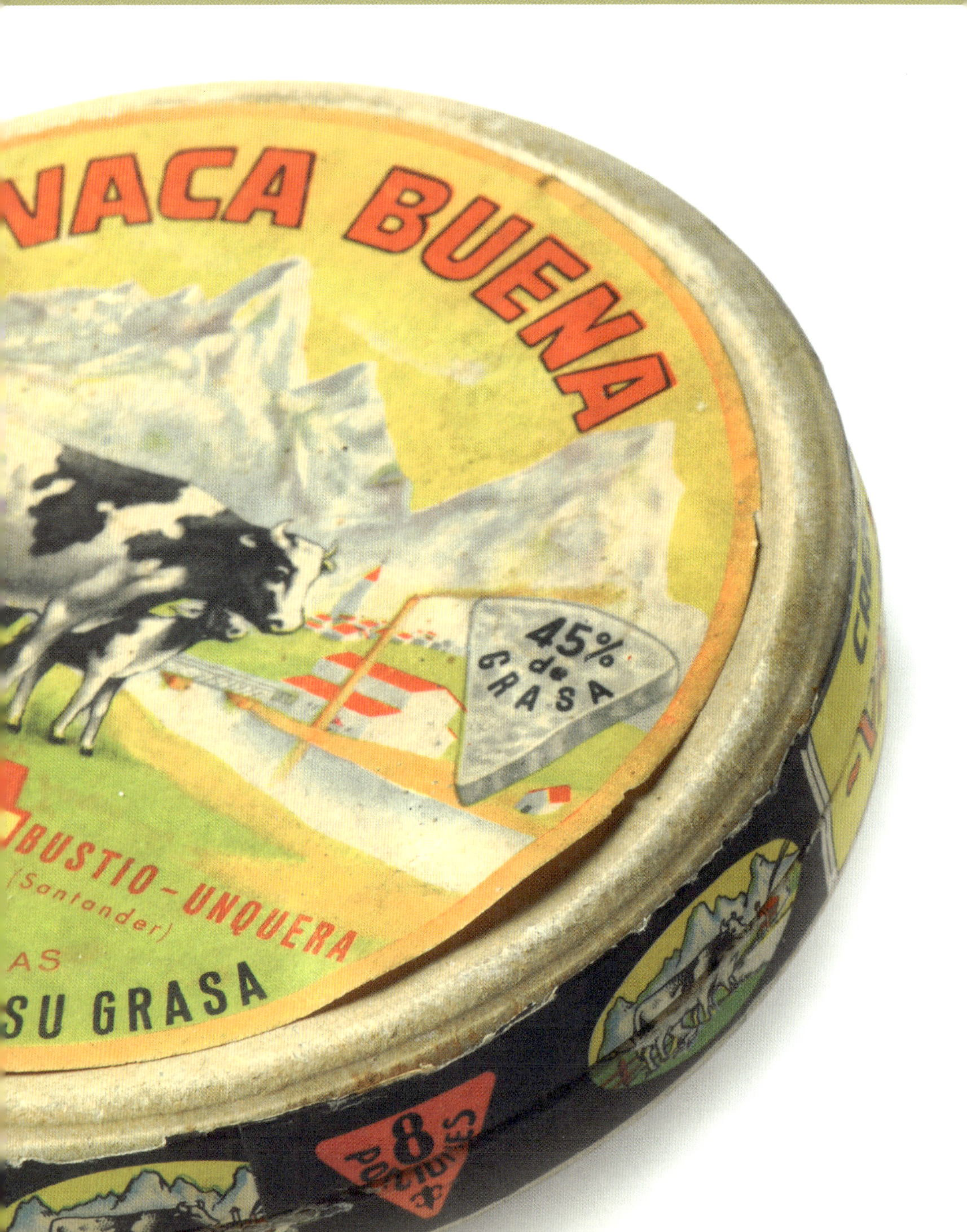
VACA BUENA
45% de GRASA
BUSTIO – UNQUERA
(Santander)
SU GRASA
8 PORCIONES

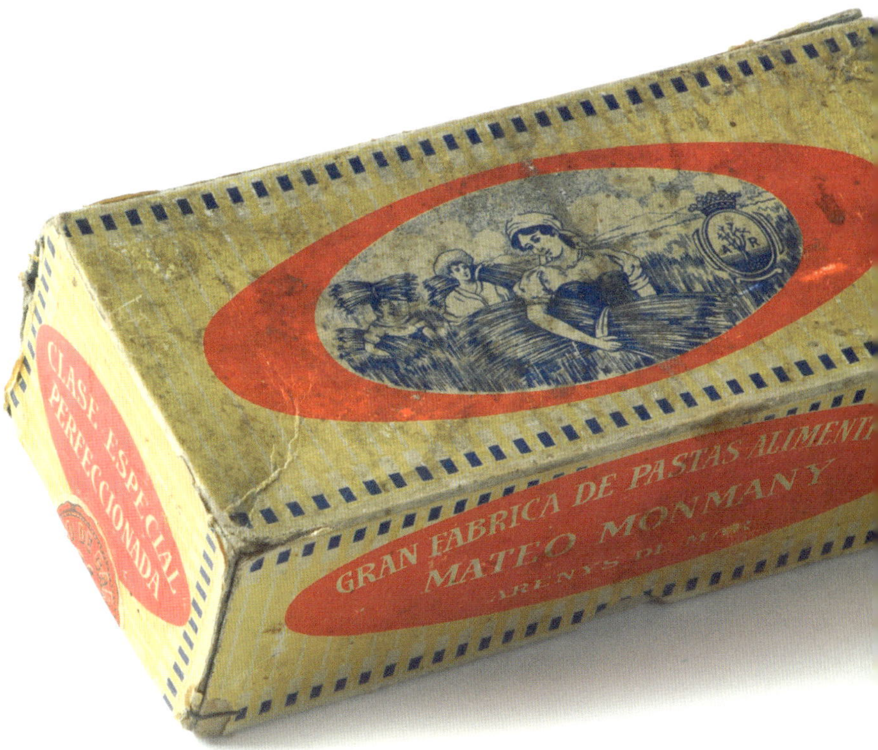
CLASE ESPECIAL
PERFECCIONADA
GRAN FABRICA DE PASTAS
MATEO MONMANY

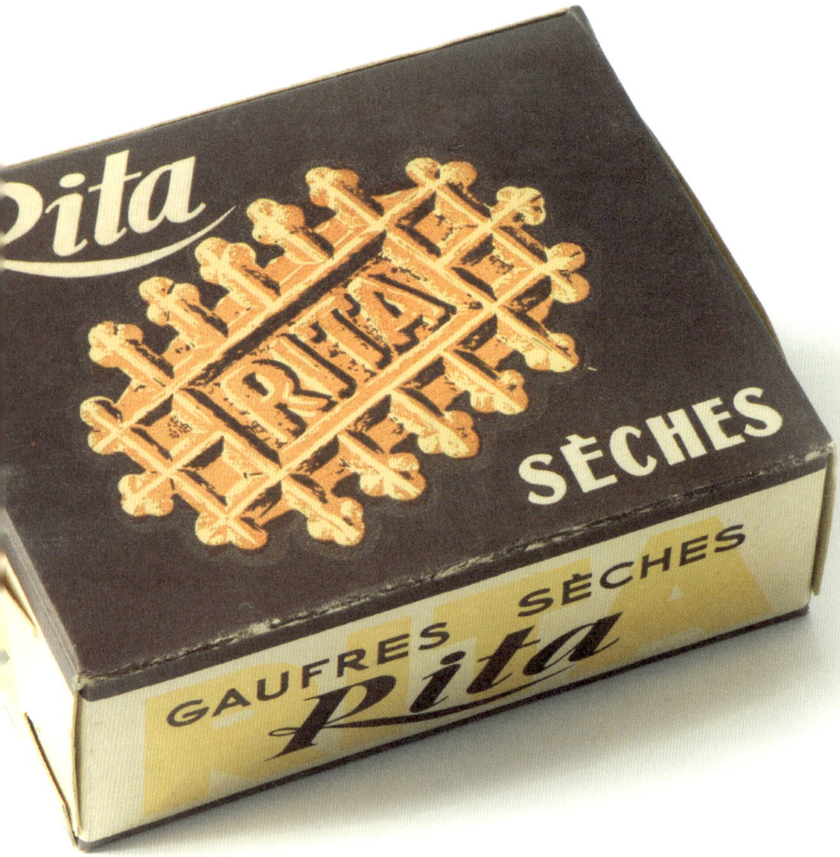
Rita
RITA
SÈCHES
GAUFRES SÈCHES
Rita

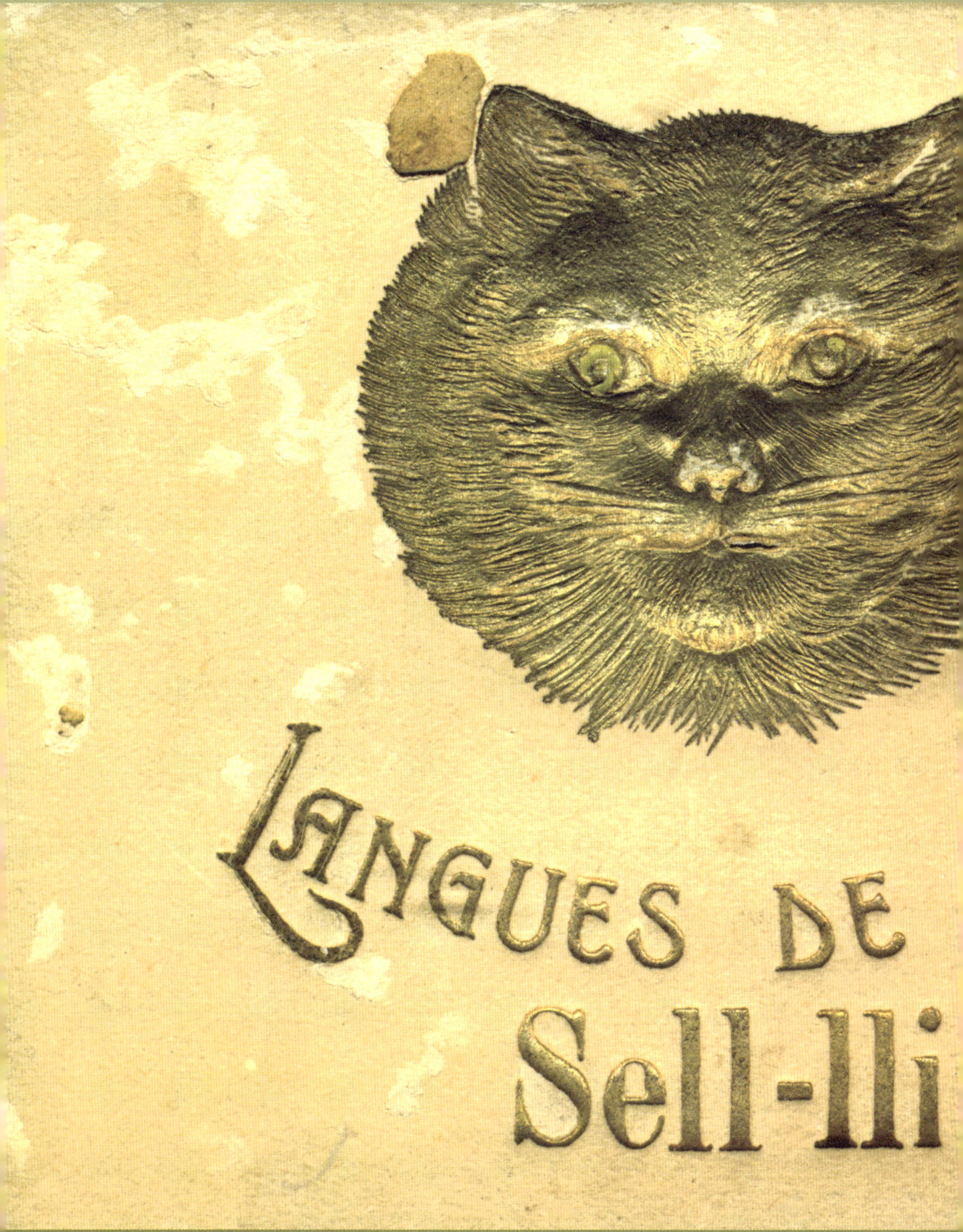
LANGUES DE
Sell-lli

OBACCO
CCOTOBACCOTOB

TOBACCO

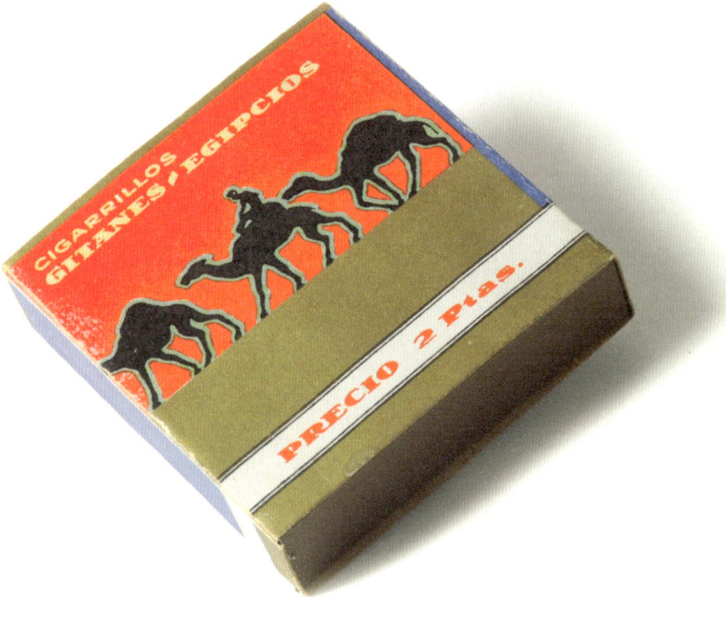
CIGARRILLOS
GITANES / EGIPCIOS
PRECIO 2 Ptas.

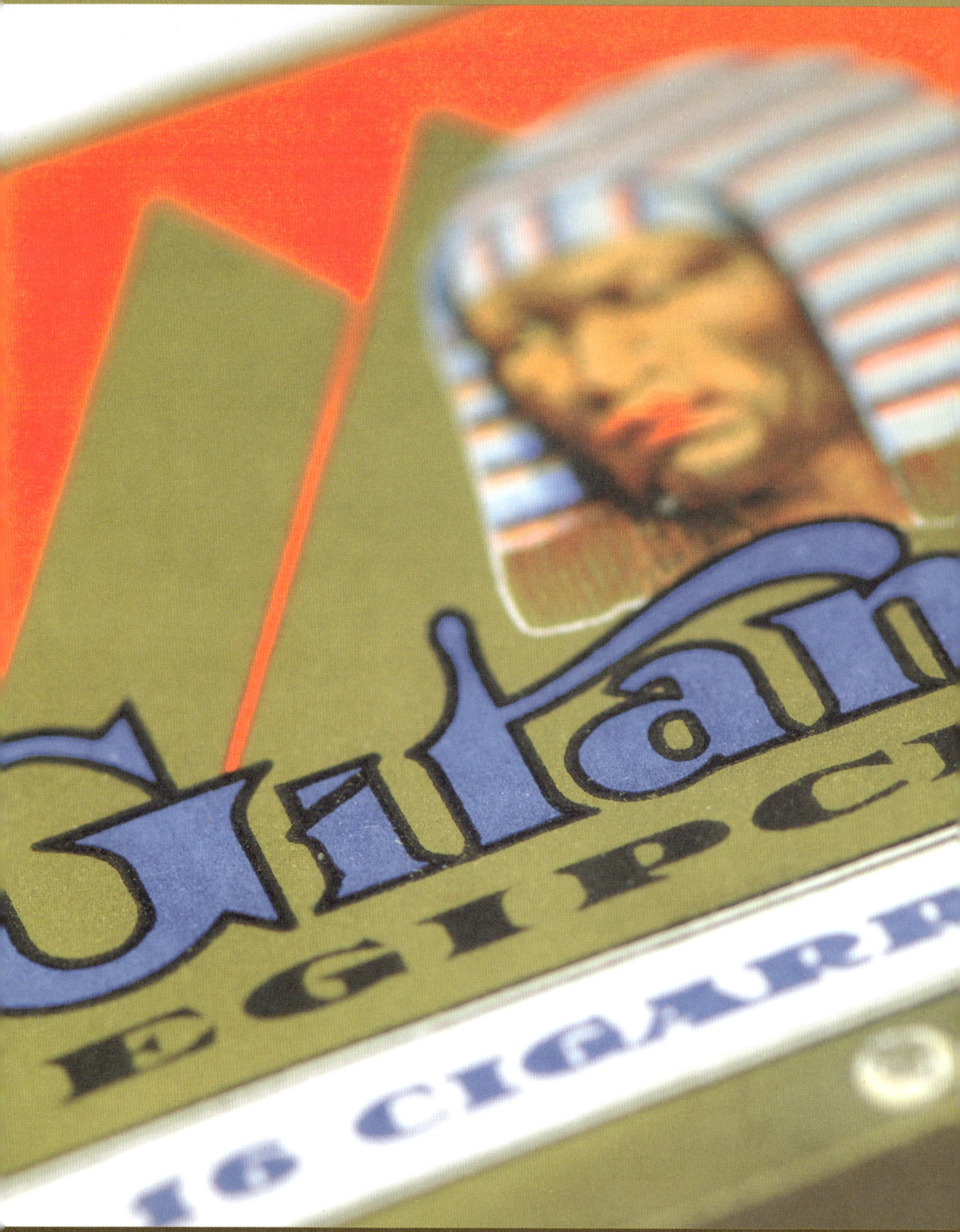

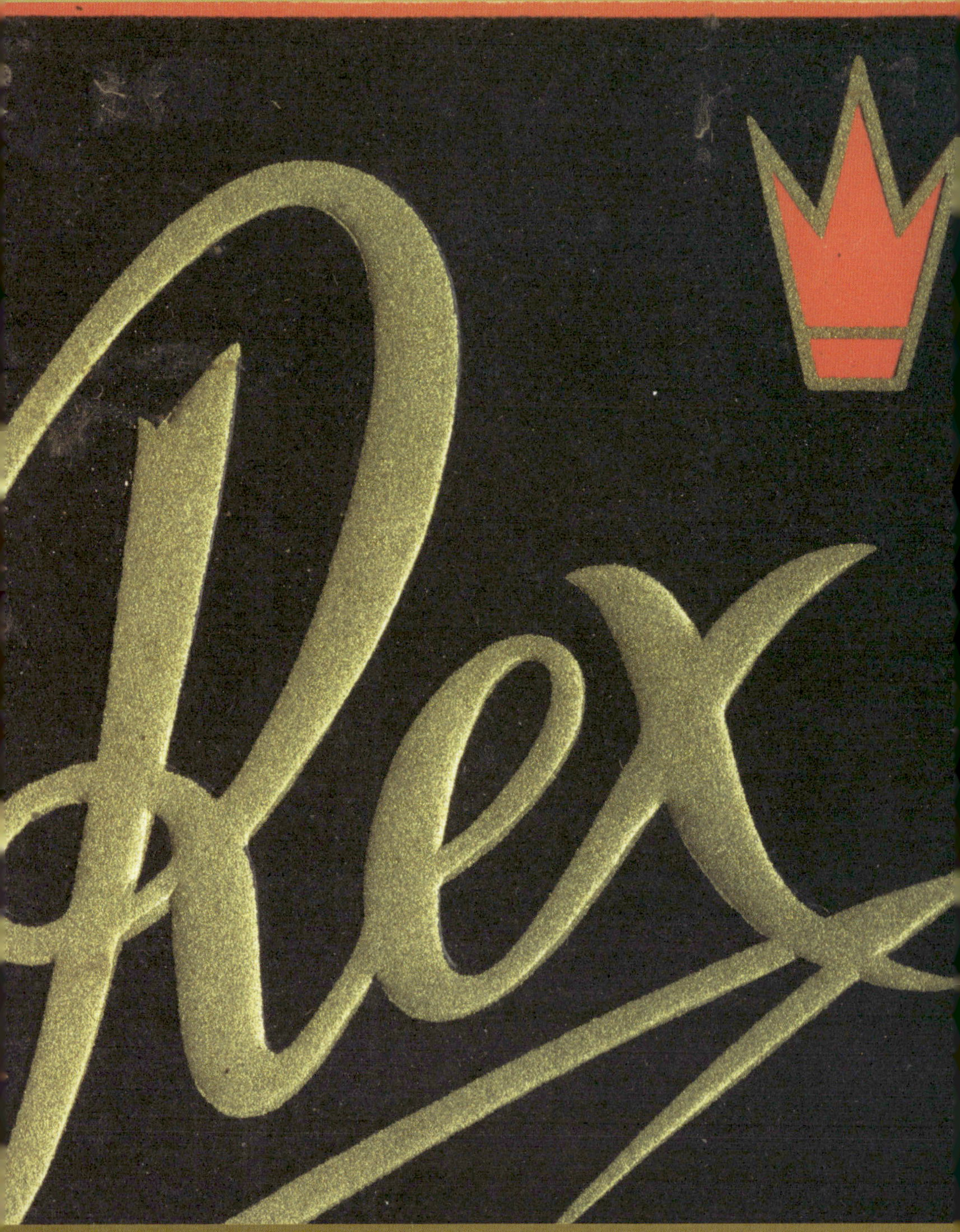
Rex

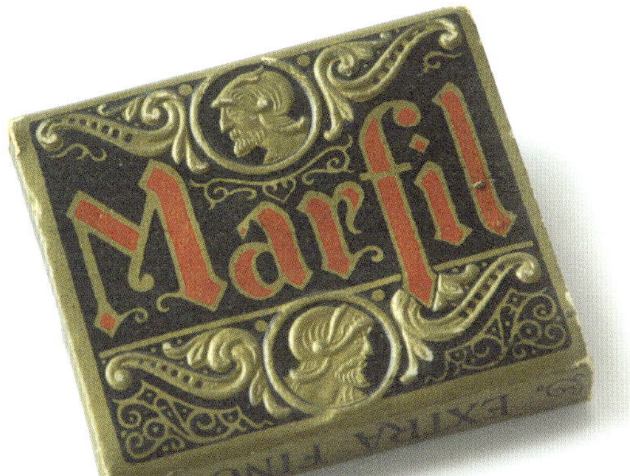
Marfil
EXTRA FINO

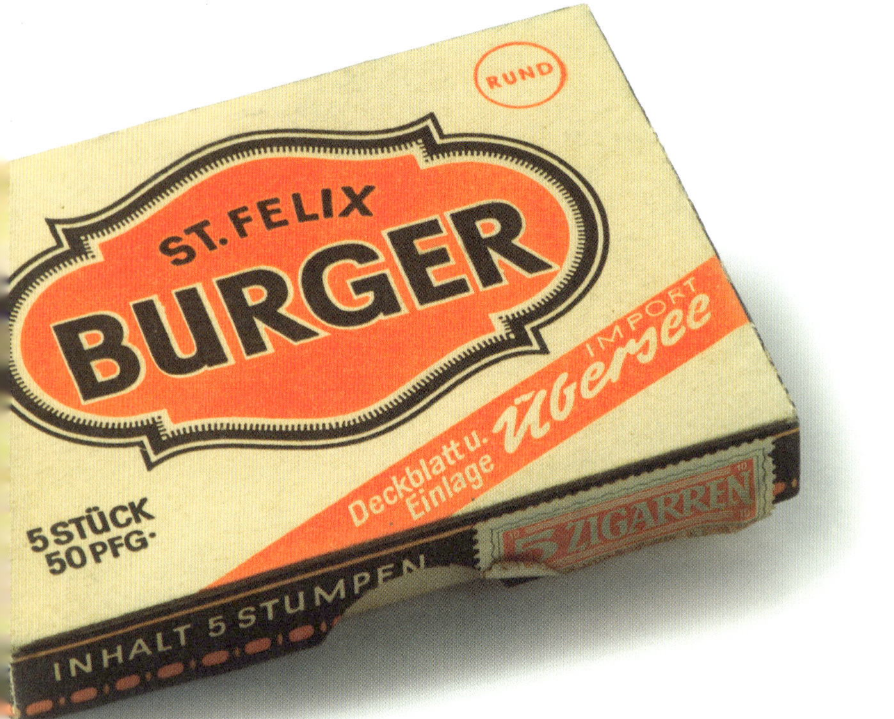
RUND
ST. FELIX
BURGER
IMPORT
Deckblatt u.
Einlage Übersee
5 STÜCK
50 PFG.
5 ZIGARREN
INHALT 5 STUMPEN

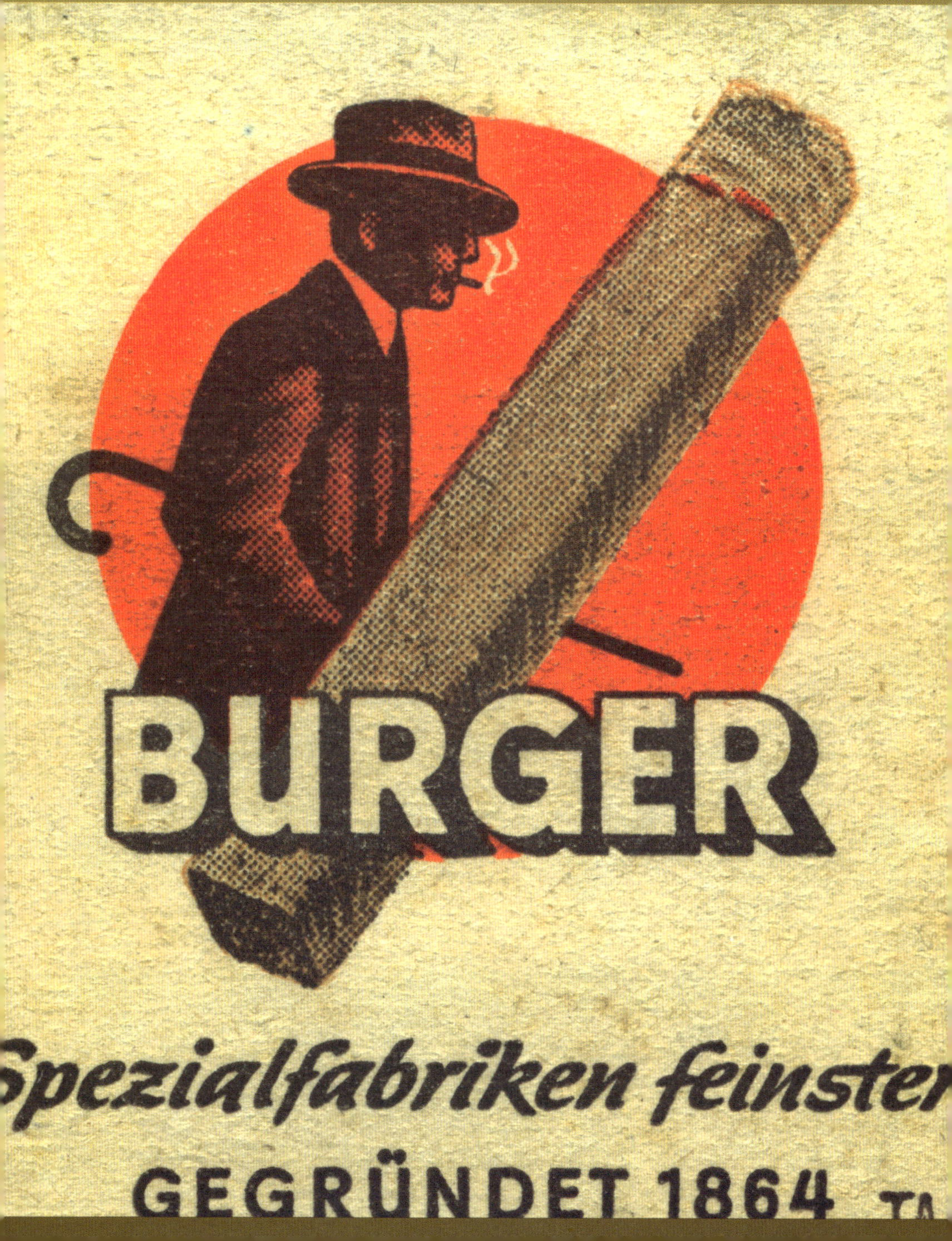
BURGER
Spezialfabriken feinster
GEGRÜNDET 1864

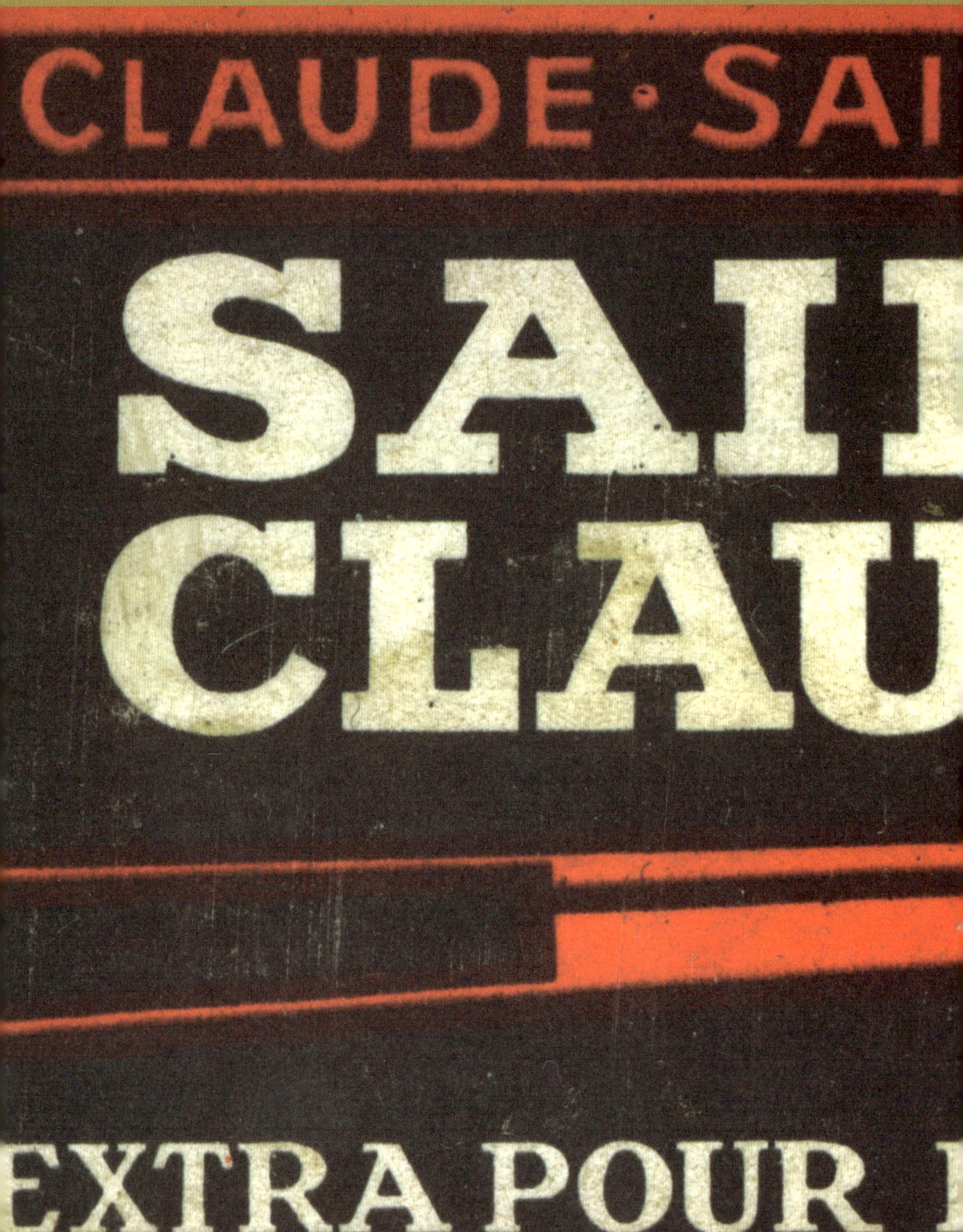
CLAUDE·SAI
SAI
CLAU
EXTRA POUR

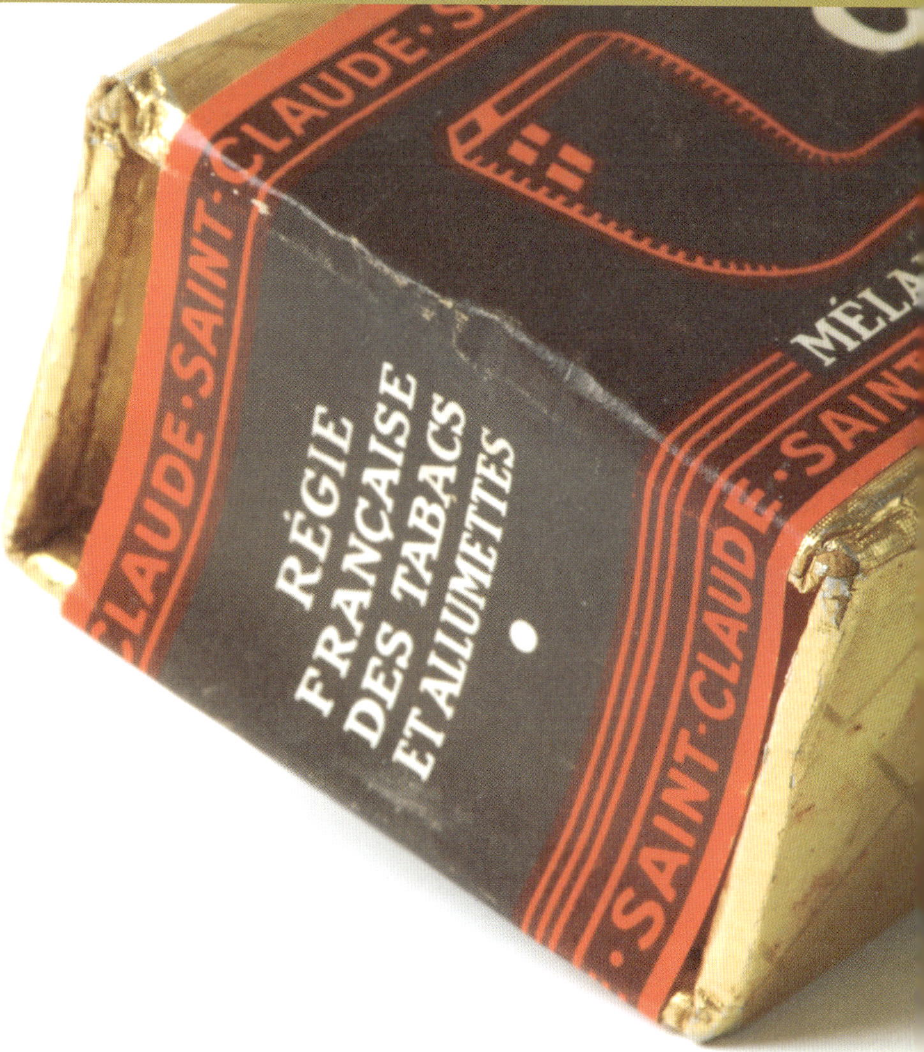
RÉGIE
FRANÇAISE
DES TABACS
ET ALLUMETTES
SAINT·CLAUDE

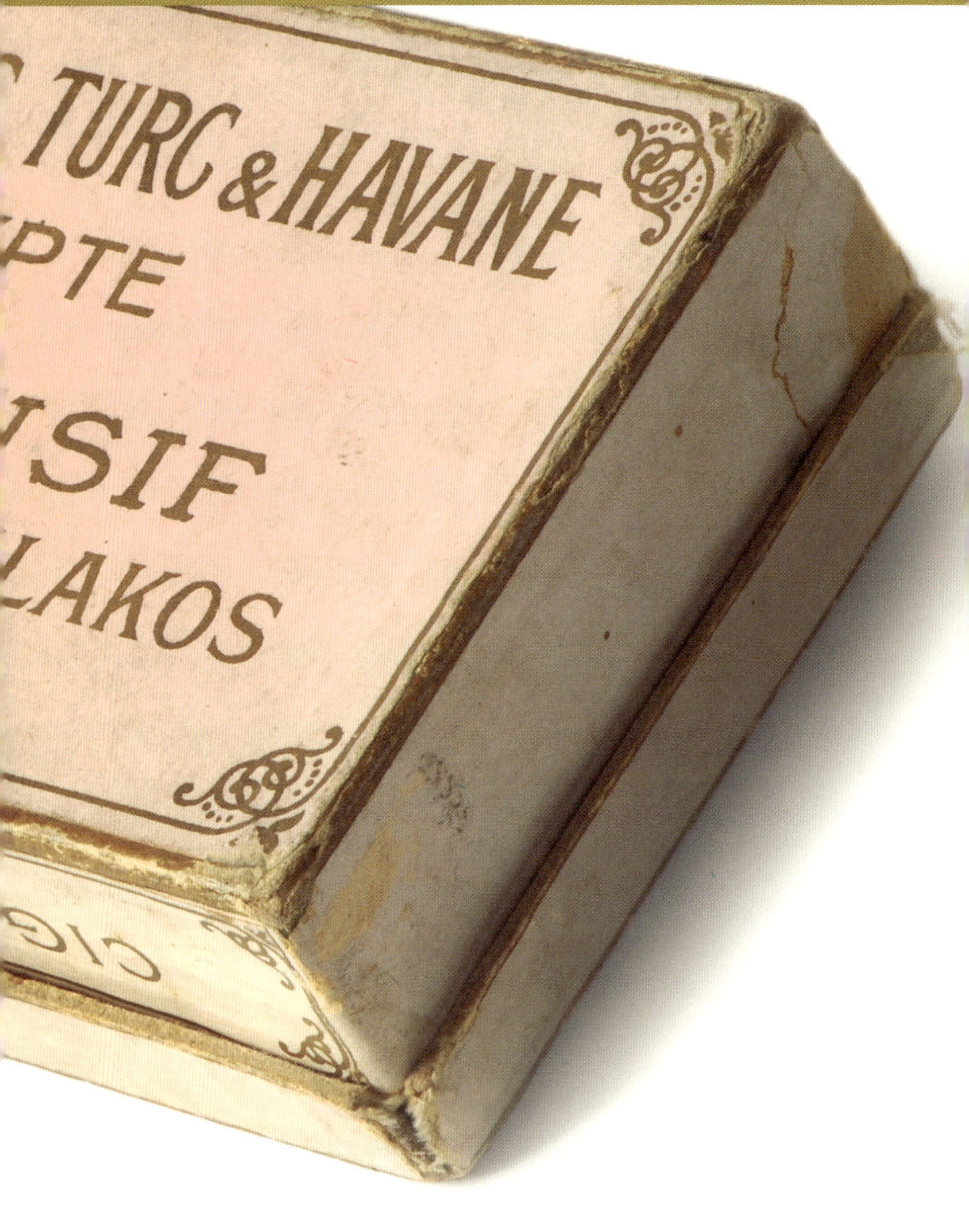
TURC & HAVANE
PTE
SIF
LAKOS

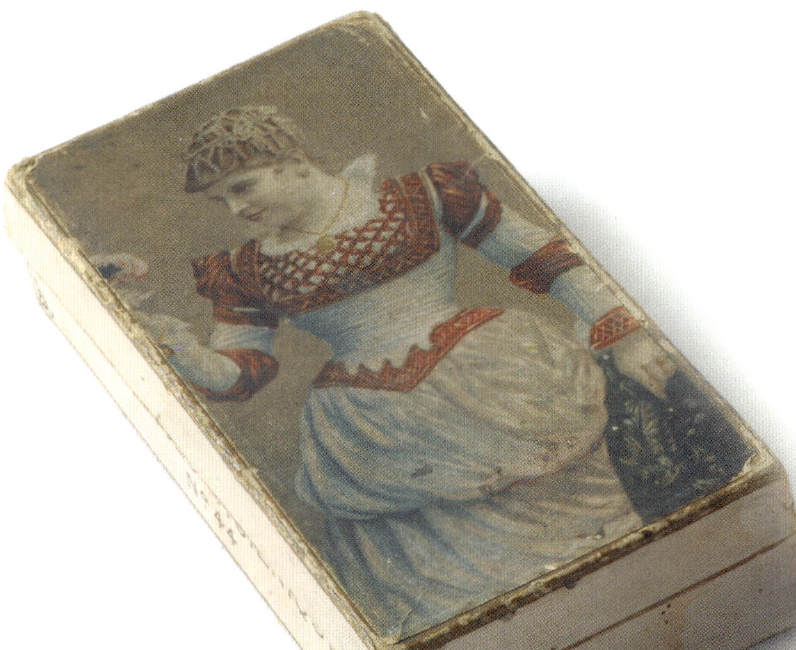

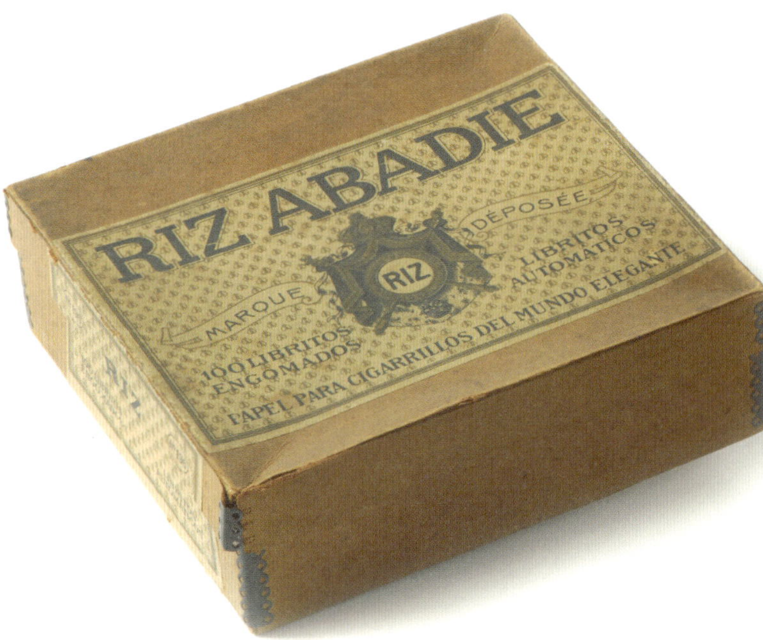
RIZ ABADIE
MARQUE
DÉPOSÉE
RIZ
100 LIBRITOS ENGOMADOS
LIBRITOS AUTOMÁTICOS
PAPEL PARA CIGARRILLOS DEL MUNDO ELEGANTE

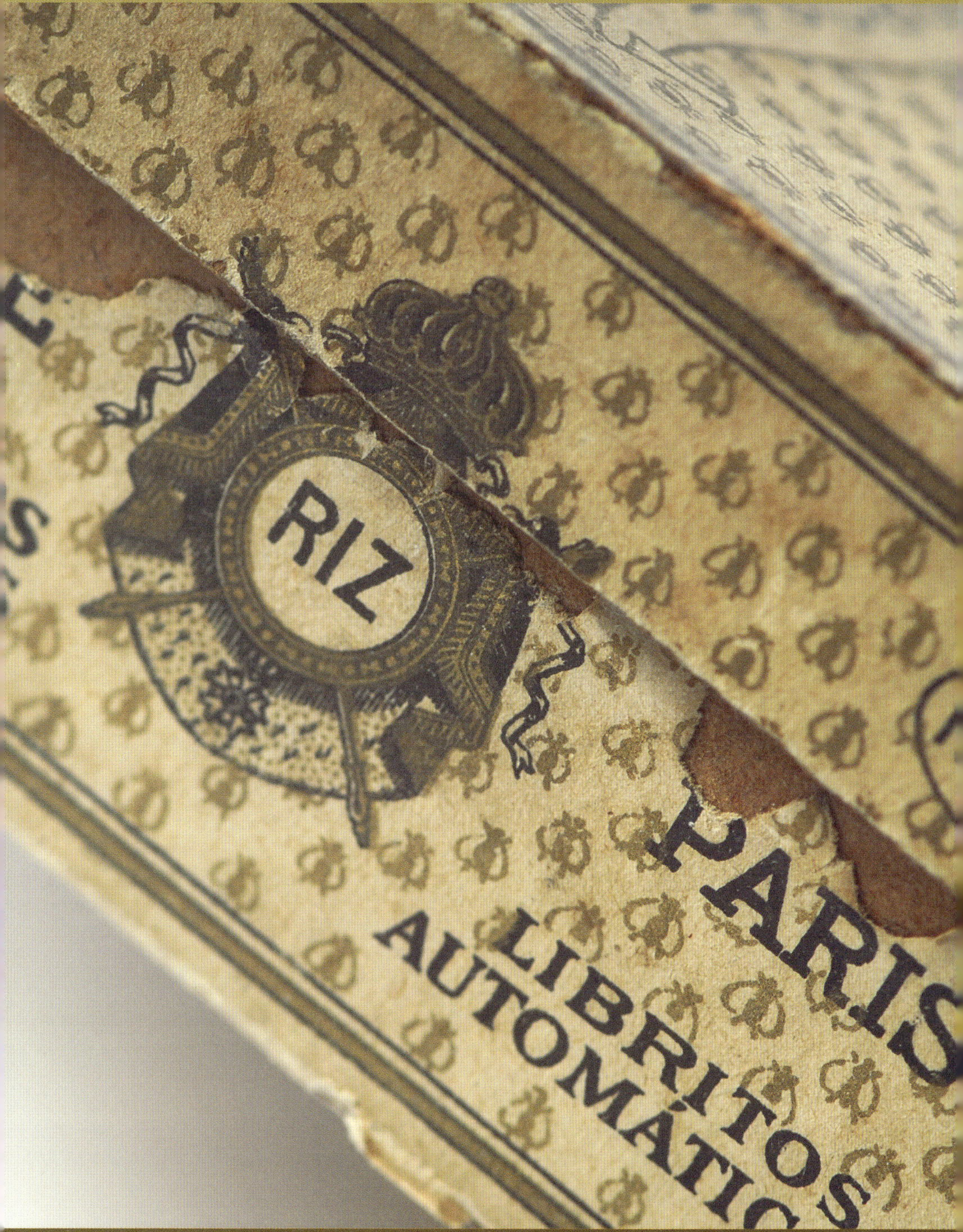
RIZ
PARIS
LIBRITOS

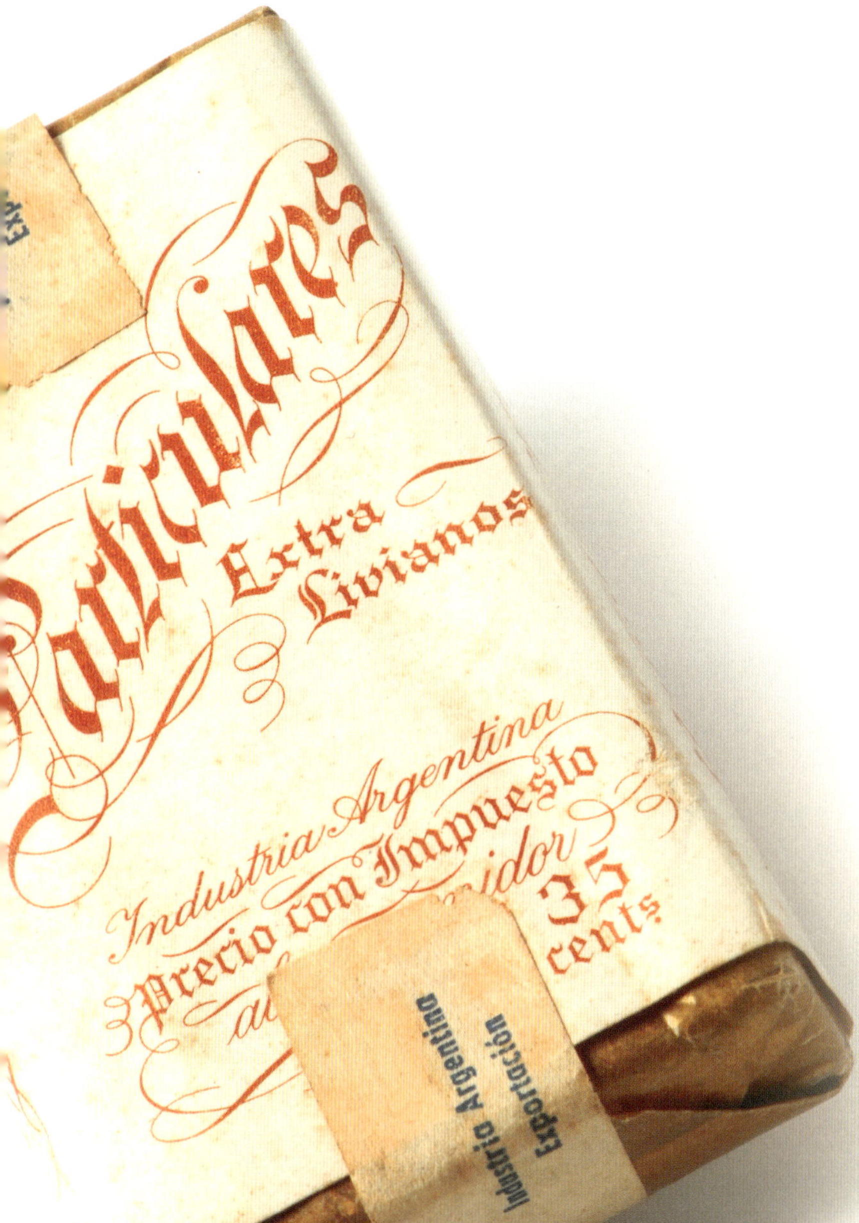
Extra
Livianos
Industria Argentina
Precio con Impuesto
35
cents
Industria Argentina
Exportación

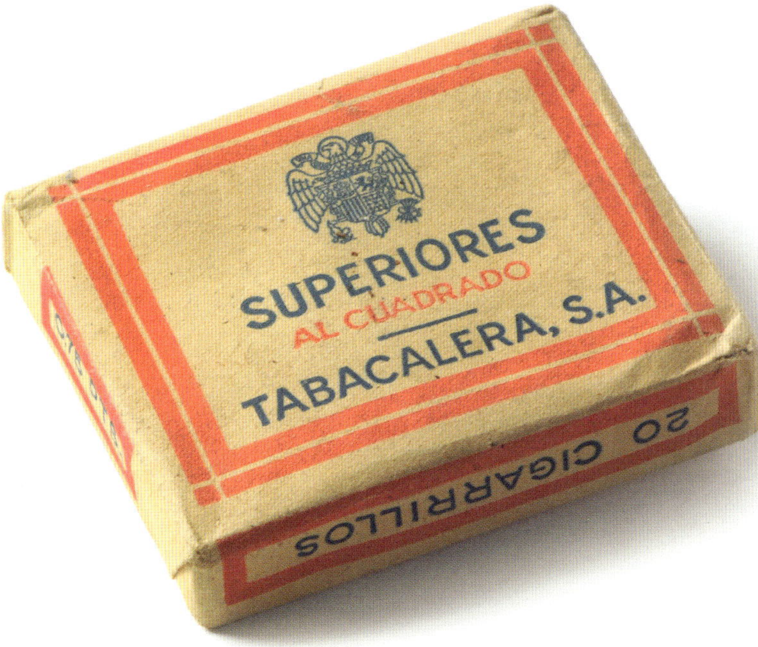
SUPERIORES
AL CUADRADO
TABACALERA, S.A.
20 CIGARRILLOS

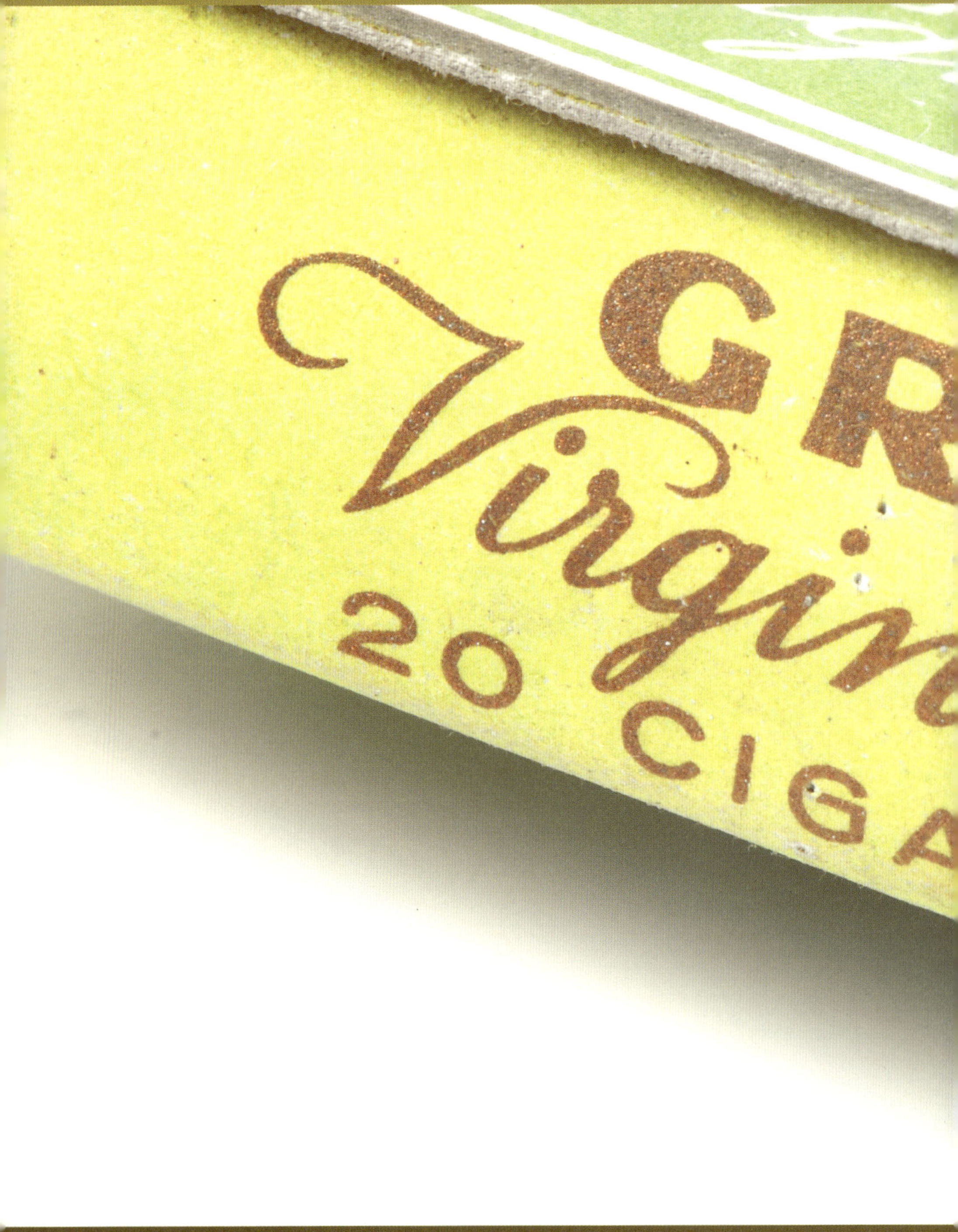
GR
Virgin
20 CIGA

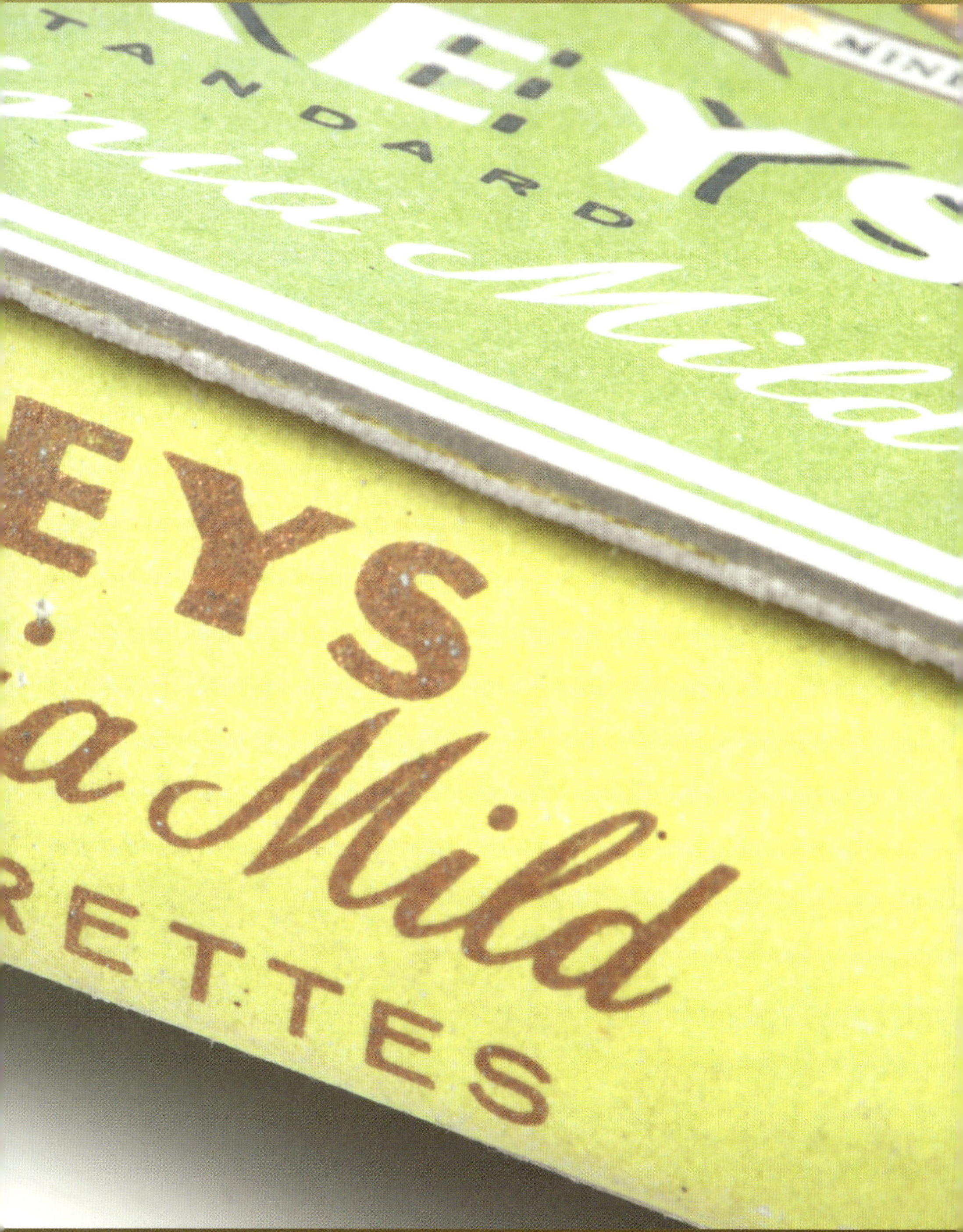
TANDARD
EYS
Mild
EYS
a Mild
RETTES

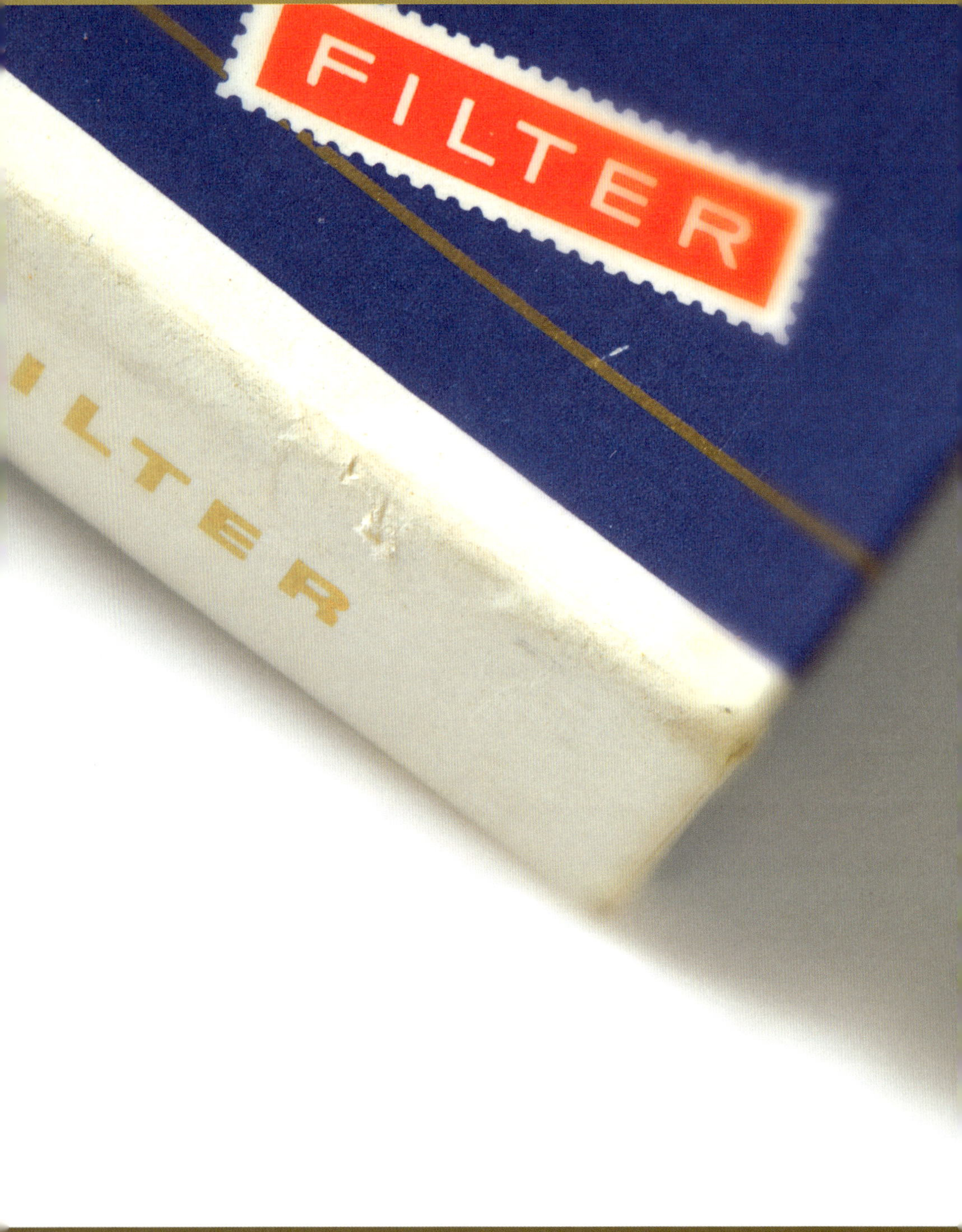
FILTER

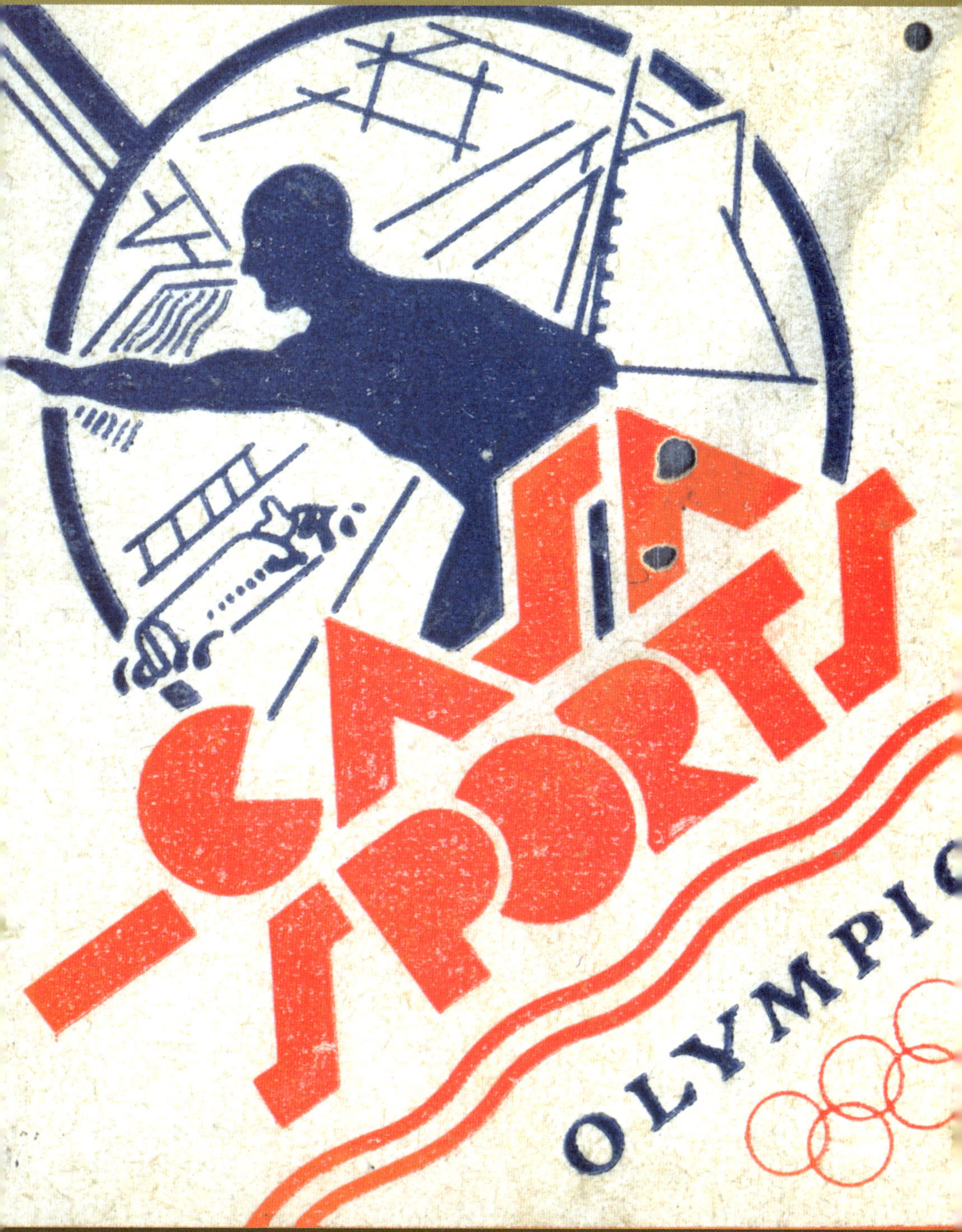
CASA
SPORTS
OLYMPIC

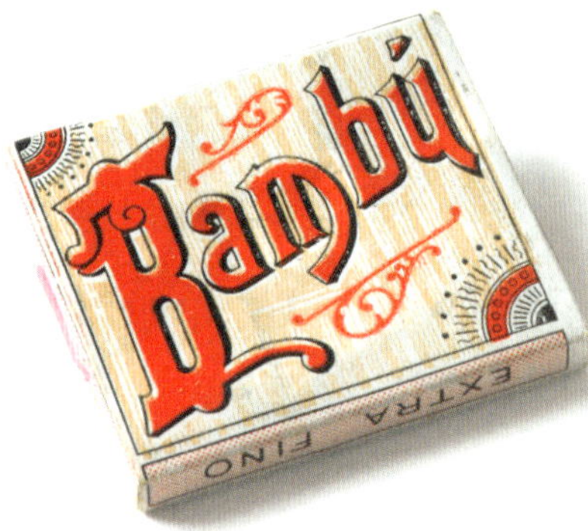
Bambú
EXTRA FINO

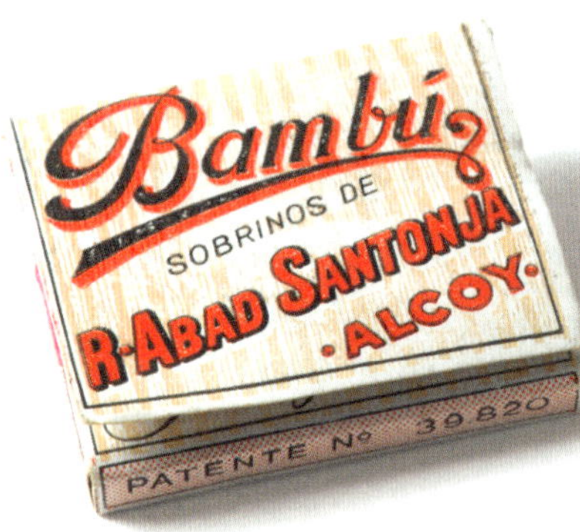
Bambú
SOBRINOS DE
R·ABAD SANTONJA
·ALCOY·
PATENTE Nº 39.820

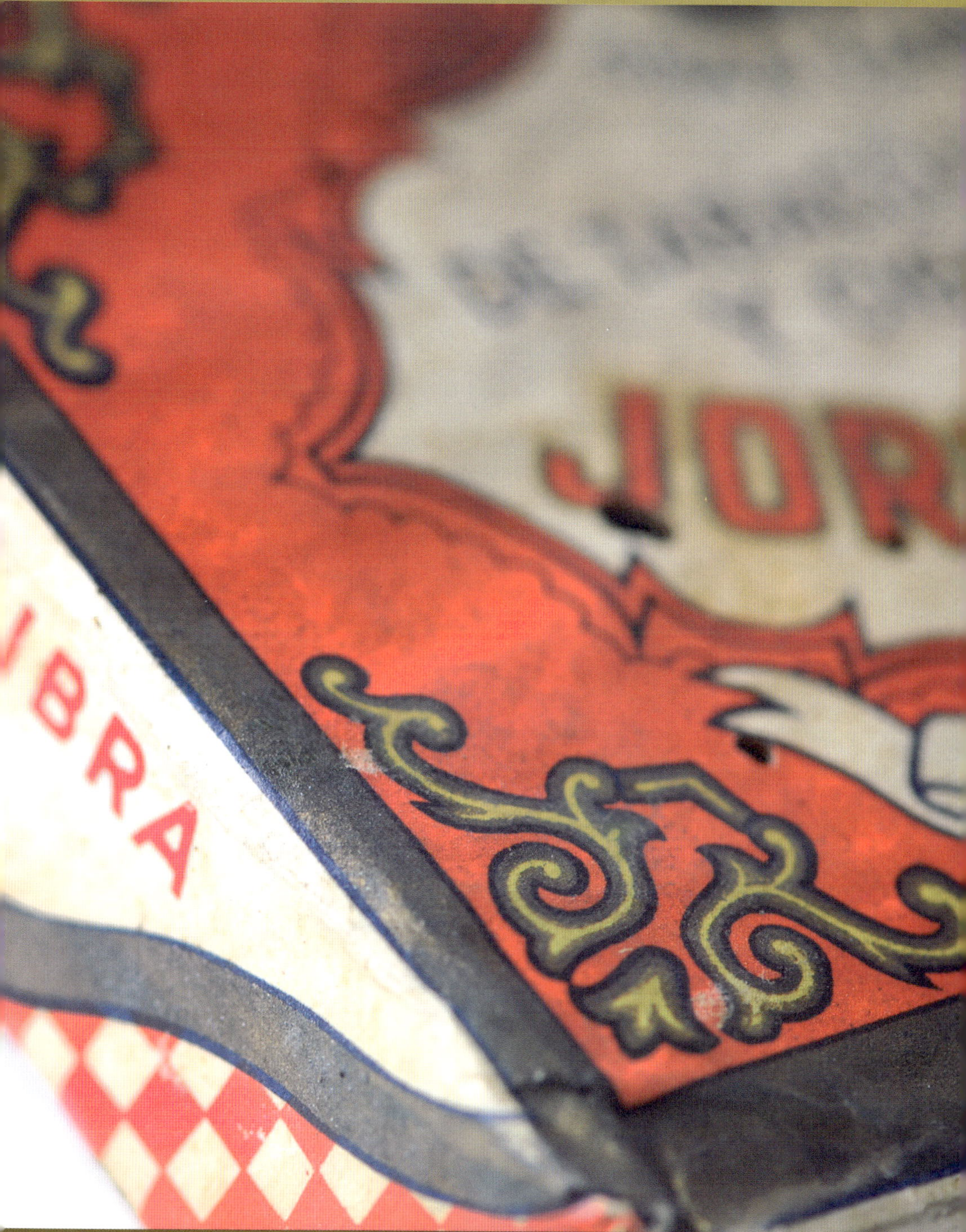

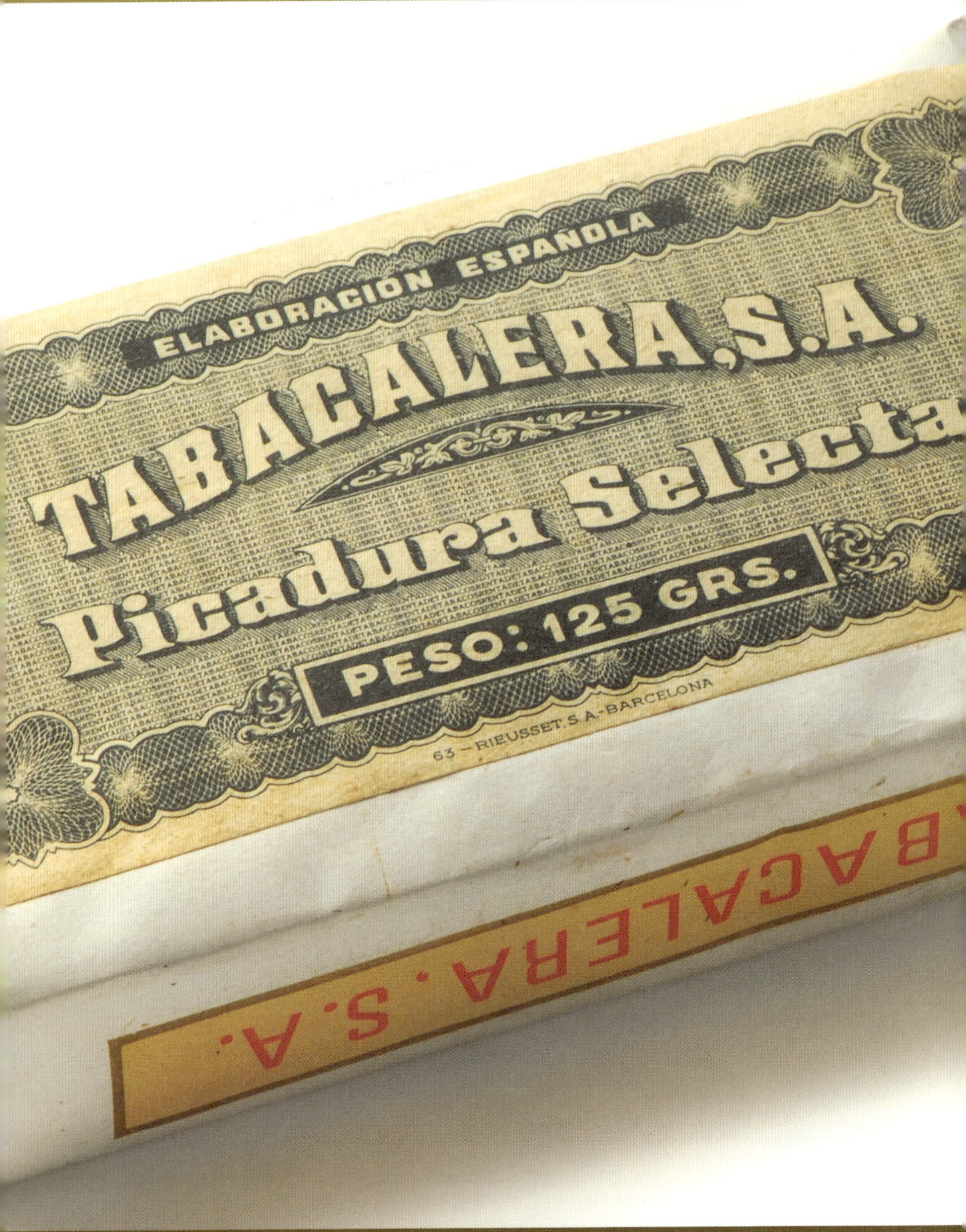
ELABORACION ESPAÑOLA
TABACALERA, S.A.
Picadura Selecta
PESO: 125 GRS.
63 – RIEUSSET, S.A.-BARCELONA
TABACALERA, S.A.

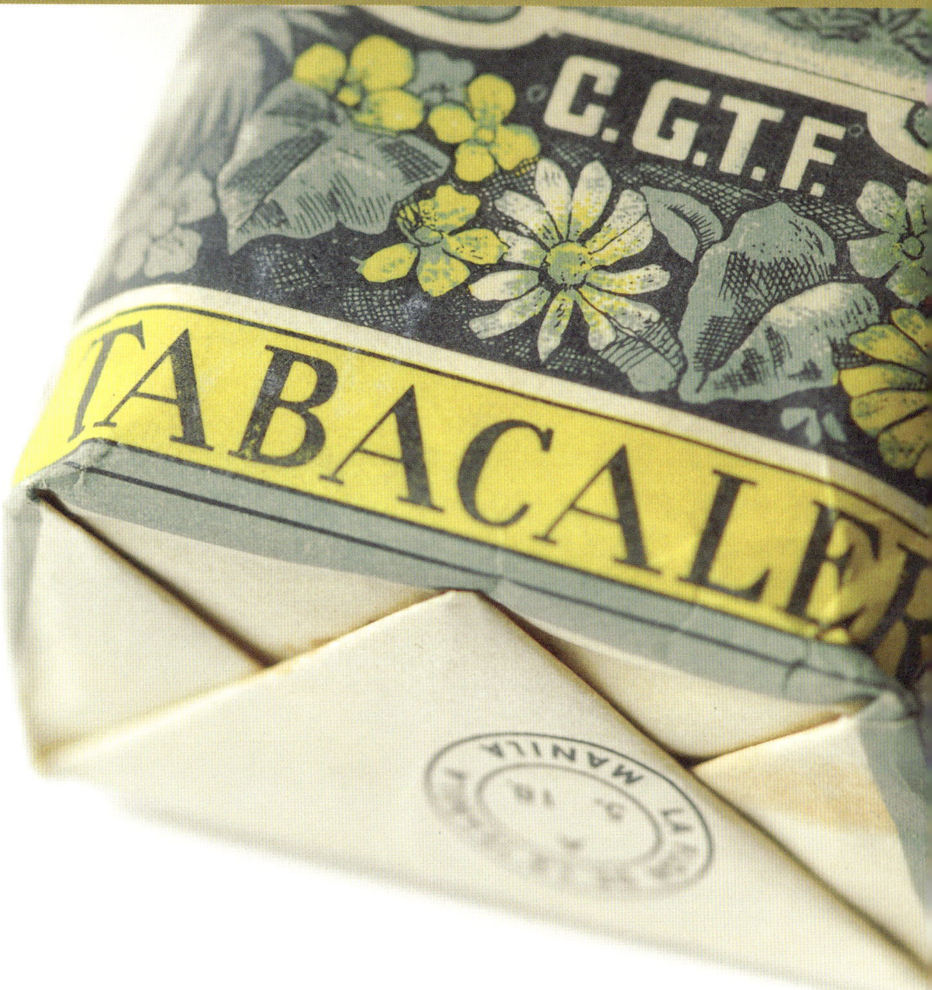
C.G.T.F.
TABACALE

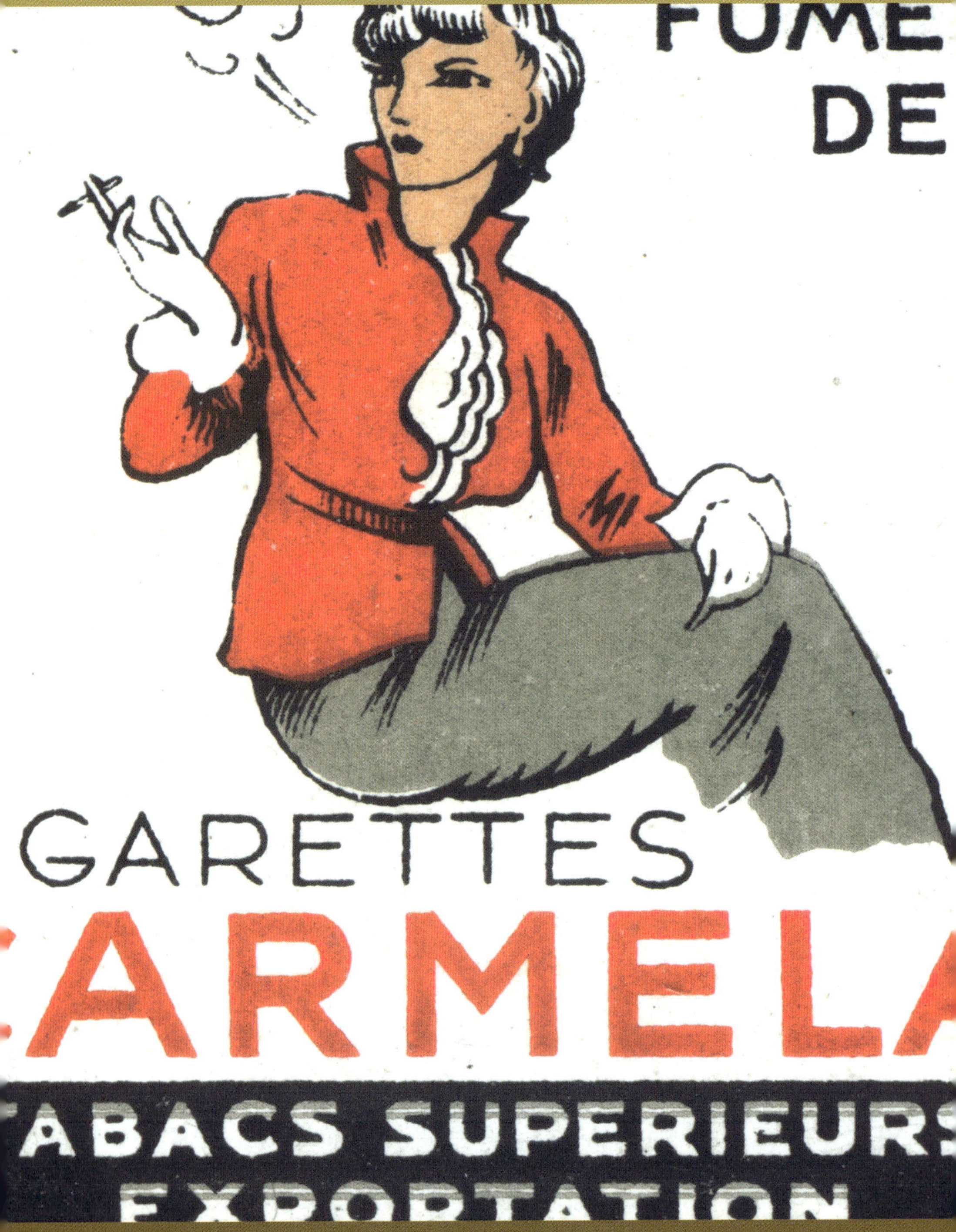
FUME
DE
GARETTES
ARMELA
ABACS SUPERIEURS
EXPORTATION

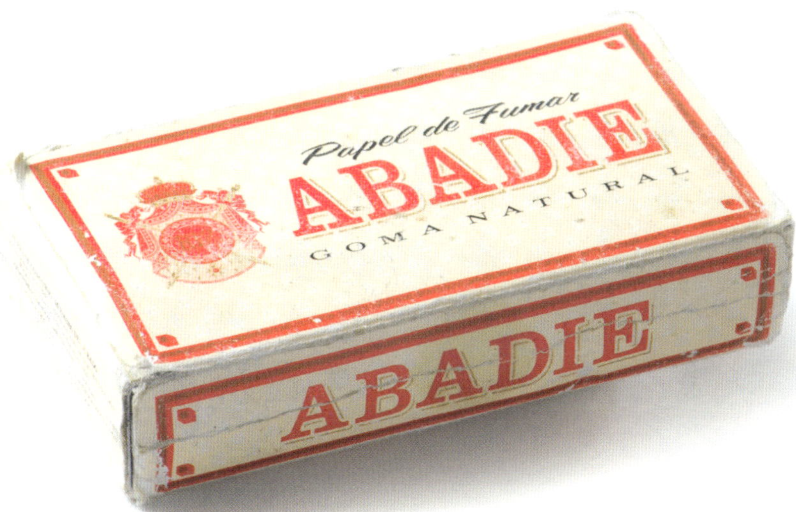
Papel de Fumar
ABADIE
GOMA NATURAL
ABADIE

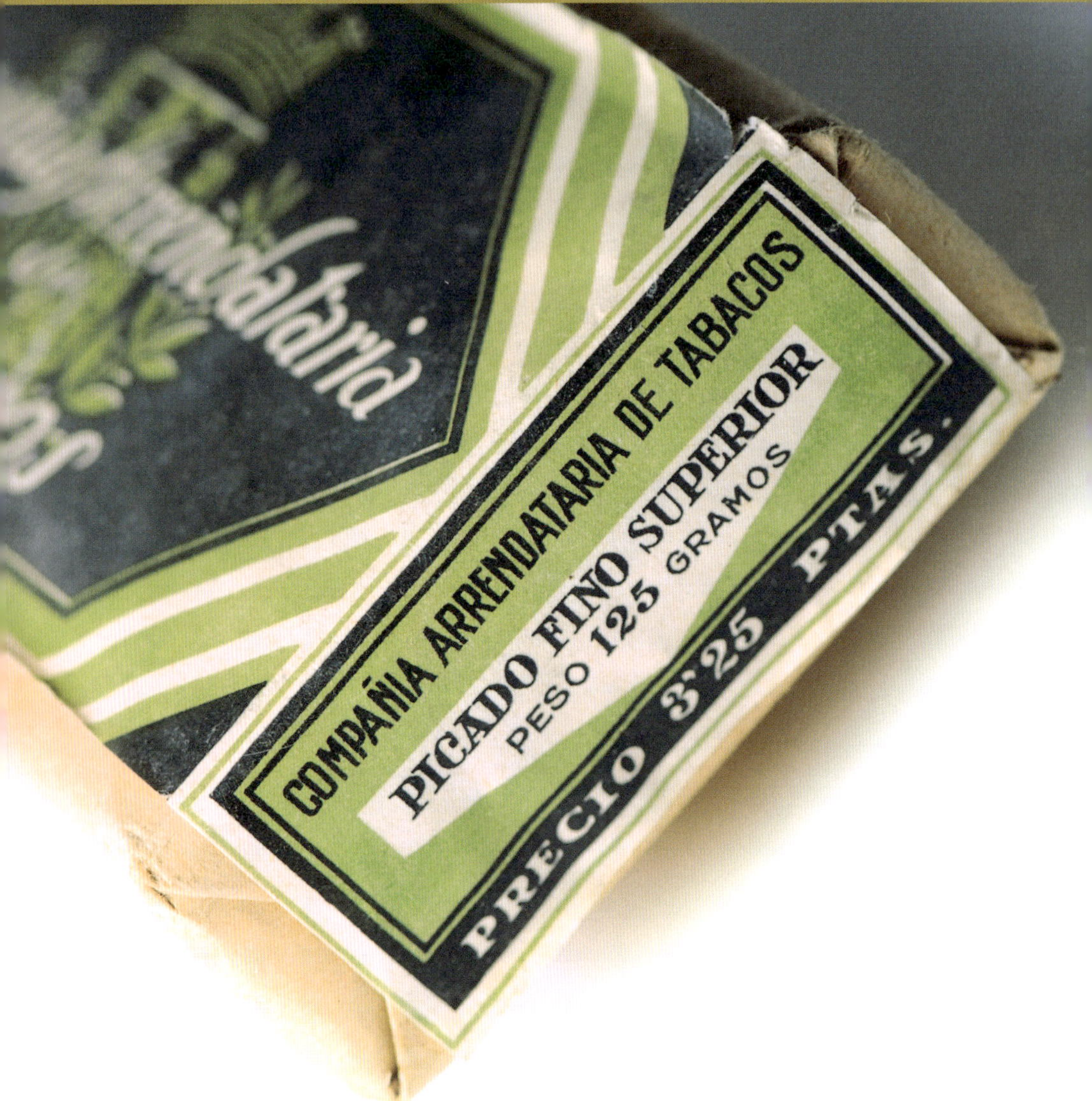
COMPAÑIA ARRENDATARIA DE TABACOS
PICADO FINO SUPERIOR
PESO 125 GRAMOS
PRECIO 3'25 PTAS.

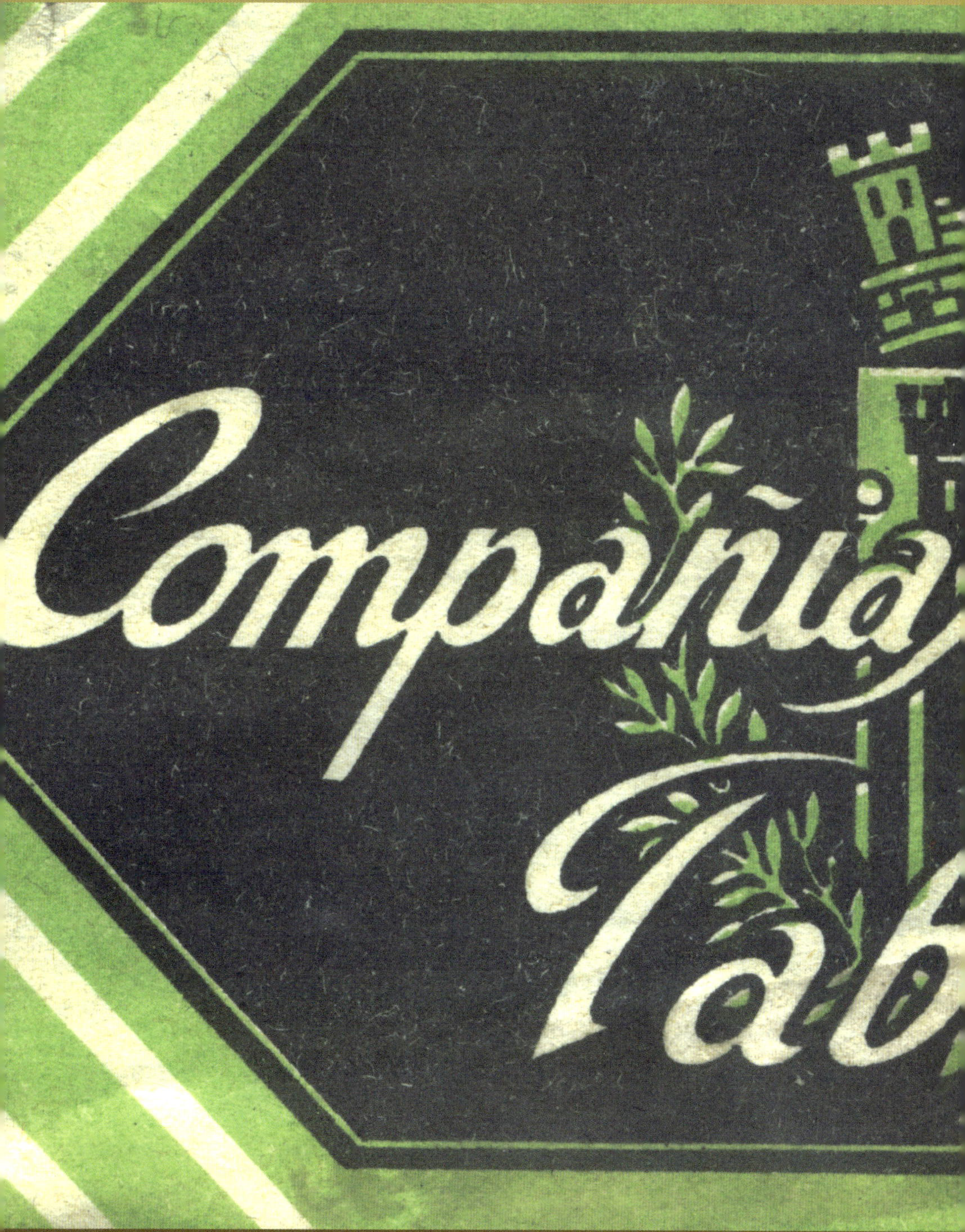
Compañía

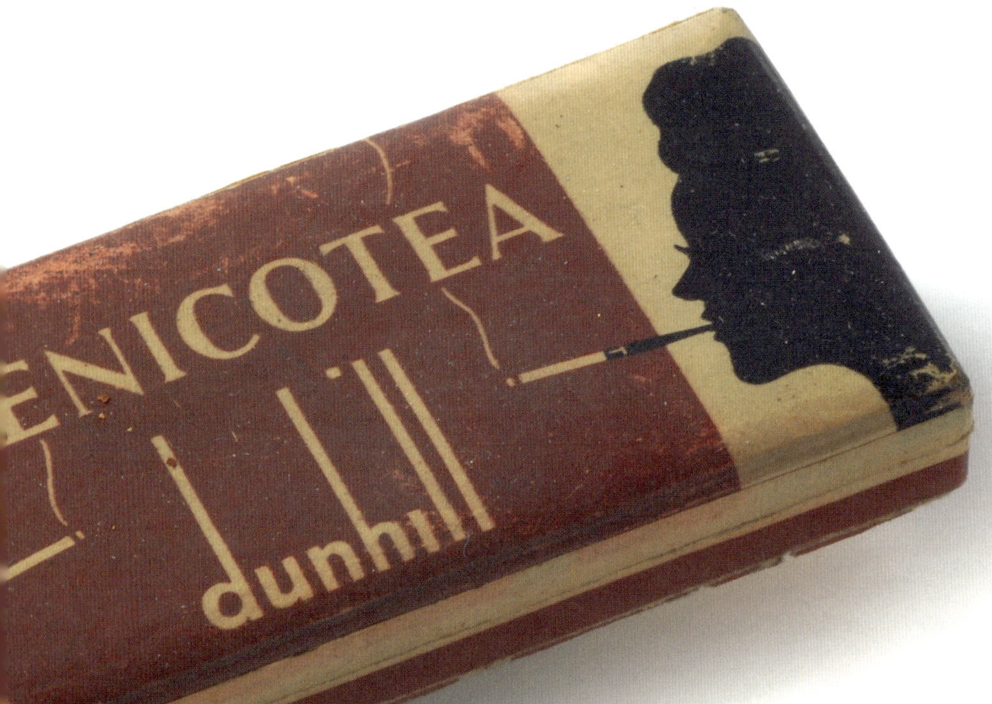
ENICOTEA
dunhill

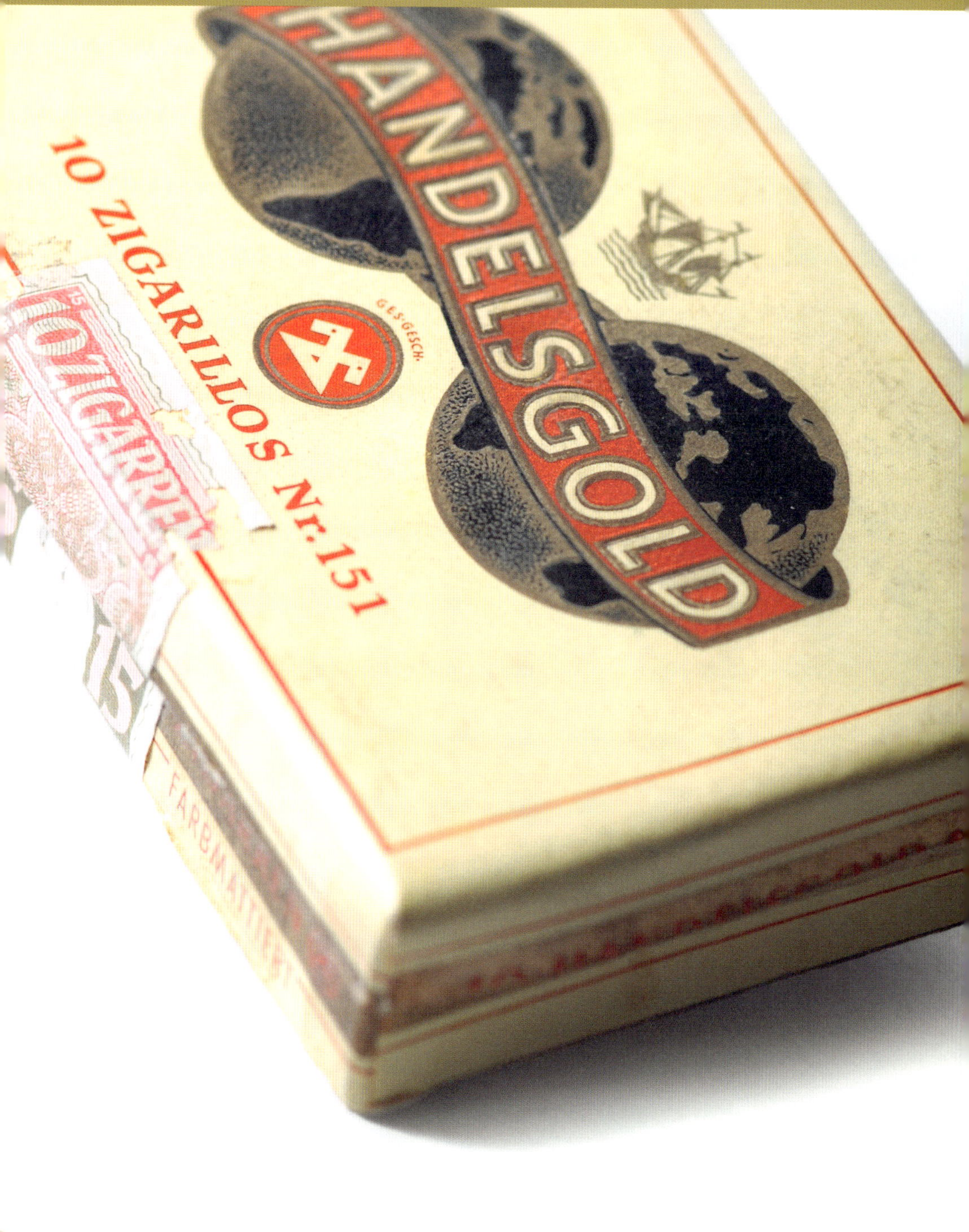
HANDELSGOLD
10 ZIGARILLOS Nr. 151
GES. GESCH.
FARBMATTIERT
15

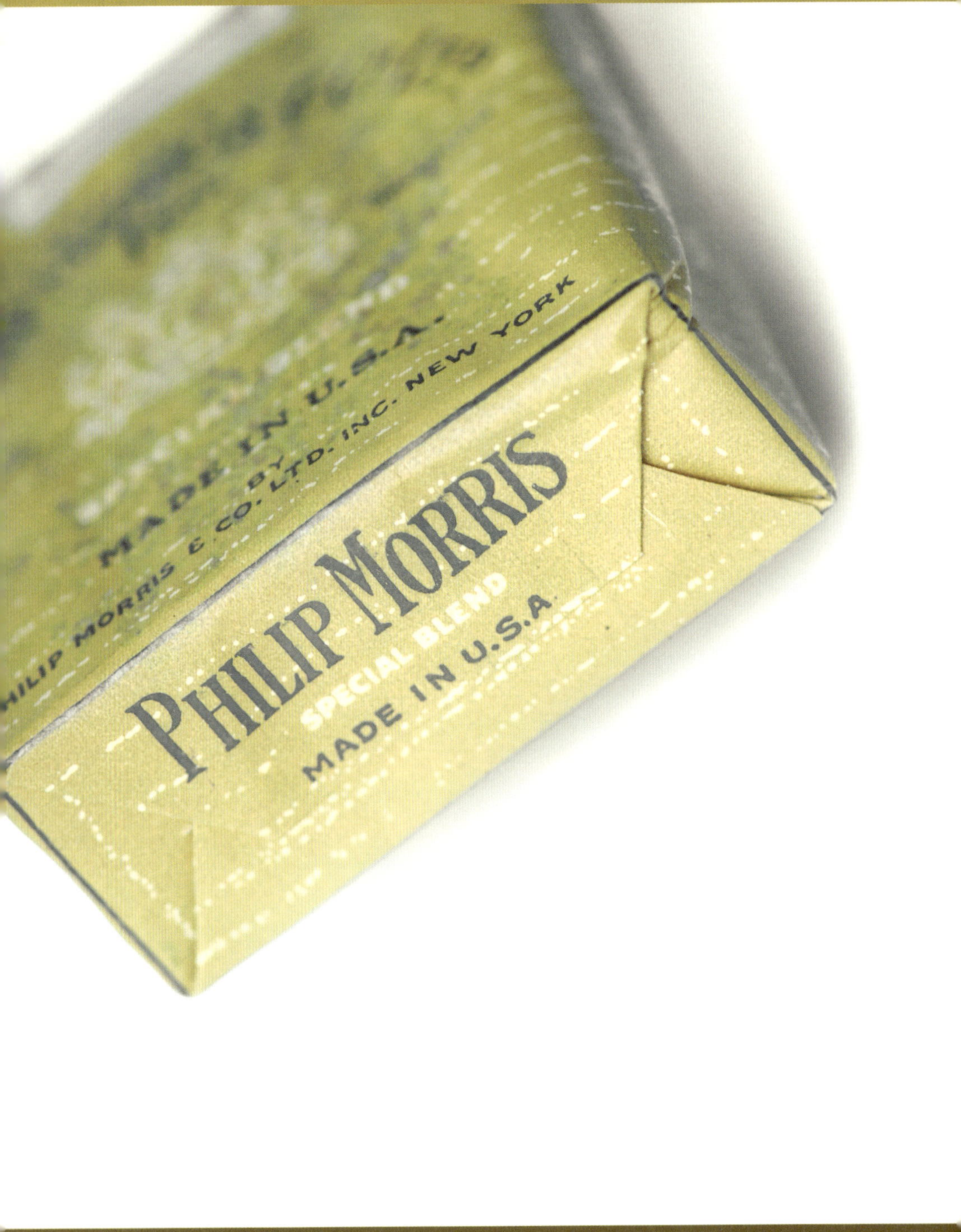
MADE IN U.S.A.
BY
PHILIP MORRIS & CO. LTD. INC. NEW YORK
PHILIP MORRIS
SPECIAL BLEND
MADE IN U.S.A

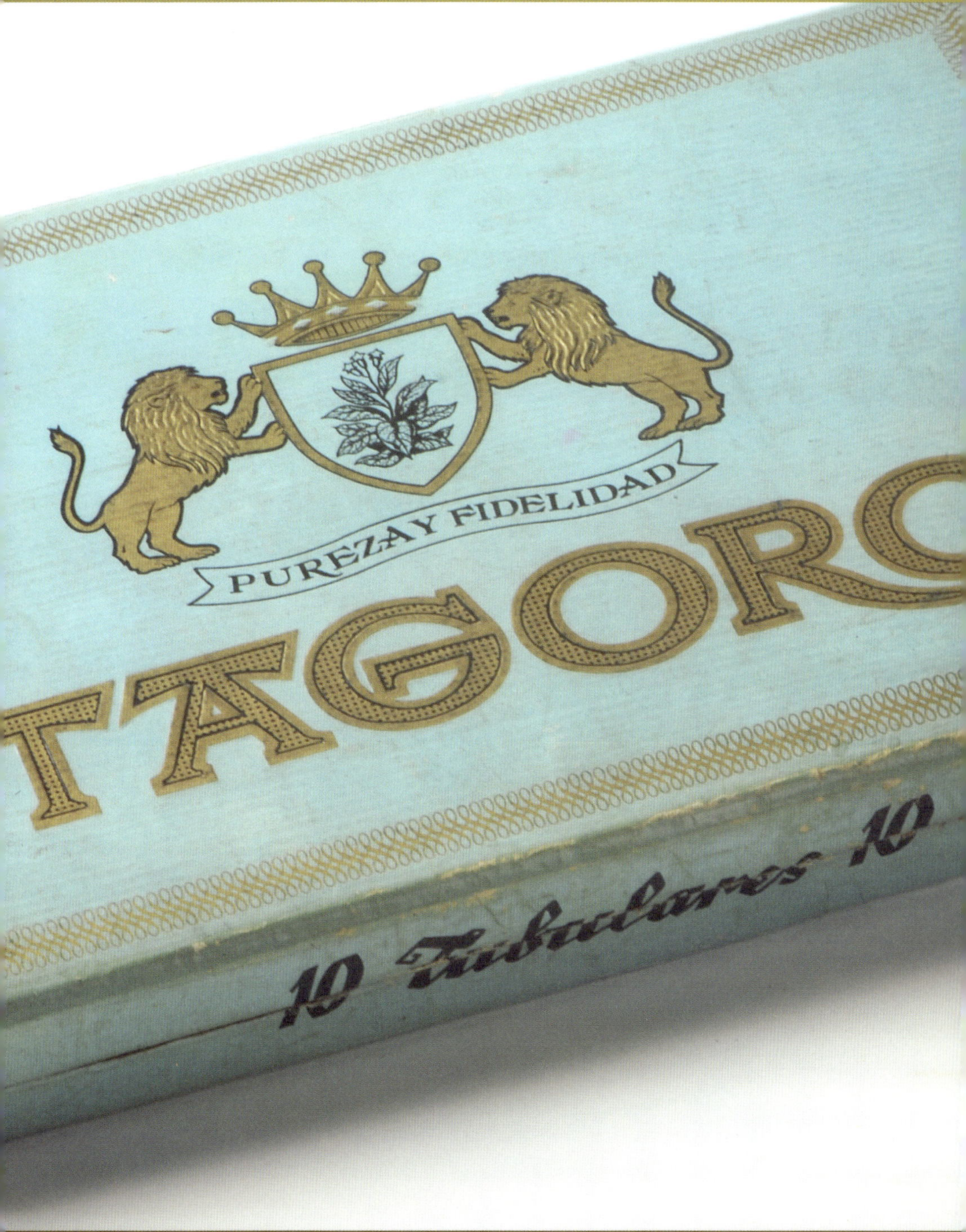
PUREZA Y FIDELIDAD
TAGORO
10 Tubulares 10

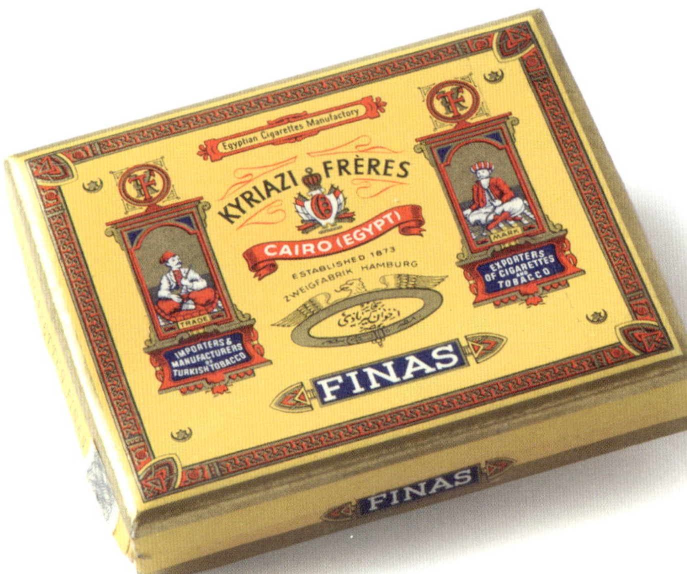
Egyptian Cigarettes Manufactory
KYRIAZI FRÈRES
CAIRO (EGYPT)
ESTABLISHED 1873
ZWEIGFABRIK HAMBURG
TRADE
MARK
IMPORTERS & MANUFACTURERS OF TURKISH TOBACCO
EXPORTERS OF CIGARETTES AND TOBACCO
FINAS
FINAS

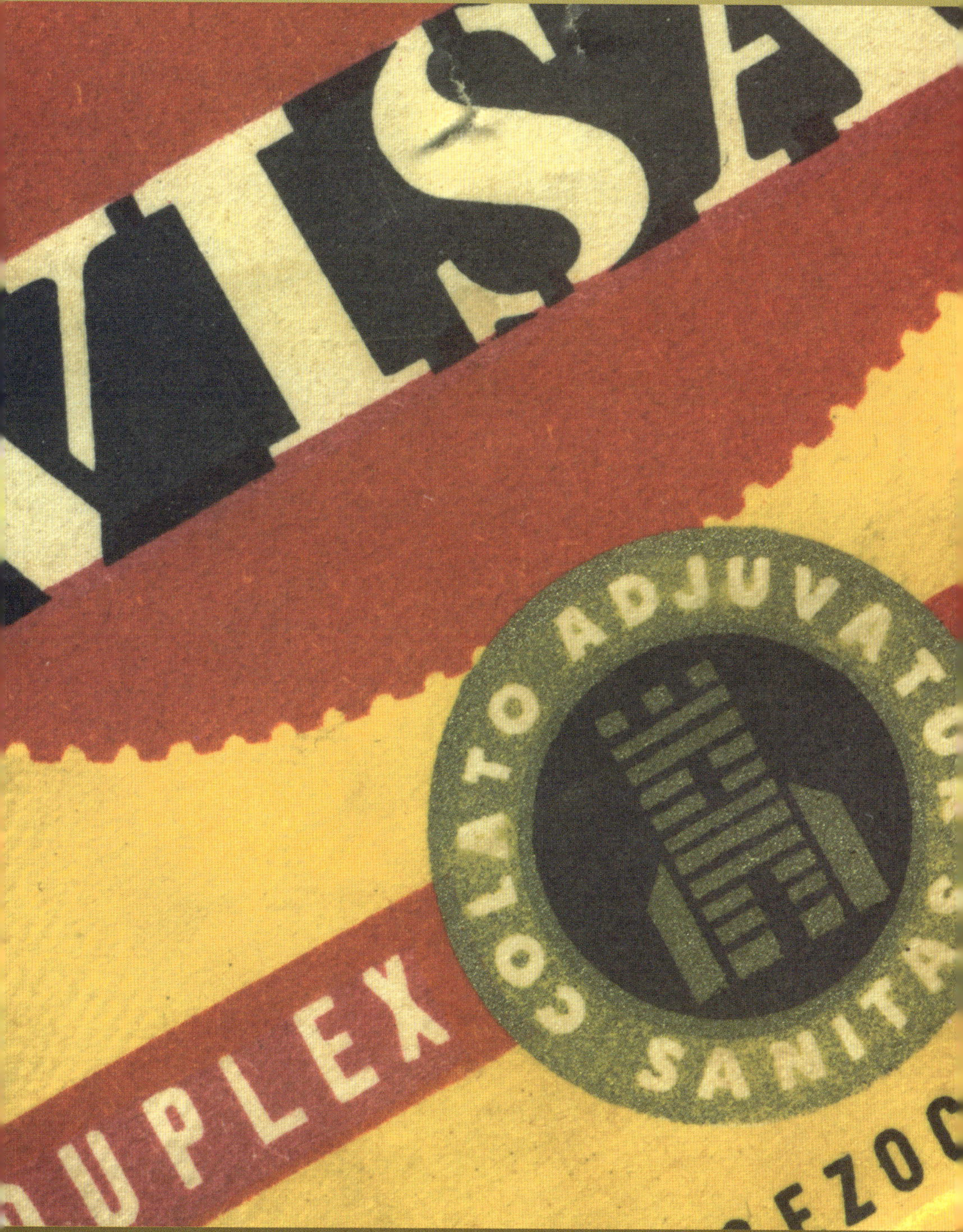
DUPLEX
COLATO ADJUVAT
SANITA

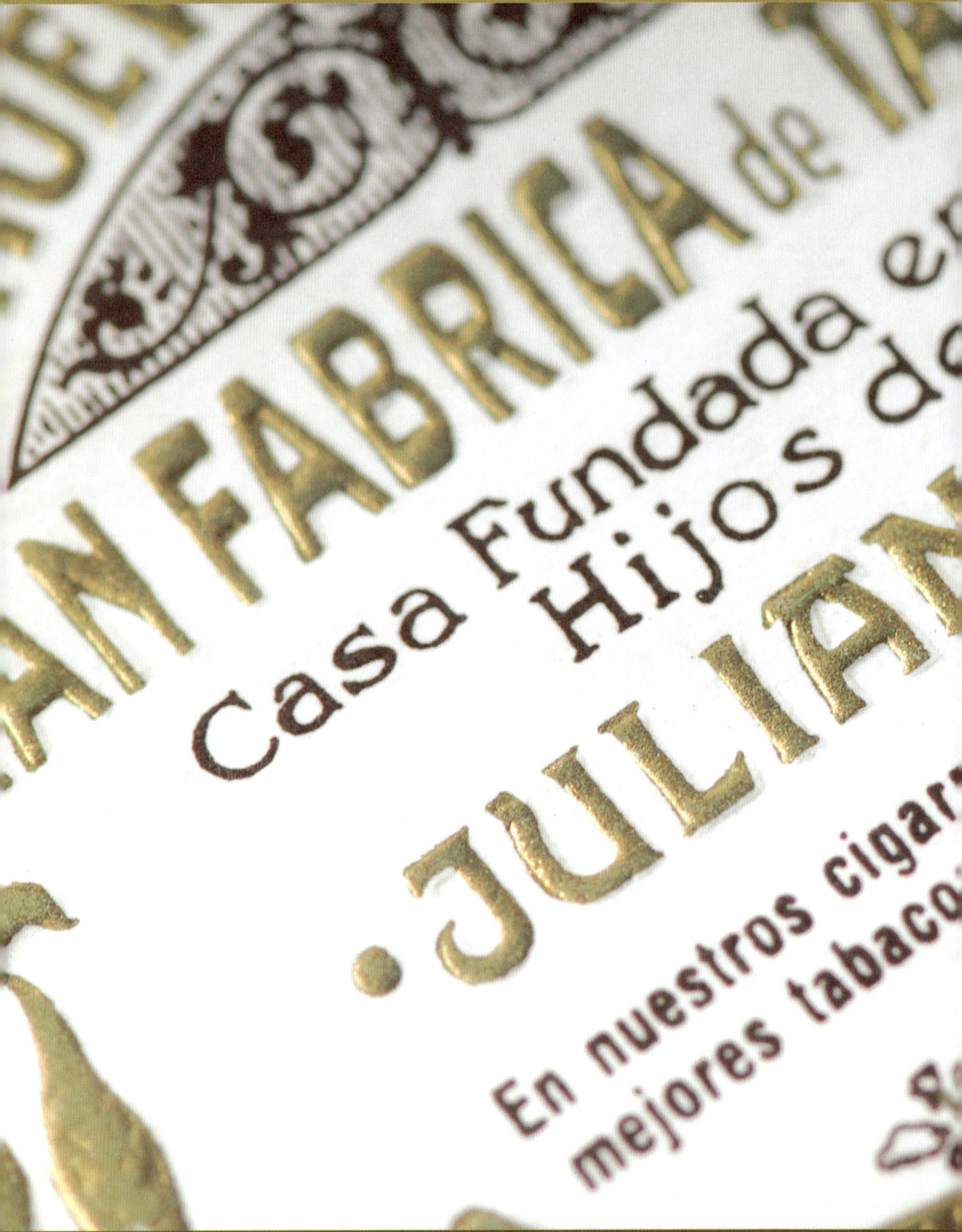
FABRICA de
Casa Fundada
Hijos
En nuestros
mejores

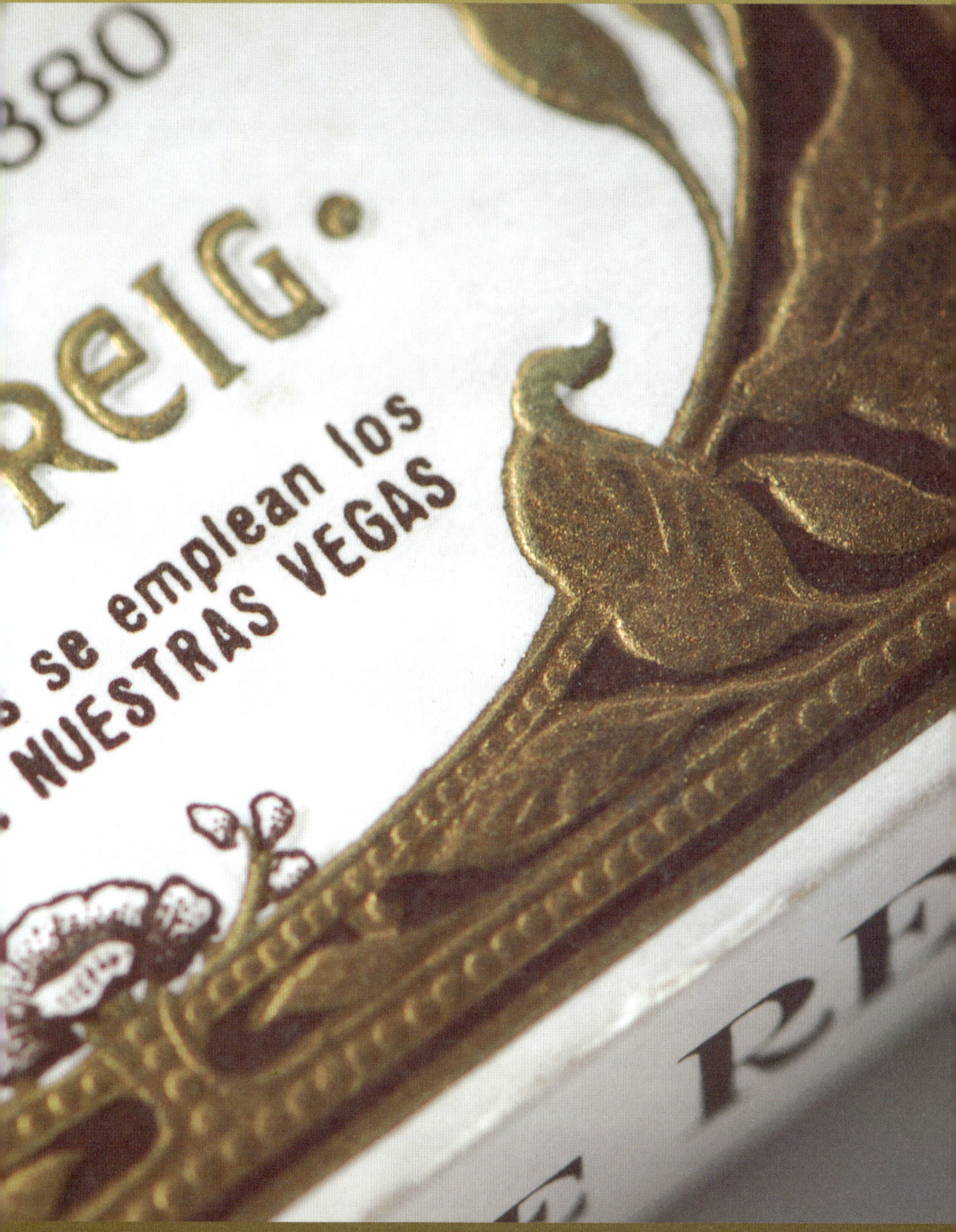
380
REIG
se emplean los
NUESTRAS VEGAS

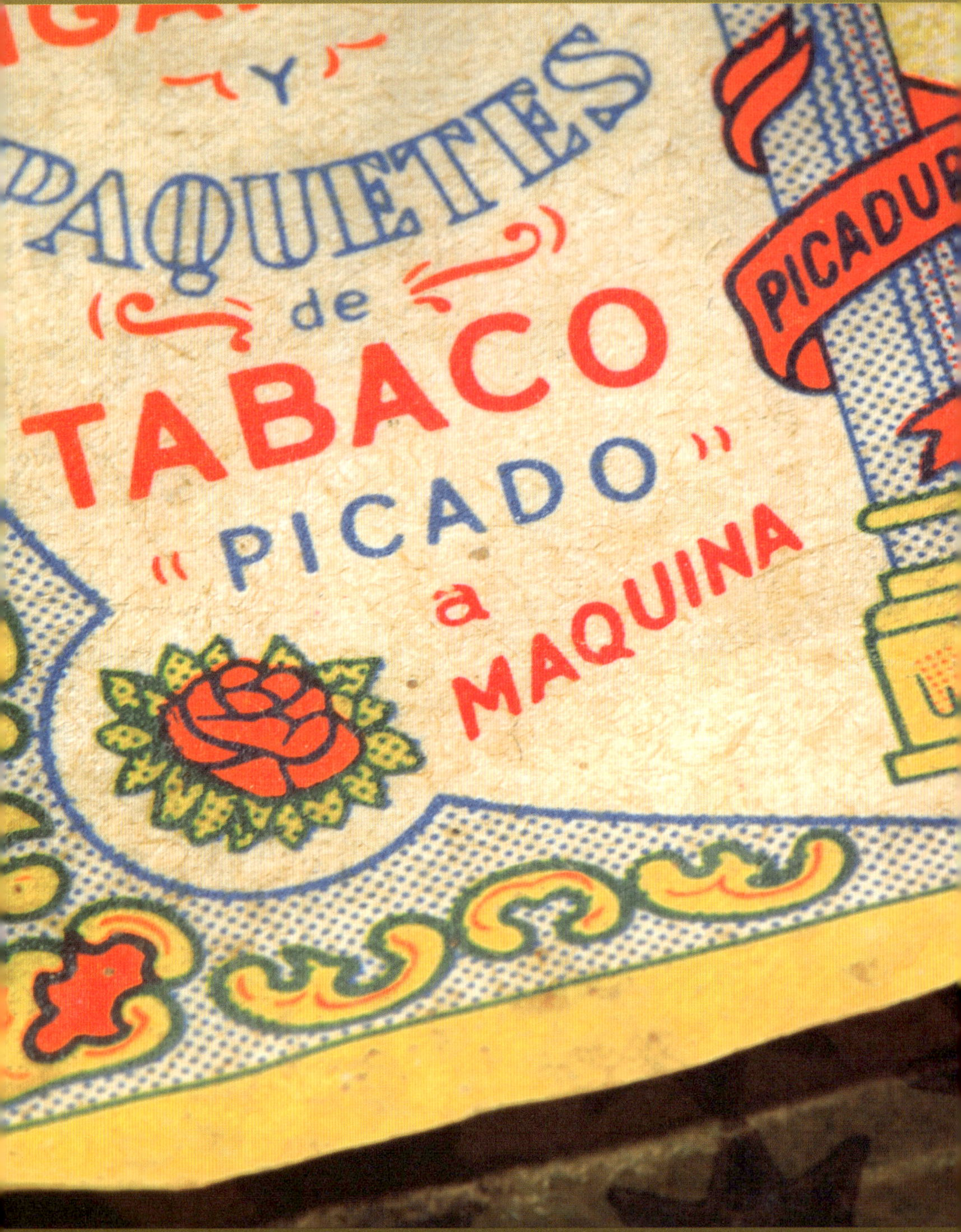
Y
PAQUETES
de
TABACO
"PICADO"
a
MAQUINA

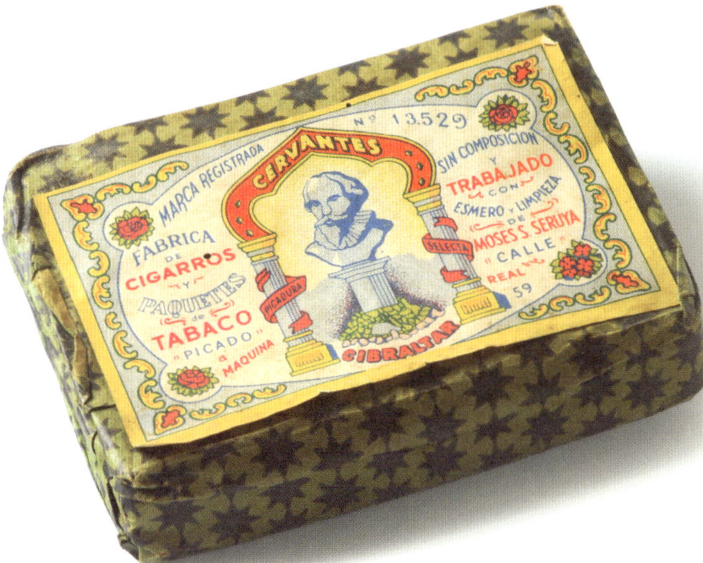
Nº 13529
CERVANTES
MARCA REGISTRADA
FABRICA DE CIGARROS Y PAQUETES de TABACO "PICADO" a MAQUINA
SIN COMPOSICION Y TRABAJADO CON ESMERO Y LIMPIEZA DE MOSES S. SERUYA "CALLE" REAL 59
SELECTA
PICADURA
GIBRALTAR

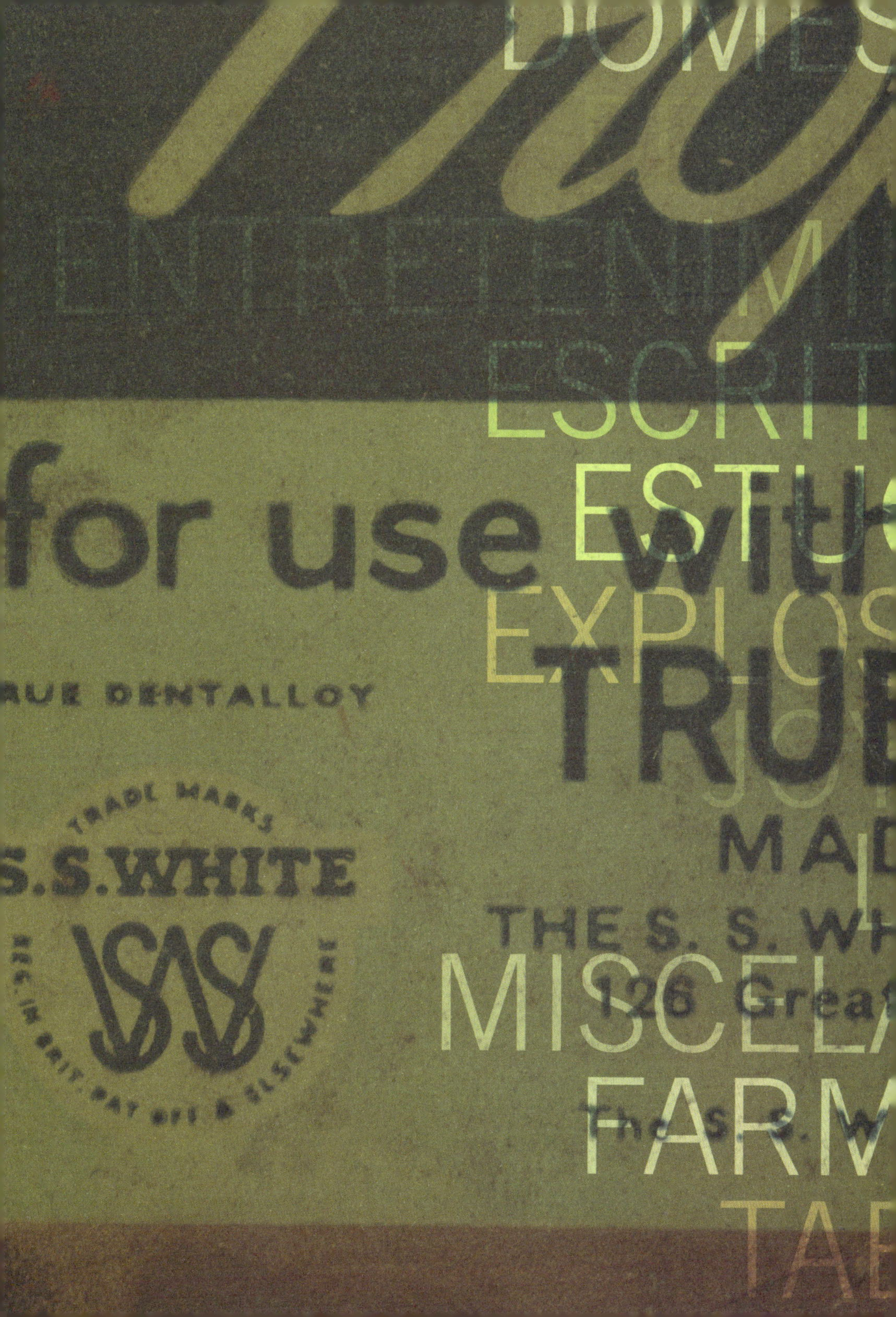
DOMES
ENTRETENIM
ESCRIT
ESTU
for use with
EXPLOS
RUE DENTALLOY
TRU
TRADE MARKS
S.S.WHITE
REG. IN BRIT. PAT OFF & ELSEWHERE
MAD
THE S. S. WH
MISCELA
126 Great
FARM
The S. S. W
TAB

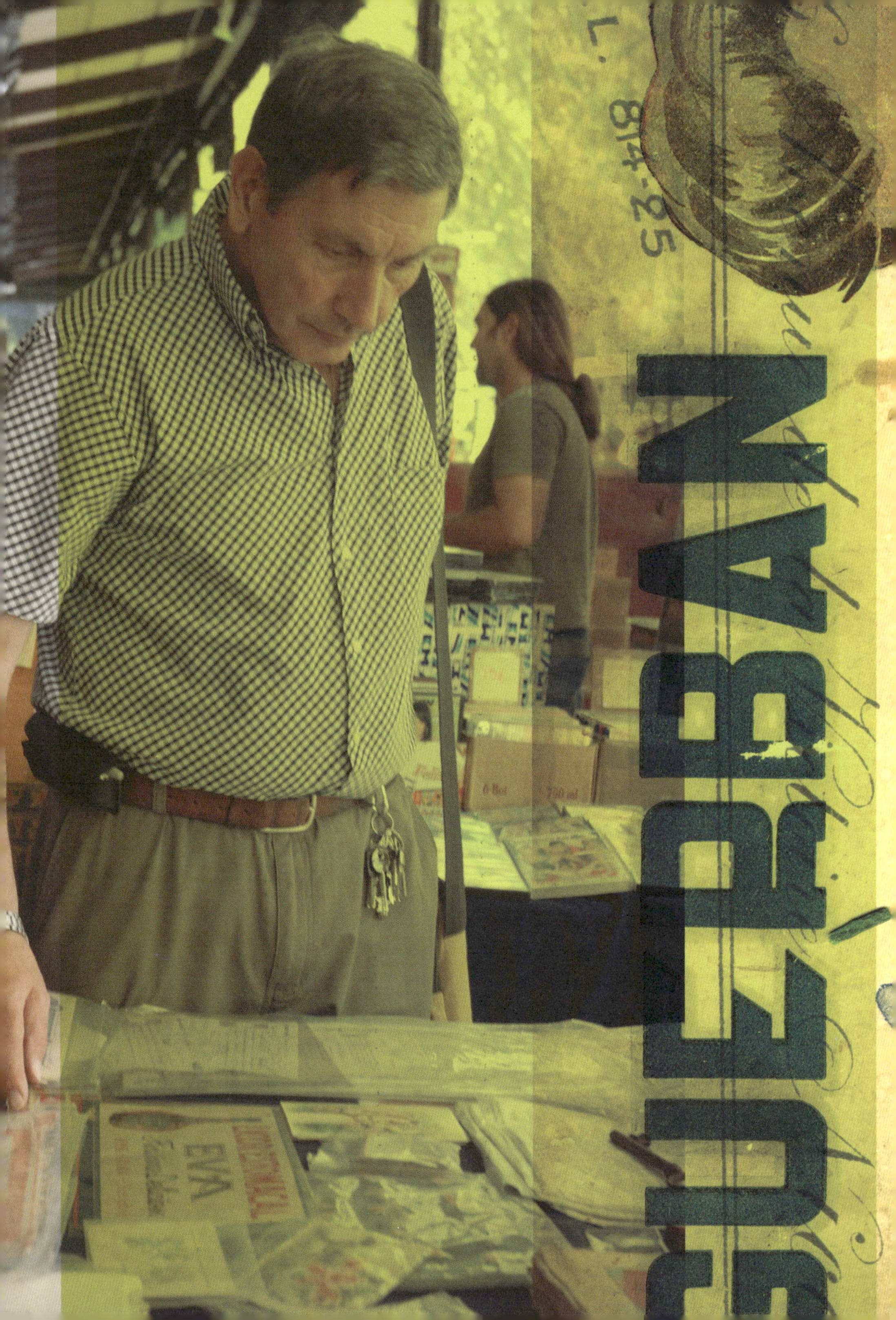

MR. SERRA

The participation of Francesc Serra in this project has been essential. He is the one who has been collecting packs for a long time and has eventually classified them to create Antique Packaging.

There is no way he will reveal us where he has found the boxes and how. But never mind, they are really beautiful and technically and aesthetically valuable. This way, little by little, with a real romantic vocation, tireless curiosity and unbreakable spirit of research, he has created a unique collection which has become an important part of Tribu-3.

That is why Antique Packaging owes a lot to Francesc Serra. And Tribu-3 would not be a complete company of creative packaging without this collection of boxes and packaging.

There is no doubt about it.

Serra

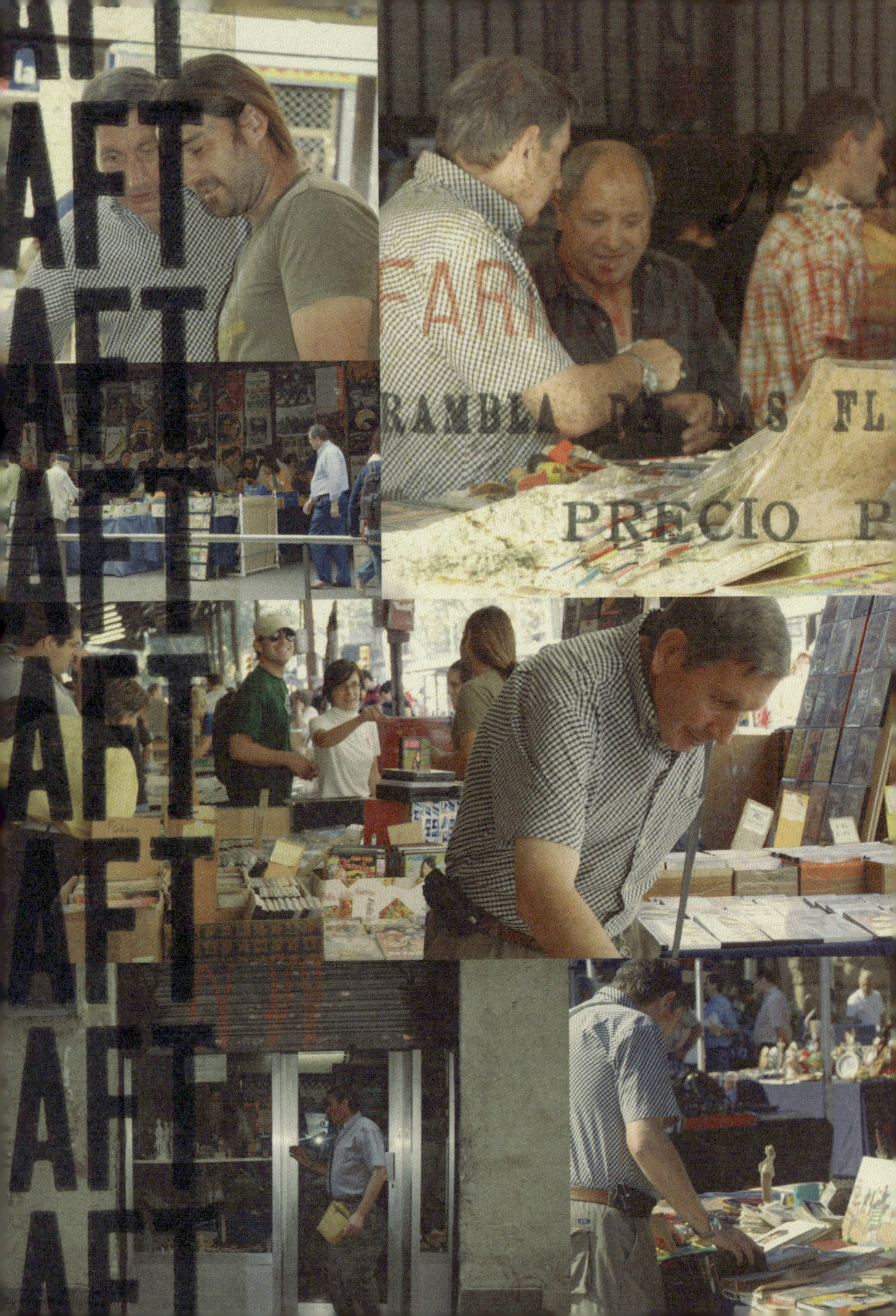
PRECIO P

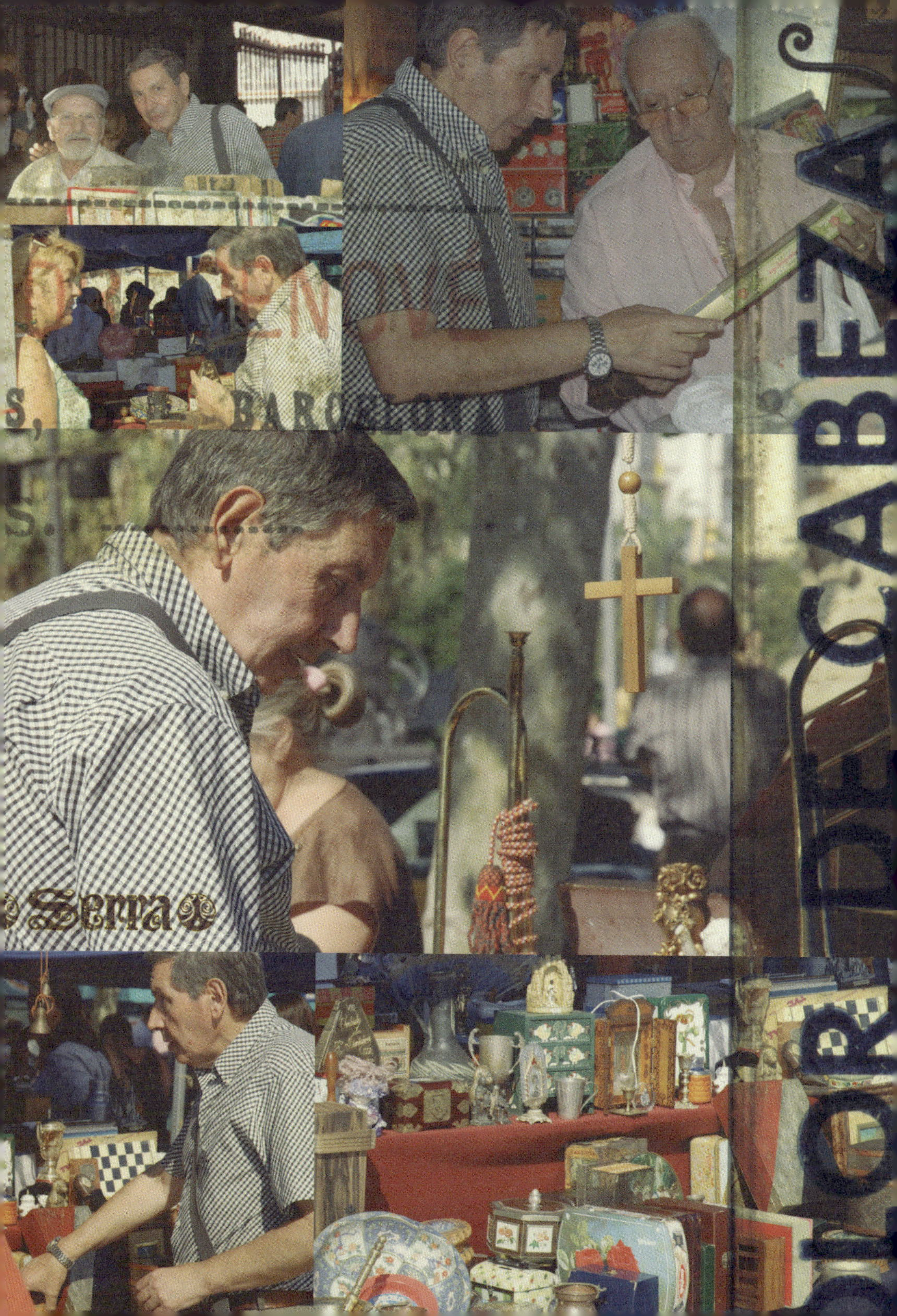
Serra
DE CABEZA

This book was finished in March, 2008.